Bioadsorbents for the Remediation of Environmental Pollutants

Dr. P. Senthil Kumar
Centre for Pollution Control and Environmental Engineering, Pondicherry University, Puducherry, India

Dr. B. Senthil Rathi
Department of Chemical Engineering, Sri Sivasubramaniya Nadar College of Engineering, Kalavakkam, Tamil Nadu, India

WILEY

Published by John Wiley & Sons, Inc., Hoboken, New Jersey.

Library of Congress Cataloging-in-Publication Data has been applied for:

Paperback ISBN: 9781394312160
ePDF ISBN: 9781394312184
ePub ISBN: 9781394312177
oBook ISBN: 9781394312191

Cover Design: WIley
Cover Image: © LuckyStep/stock.adobe.com

Contents

Preface

The rapid industrialization and urbanization of the modern world have resulted in a significant increase in environmental pollutants, ranging from heavy metals and dyes to emerging contaminants such as pharmaceuticals and microplastics. Conventional remediation methods, though effective to some extent, often involve high costs, energy-intensive processes, or the generation of secondary pollutants. This has led to a growing interest in sustainable, cost-effective, and environmentally friendly alternatives for pollution mitigation.

Bioadsorbents, derived from agricultural residues, forestry waste, algal and microbial biomass, offer a promising solution to these challenges. These materials not only provide a low-cost option for pollutant removal but also contribute to waste valorization, promoting circular economy principles. The versatility, renewability, and eco-friendliness of bioadsorbents make them a crucial research focus in environmental engineering and pollution control.

This book, *Bioadsorbents for the Remediation of Environmental Pollutants*, aims to provide a comprehensive and up-to-date resource on the subject. It systematically covers the fundamentals of adsorption, the classification and sources of bioadsorbents, and their environmental significance. The chapters explore biomass-derived adsorbents, novel synthesis techniques including nanotechnology-based approaches, and advanced characterization methods. The applications of bioadsorbents in water and wastewater treatment, heavy metal and dye removal, emerging contaminants remediation, and soil and air purification are discussed in detail, supported by case studies and comparative analyses.

In addition, the book addresses the economic and environmental aspects of bioadsorbents, highlighting cost factors, scale-up challenges, and sustainability considerations. Emerging trends, future prospects, and research gaps are identified to guide scientists, engineers, and policymakers in the development and implementation of bioadsorbent technologies.

This book is intended for graduate students, researchers, academicians, and industry professionals working in environmental science, chemical engineering, and sustainable technologies. It is our hope that this work will not only serve as an academic resource but also inspire further research and practical applications that contribute to cleaner and healthier environments.

1

Introduction

1.1 Introduction

Adsorption is a phenomenon related to surfaces. It refers to the process by which particles from a liquid or gas accumulate at the interface between a liquid and a solid. Adsorption plays a crucial role in biological, environmental, and industrial processes, including catalysis, separation, and purification. It is especially important in water and air treatment, chemical production, and various biological applications. Although initially developed empirically, its practical utility has significantly increased due to advances in theoretical models and nanomaterials (Dąbrowski, 2001). Adsorption involves the movement of guest molecules from the surrounding fluid to the surface of a solid or liquid. According to Pourhakkak et al. (2021), its efficiency, ease of use, and low cost make adsorption preferable to other techniques such as membrane filtration and ion exchange.

Adsorbents can be biological, organic, or mineral in nature. From an industrial perspective, activated carbon (AC) is the most conventional adsorbent due to its exceptional surface area and adsorption capabilities. Activated carbon is widely used to purify drinking water by removing pollutants from groundwater, rivers, lakes, reservoirs, and wastewater. Despite its efficacy, the high production and regeneration costs of AC limit its widespread use, especially in resource-limited settings, prompting the search for more affordable alternatives (Crini et al., 2019). Low-cost adsorbents, sometimes referred to as "green adsorbents," are derived from readily available and environmentally friendly sources such as waste materials, residues, and agricultural byproducts. These include fruit and vegetable peels, food industry byproducts, and biomass, which can be pyrolyzed to produce more effective adsorbents like AC. These materials are attractive for wastewater treatment due to their abundance, low manufacturing costs, and environmental sustainability, even though they may have lower adsorption capacities compared to advanced synthetic adsorbents such as complex composites or modified chitosans. Their use promotes environmentally friendly methods and offers a cost-effective alternative for large-scale pollution management (Kyzas and Kostoglou, 2014).

Carbon's most common allotropic form is graphite, although fullerenes and diamond-like carbons also exist. Graphitic and non-graphitic carbons are differentiated by their crystalline structures, which are further classified as graphitizable or non-graphitizable. Porous carbon adsorbents, such as powdered and granular ACs, contain heteroatoms like oxygen and hydrogen within their structure. These heteroatoms form surface functional groups that influence the adsorbents' chemical properties. The physical and chemical characteristics of carbon adsorbents significantly affect their adsorption behavior and efficiency in various applications (Menéndez-Díaz and Martín-Gullón, 2006).

Bioadsorbents derived from waste biomass, particularly fruit wastes, are gaining popularity for wastewater treatment due to their affordability, sustainability, and environmental friendliness. By recycling waste into valuable materials for environmental purification, these adsorbents support a circular economy. When ACs are combined with nanoparticles, the resulting material exhibits a higher surface area and an increased number of active sites, enhancing adsorption capacity and kinetics. Praseodymium-loaded carbon nanotubes (CNTs) and cellulose@CNT composites show promise for both dye adsorption and heavy metal sequestration. To fully understand the role of bioadsorbents and optimize their use in wastewater treatment, characterization techniques, adsorption isotherms, and kinetic models are essential (Solangi et al., 2021).

Heavy metals can be effectively removed from wastewater using inexpensive agricultural wastes such as rice husk, coconut shell, and sugarcane bagasse. To improve treatment efficiency, a new generation of biosorbents has been developed with superior adsorption performance and enhanced reusability (Mathew et al., 2016). For example, cellulose-based bioadsorbents have demonstrated consistent adsorption capacity and effective dye removal over at least three reuse cycles, even in complex systems containing binary dyes, salts, and surfactants. This highlights their potential to efficiently extract dyes from industrial textile wastewater (Liu et al., 2015). Aguilar-Rosero et al. (2022) reviewed the development of bioadsorbents from a variety of agricultural wastes, including fruits, vegetables, grains, seeds, and fodder. Their study investigates the effectiveness of over 60 bioadsorbents in removing heavy metals, industrial pollutants, nutrients, and pharmaceuticals.

This chapter explores adsorption as a fundamental separation technique, with a focus on bioadsorbents derived from low-cost, sustainable biomass sources. It emphasizes their advantages over synthetic adsorbents in environmental remediation, particularly in wastewater treatment and pollution control. Despite significant progress, challenges remain in improving adsorption capacities, regeneration, and scalability. Addressing these issues through innovative material development and process optimization will promote sustainable waste valorization and environmentally friendly treatment methods, driving advancements in bioadsorbent applications.

1.2 Overview of Adsorption and Adsorbents

Adsorption is a surface phenomenon in which atoms, ions, or molecules of a gas or liquid accumulate on a solid surface due to chemical bonding or physical forces such as van der Waals interactions. Mhemeed (2018) provided a comprehensive description of this process, highlighting its thermodynamic aspects, including decreases in enthalpy (ΔH), entropy (ΔS), and free energy (ΔG) during adsorption. The study discussed porous adsorbents like charcoal, silica gel, and zeolite, and distinguished between unimolecular and multimolecular adsorption. It also clarified the difference between adsorption and absorption, noting that both processes often occur simultaneously in sorption, which may be endothermic.

Adsorption is a well-established method for removing heavy metals from water and wastewater. Wang and Guo (2023) thoroughly reviewed adsorption kinetics and isotherm models, emphasizing their importance in understanding adsorption mechanisms and optimizing system design. Their study assessed the adsorption capacities of heavy metals such as Cu, Cd, Zn, Ni, and Cr using biosorbents (e.g., algae, agricultural waste biochar, bacteria) and abiotic adsorbents (e.g., metal–organic frameworks (MOFs), polymers, and clays).

Nanomaterial-based adsorbents (NAs), including metal oxides, graphene sheets, and CNTs, have shown great promise in wastewater treatment due to their strong adsorption properties, high surface area, and tunable surface chemistry. Awad et al. (2020) investigated the adsorption performance of NAs for a variety of organic contaminants, including dyes, phenolic compounds, pharmaceuticals, and antibiotics. Their study focused on how adsorption mechanisms, solution chemistry, and adsorbent properties influence removal efficiency. Raji et al. (2023) examined various adsorbents for heavy metal removal from wastewater, highlighting alternatives derived from plants, biological sources, and nanomaterials. While algae and fungi are environmentally friendly bioadsorbents, their effectiveness varies depending on biomass type and pretreatment methods. Although nanomaterials offer high reactivity and reusability, they also pose environmental risks. Despite their need for surface modification to enhance performance, low-cost plant-based adsorbents align well with green chemistry principles. Challenges remain in scaling up modified adsorbents due to high costs and potential pollution concerns, as well as in predicting biosorption behavior in complex systems.

1.2.1 Definition and Significance of Adsorption

Adsorption is a surface phenomenon in which molecules from a fluid phase (gas or liquid) accumulate on a solid surface and interact chemically or physically to form an interfacial layer. Desorption is the reverse process, which is generally reversible. Adsorption plays a vital role in environmental cleanup, catalysis, and water purification. This overview briefly explains key adsorption isotherm models (e.g., Langmuir, Freundlich), kinetics (e.g., pseudo-first-order and pseudo-second-order), and thermodynamic parameters ($\Delta G, \Delta H, \Delta S$) that influence the process. It also discusses the two main types of adsorption: physisorption and chemisorption. Understanding these fundamentals is critical for optimizing adsorption-based technologies (Mhemeed, 2018).

The term "biosorption" refers to the passive binding and removal of chemicals—whether soluble or insoluble, organic or inorganic—by biological materials. Mechanisms involved include adsorption, ion exchange, surface complexation, and precipitation. As a physicochemical process occurring in both living and dead biomass, biosorption offers an effective, affordable, and environmentally responsible approach to remove radionuclides, hazardous metals, and other contaminants. Biomass sources such as seaweeds, fungi, algae, bacteria, and agricultural wastes have shown considerable promise. However, despite encouraging laboratory results, industrial-scale adoption remains limited. Biosorption continues to be a valuable area of research for environmental biotechnology and pollution management (Gadd, 2009).

The inconsistent use of adsorption definitions—such as actual, absolute, excess, net, and column adsorption—often causes confusion in gas adsorption studies. Turner et al. (2024) clarified these distinctions by applying fundamental concepts of volume, porosity, and density in packed columns, and derived interconversion relationships among them. They developed dynamic mass balance models tailored to each form of adsorption, emphasizing the importance of aligning the mass balance model with the appropriate isotherm type. Although excess adsorption isotherms are widely used, they can yield inaccurate results when applied as actual or absolute models, especially under varying pressures and mass transfer conditions.

1.2.2 Classification of Adsorbents

Adsorbents are solid materials characterized by a significant volume of micro- and mesopores. They are broadly classified into three main categories: organic, inorganic, and composite adsorbents. Organic adsorbents—comprising carbon-based materials such as agricultural residues, biochar, synthetic polymers, and biopolymers like chitosan—dominate the field, accounting for 68.3% of studies. Inorganic adsorbents, which make up about 20.8%, include minerals like silica, metal oxides, clays, sediments, and soils, both natural and synthetic. Composite adsorbents, hybrids of organic and inorganic components, represent 10.9% and have shown promising results in removing dyes, heavy metals, and emerging contaminants (Quintero-Jaramillo et al., 2021).

Due to their affordability and availability, a wide range of adsorbents—including natural materials, agricultural wastes, and industrial byproducts—have been explored for effective water and wastewater treatment. Nano-adsorbents, such as metal and metal oxide nanoparticles, spinel ferrite nanomaterials, carbon-based nanomaterials, and nanocomposites, alongside biomass-derived adsorbents, offer enhanced adsorption capacities owing to their high surface area and reactivity. Advanced materials like dendritic polymers and geopolymer cement are also gaining attention for contaminant removal. However, since each adsorbent type has unique advantages and limitations, further optimization of process parameters and comprehensive research are essential to maximize efficiency (Hajjizadeh et al., 2020).

Polymer-based adsorbents are organic materials notable for their high specific surface areas, microporous structures, and interconnected three-dimensional pore networks. These include membranes, covalent organic polymers, molecularly imprinted polymers, ion exchange resins, and microporous organic polymers. Polymeric composites—blends of polymers with metals, ceramics, or clays—offer flexible architectures and have proven highly effective. Thanks to their specific porosity and surface chemistry, polymer-based adsorbents outperform many conventional adsorbents in removing dyes from wastewater (Moradi and Sharma, 2021).

Owing to their large surface areas and strong affinity for hazardous chemicals, polymeric adsorbents have attracted significant interest. This chapter reviews various types including MOFs, polymers of intrinsic microporosity, covalent organic frameworks, hyper cross linked polymers (HCPs), conjugated microporous polymers, and porous aromatic frameworks. It also highlights their common precursors and synthesis methods. Among these, HCPs stand out due to their exceptional porosity, high adsorption capacity, and low fabrication costs. The choice of monomers and synthesis conditions plays a crucial role in determining the structural properties of these adsorbents (Masoumi, 2024).

1.3 Concept of Bioadsorbents

Significant environmental pollution arises from high concentrations of dyes such as methylene blue (MB) in industrial wastewater, especially from the rapidly growing textile sector. While AC is effective for MB removal, its high cost limits widespread use. Consequently, adsorbents derived from biowaste—originating from wood, animal, and agricultural residues—have emerged as viable, affordable, and renewable alternatives with considerable sorption capacities. Recent research focuses on their properties, regeneration potential, processing methods, and economic feasibility. These advances pave the way for large-scale wastewater treatment applications using biowaste materials and underscore the urgent need for cost-effective adsorbents (Hamad and Idrus, 2022).

Growing environmental concerns have accelerated the development of eco-friendly bioadsorbents for wastewater remediation. Natural biopolymers, especially polysaccharides like alginate, cellulose, starch, chitosan, and natural gums, are favored due to their biodegradability, renewability, availability, and low cost. These bioadsorbents have successfully removed a variety of contaminants from aqueous solutions. Recent studies highlight their sustainable advantages over conventional materials by focusing on their synthesis, adsorption mechanisms, and recyclability. Such developments underscore the promising potential of polysaccharide-based bioadsorbents as economical and environmentally friendly solutions for wastewater treatment (Fakhri et al., 2023).

1.3.1 What Are Bioadsorbents?

Bioadsorbents are naturally occurring materials used to remove contaminants from soil or water through adsorption. They are often derived from waste products of the food, forestry, or agricultural industries. These materials are rich in functional groups such as —OH, —COOH, and $—NH_2$, which interact with pollutants like heavy metals, dyes, and pharmaceuticals. For instance, a recent study demonstrated that antibiotics including amoxicillin (AMX), ciprofloxacin, and trimethoprim were effectively adsorbed by pine bark, oak ash, and mussel shells, achieving near-complete removal under batch conditions (Míguez-González et al., 2023). Due to their effectiveness, affordability, and renewability, bioadsorbents are promising tools for environmentally sustainable remediation.

Fruit peels—such as those from banana, orange, mango, watermelon, and lemon—are abundant and cost-effective bioadsorbents for water contaminants. Rich in functional groups, these peels efficiently bind pollutant cations, especially heavy metals and dyes. However, physicochemical modifications are often necessary to enhance their adsorption capacity. Studies have also demonstrated their reusability over multiple adsorption cycles. Factors such as pollutant type, pH, temperature, and adsorbent surface properties significantly influence adsorption efficiency. Further research is needed to optimize biosorption parameters, integrate hybrid purification techniques, and sustainably manage residual wastes (Nyika and Dinka, 2023).

Recent advances in bioadsorbents have improved heavy metal removal from wastewater and broadened their application scope. Various materials—including lignin, bacteria, fungi, magnetic bioadsorbents, nanocellulose, and agricultural or animal wastes—have been investigated for their unique adsorption properties. Key advantages of these bioadsorbents include reusability, eco-friendliness, high selectivity, and specificity toward target metals. Their diverse sources and functional traits allow tailored approaches for treating complex wastewater streams, delivering cost-effective and environmentally sound remediation solutions. These developments highlight the growing potential of bioadsorbents in enhancing wastewater treatment technologies (Hussain et al., 2024). Table 1.1 provides a comparative analysis of bioadsorbents, detailing their sources, compositions, and adsorption capabilities.

1.3.2 Differentiation from Synthetic Adsorbents

Natural bioadsorbents often provide performance comparable to or better than synthetic adsorbents like AC and chromium-doped metal oxides, but at a lower cost and with less environmental impact. Unmodified materials such as peanut shells, neem leaves, and coffee waste have demonstrated remarkable removal efficiencies for heavy metals including Cd(II), Pb(II), and Cu(II), rivaling those of synthetic alternatives. Additionally, plant-derived materials like mint waste and casuarina pine effectively adsorb organic dyes. These natural adsorbents are attractive options for environmentally friendly water treatment solutions that align with global sustainability goals due to their high sorption capacity, renewability, and reusability (Meftah et al., 2025).

Bioadsorbents made from natural or waste materials—such as bagasse, red mud, clays, and agricultural residues—are highly valued for their affordability, availability, and eco-friendliness. In contrast, synthetic adsorbents like synthetic resins and AC, while highly efficient, are expensive to manufacture. Although bioadsorbents may require surface modifications to enhance their adsorption capacity, their sustainability and abundance make them ideal for large-scale wastewater treatment applications. Despite offering greater consistency and tunability, synthetic adsorbents can have negative environmental impacts and lead to higher operational costs (Lin and Juang, 2009). Consequently, bioadsorbents are increasingly recognized as a viable and environmentally responsible alternative.

Table 1.1 Comparative analysis of bioadsorbents: source, composition, and adsorption capabilities.

Biomass source	Bioadsorbent type	Main components	Target contaminants	Advantages	Limitations	Reference
Orange peel	Dried powder, modified form	Pectin, cellulose	Pb^{2+}, Ni^{2+}, dyes	Renewable, good binding sites	Poor regeneration	Ajmal et al. (2000)
Banana peel	Dried powder, treated peel	Pectin, cellulose, lignin	Dyes, Pb^{2+}, Cu^{2+}	Waste valorization, eco-safe	Degrades over time	Annadurai et al. (2002)
Peat moss	Raw and acid-treated	Lignin, humic acid	Hg^{2+}, dyes	High sorption capacity	Biodegradable; needs stabilization	Brown et al. (2000)
Sawdust	Raw, acid-treated, modified	Cellulose, hemicellulose, lignin	Cr^{6+}, Cu^{2+}, Zn^{2+}	Readily available, eco-friendly	Low mechanical strength	Demirbas (2008)
Sugarcane bagasse	Carbonized, chemically treated	Cellulose, lignin	Textile dyes, Pb^{2+}, Ni^{2+}	Porous, cost-effective	Limited regeneration cycles	Mubarak et al. (2024)
Rice husk	Activated carbon, raw husk	Cellulose, lignin, silica	Heavy metals (Pb^{2+}, Cd^{2+}), dyes	Abundant, low-cost, renewable	Requires activation for better efficiency	Foo and Hameed (2009)
Tea waste	Raw and carbonized	Polyphenols, cellulose	Cu^{2+}, Cd^{2+}, Cr^{3+}	Cost-effective, widely available	Limited adsorption under acidic pH	Malkoc and Nuhoglu (2005)
Corncob	Raw, acid-treated	Hemicellulose, cellulose	Dyes, arsenic	Abundant, biodegradable	Requires pretreatment	Akindolie and Choi (2023)

Groundnut shell	Carbonized, chemically treated	Cellulose, hemicellulose, lignin	Dyes, Cu^{2+}	Agricultural waste reuse	Pretreatment needed	Kali et al. (2024)
Algae (e.g., *Chlorella*)	Dried biomass, alginate beads	Polysaccharides, proteins	Cd^{2+}, Pb^{2+}, Ni^{2+}	High biosorption capacity	Affected by environmental conditions	Romera et al. (2007)
Neem leaves	Dried, powdered	Lignin, cellulose, flavonoids	Cr^{6+}, dyes	Antimicrobial, eco-friendly	Seasonal availability	Hossain et al. (2024)
Coconut shell	Activated carbon	Lignin, cellulose	Hg^{2+}, phenols, organics	High surface area, durable	Costly activation process	Tan et al. (2008)
Palm kernel shell	Activated carbon	Lignin, cellulose	Phenols, heavy metals	High carbon content, hard structure	Processing cost	Uchegbulam et al. (2022)
Fungal biomass (*Aspergillus*)	Live/dead biomass	Chitin, protein	Cr^{6+}, Cd^{2+}	Fast adsorption, reusable	Needs cultivation facilities	Volesky (2001)
Eggshell	Powdered or calcined form	Calcium carbonate	Fluoride, Pb^{2+}	Strong binding, easy availability	Limited surface area	Wu et al. (2025)

1.4 Sources and Types of Bioadsorbents

Bioadsorbents derived from biomass wastes—such as fruit peels and agricultural residues—are gaining increasing popularity due to their high adsorption capacities, low cost, environmental friendliness, and renewability. Examples include peels from bananas, oranges, pomegranates, and mangoes, which have demonstrated effectiveness in removing dyes, heavy metals, pharmaceuticals, and various organic pollutants. The adsorption performance is further enhanced by AC produced from oak or eucalyptus shells, especially when functionalized with nanoparticles. Additionally, iron oxide-functionalized carbons and magnetic biochar derived from marine algae have shown promising results. By promoting waste valorization, these bioadsorbents contribute to a circular economy, providing dual environmental benefits: reducing waste and purifying water (Solangi et al., 2021). Kumar and Balasubramani (2025) emphasize their suitability for removing dyes, heavy metals, and organic contaminants from wastewater, attributing this to their porous structures and abundant functional groups that enhance adsorption efficiency. Figure 1.1 illustrates the sources of bioadsorbents, including biomass categories such as agricultural waste, forestry residues, algae, and microbes.

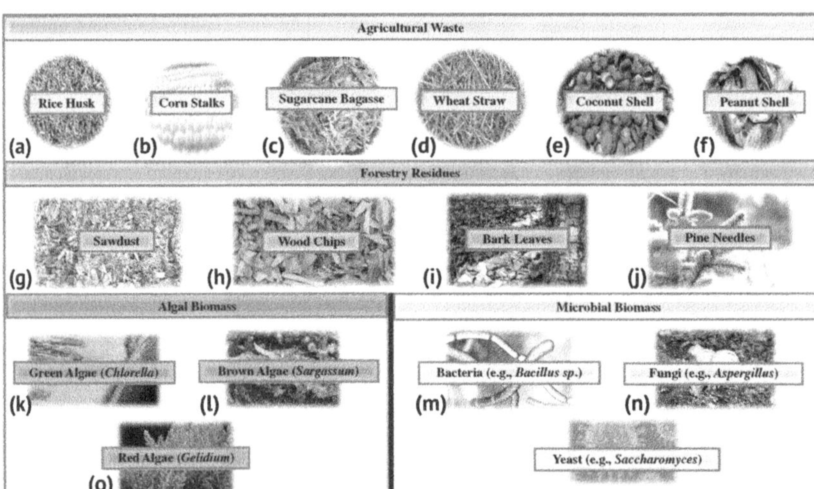

Figure 1.1 Sources of bioadsorbents. *Source:* (c): Thamizhpparithi Maari/Wikimedia Commons/CC BY-SA 3.0, (d): Volodymyr Chaban/Adobe Stock, (e): verinize/Adobe Stock Photos, (f): Delia/Adobe Stock Photos, (g): _Alicja_/Pixabay, (h): spr0ut7 design/Adobe Stock Photos, (i): M.Fawaid.M/Adobe Stock Photos, (j): gosiak1980/Pixabay, (m): Dr_Microbe/Adobe Stock Photos, (o): damedias/Adobe Stock.

1.4.1 Agricultural Waste

Global food and agricultural industries generate vast quantities of waste, which, if improperly managed, pose significant environmental hazards. A sustainable approach to both reducing this waste and removing heavy metals from wastewater involves repurposing such wastes as bioadsorbents. Agricultural byproducts like coconut coir, rice husk, sugarcane bagasse, and watermelon rind are rich in cellulose, lignin, and functional groups such as hydroxyl (—OH), amino ($—NH_2$), and carbonyl (—CO) groups, which effectively bind metal ions. For instance, sugarcane bagasse has demonstrated the ability to remove up to 97% of chromium from aqueous solutions, while coconut coir exhibits an adsorption capacity of 263 mg/g for lead ions. These abundant and low-cost materials promote waste valorization while enabling eco-friendly water treatment (Mathew et al., 2023).

Activated carbon can also be synthesized from fruit wastes through methods such as pyrolysis, hydrothermal carbonization, and chemical activation using agents like KOH and $ZnCl_2$. These processes significantly enhance surface area and porosity, thereby improving the capacity of ACs to adsorb dyes, heavy metals, and other pollutants. For example, ACs derived from banana peels, coconut shells, and lapsi seeds have proven effective in wastewater contaminant removal. Utilizing fruit waste not only mitigates environmental degradation but also provides a sustainable and cost-effective means of water purification (Solangi et al., 2021).

Repurposing agricultural waste as biosorbents offers an economical and environmentally responsible solution for eliminating hazardous heavy metals from wastewater. These biomaterials possess irregular structures laden with metal-binding functional groups such as amine, carboxyl, and hydroxyl groups. Modified and composite biosorbents further improve adsorption efficiency and enable the removal of multiple metal ions simultaneously. Key parameters influencing biosorption include particle size, temperature, adsorbent dosage, and pH. Kinetic and isotherm studies commonly indicate that the biosorption process follows pseudo-second-order kinetics and fits the Langmuir adsorption model. However, regeneration studies highlight the pressing need for universal eluents effective across different biosorbents. Continued research into green conversion technologies and regeneration strategies remains essential to optimize sustainability. Overall, biosorbents derived from agricultural wastes present a highly effective and sustainable method for environmental remediation (Harshala and Wagh, 2022).

Biochar and AC produced from agricultural residues such as peat, rice straw, beer dregs, palm kernel husks, sugarcane bagasse, nut shells, coffee, tea, and wood are valued due to their low ash content and high carbon content. Biochar is produced via pyrolysis of plant biomass without chemical treatment and has shown high efficacy in heavy metal removal from wastewater. Pyrolysis conditions, especially temperature, critically affect biochar's adsorption performance. For example, biochar generated from Camellia seed husks at 300 °C exhibited Pb(II) and Cd(II) removal efficiencies of 80.61 and 53.57%, respectively, while increasing the temperature to 700 °C improved removal rates to 89.03 and 75.84%. Similarly, biochars derived from coconut and peanut shells removed Cd(II) at rates of 80 and 99.9%, respectively. Biochar made from banana peels demonstrated an impressive Cu(II) adsorption capacity of 351 mg/g. These findings underscore the potential of agricultural waste-derived biochar as a sustainable and efficient biosorbent for wastewater treatment (Meftah et al., 2024).

1.4.2 Forestry Residues

Forestry wastes such as bark, sawdust, and other residues are gaining popularity as sustainable and cost-effective sorbents for environmental remediation. Numerous studies from Galicia (NW Spain) have demonstrated the effectiveness of both raw and processed forestry byproducts in removing heavy metals, phosphate, fluoride, and emerging contaminants from soils and water bodies. These materials can significantly reduce pollutant concentrations, highlighting their potential for reuse in both initial treatment and regeneration processes. This underscores their growing importance as valuable resources for pollution management and resource recovery (Núñez-Delgado et al., 2020).

The Camphor tree (*Cinnamomum camphora*), widely cultivated for its essential oil used in the flavor industry, generates large amounts of camphor leaf waste through extensive harvesting in plantations and coppice models. This biomass residue, rich in cellulose, hemicellulose, and lignin, contains abundant hydroxyl groups, making it a promising, low-cost bioadsorbent. Although preliminary studies show encouraging results for Pb(II) removal, its potential for adsorbing dyes and other heavy metals requires further investigation (Xu et al., 2020).

Other forestry residues such as wood ash, pine needles, and eucalyptus leaves have demonstrated significant potential as inexpensive bioadsorbents for removing emerging pollutants like AMX from contaminated environments. Given that over 80% of AMX is excreted unmetabolized, its accumulation in sewage sludge poses a notable environmental risk when sludge is applied to soils. Recent studies have shown that wood ash exhibits superior adsorption efficiency, exceeding 80%, compared to other forestry residues. The adsorption capacity is attributed to components such as organic carbon, exchangeable cations, and amorphous aluminum and iron oxides, emphasizing the critical role of these materials in sustainable wastewater treatment (Cela-Dablanca et al., 2022).

Pine and oak bark are forestry waste products that serve as low-cost, eco-friendly bioadsorbents for removing emerging pollutants from the environment. Utilizing these byproducts supports sustainable waste management and promotes circular economy principles. A recent study investigated the adsorption of three common antibiotics—ciprofloxacin, clarithromycin, and trimethoprim—onto pine and oak bark across varying pH levels. Pine bark exhibited the highest adsorption potential, with Freundlich affinity coefficients ranging from 126.6 to 2979.1 Ln μmol^{1-n} kg^{-1}, whereas oak bark showed values between 283.9 and 806.9 Ln μmol^{1-n} kg^{-1}. Adsorption efficiency was notably influenced by pH, particularly for clarithromycin, which demonstrated the greatest mobility. These findings, reported by Rodríguez-López et al. (2023), emphasize the promise of forest-derived materials in mitigating pharmaceutical contamination, advancing zero-waste initiatives, and fostering a circular economy.

Forestry residues, especially wood-based materials, are economically viable and environmentally sustainable adsorbents for environmental remediation. Physical modifications are generally preferred over chemical treatments for industrial applications due to reduced safety risks. Organic pollutants are adsorbed primarily via van der Waals forces, hydrogen bonding, and pore entrapment, while inorganic contaminants are removed through electrostatic interactions, surface complexation, ion exchange, and pore filling. The adsorption efficiency depends largely on the material's porosity and surface area. Beyond pollutant removal, modified forestry residues can enhance soil fertility and contribute to air and water purification. Moreover, spent adsorbents have potential for reuse as catalysts, fertilizers, or construction materials, supporting a holistic approach to waste valorization (Keshvardoostchokami et al., 2023).

1.4.3 Algal and Microbial Biomass

One of the major environmental challenges is the removal of heavy metals from wastewater, and adsorption using bioadsorbents has gained popularity due to its efficiency and cost-effectiveness. Algal biomass, encompassing both macroalgae and microalgae, has emerged as a promising bioadsorbent owing to its high metal-binding capacity, abundance, and eco-friendly nature. The cell walls of algae contain diverse functional groups such as amino, hydroxyl, carboxyl, and sulfate groups, which facilitate the chemical binding of heavy metal ions. Adsorption by algal biomass generally follows pseudo-second-order kinetics and conforms to Langmuir and Freundlich isotherm models, indicating efficient and spontaneous metal uptake. Additionally, algae's adaptability and renewability make them ideal candidates for sustainable wastewater treatment. Similarly, microbial biomass—including bacteria, yeast, and fungi—leverages complex cellular structures to adsorb heavy metals effectively (Ameri et al., 2020).

Living biomasses such as bacteria, yeast, fungi, and algae play a crucial role in removing pollutants from air and water via biosorption and bioaccumulation. Biosorption involves passive adsorption of contaminants onto biomass surfaces, while bioaccumulation is the active uptake of pollutants by living cells. The presence of functional groups like amine, hydroxyl, and carboxyl in these biomasses allows efficient removal of heavy metals, dyes, pharmaceuticals, and persistent organic pollutants. Immobilization of bacterial biomass on polymers like chitosan enhances pollutant removal efficiency and facilitates biomass recovery. Fungal biomass, often available as industrial waste, shows high biosorption capacity and can be reused in continuous bioreactors, improving scalability. Algal biomass, with its rich cell wall composition, efficiently adsorbs metals such as Pb, Cd, Cu, and Zn, making these biomasses economical and sustainable alternatives for environmental remediation (Zahoor, 2024).

Fungal biomass is especially valued in wastewater treatment for dye removal, due to its composition of sugars, proteins, lipids, and functional groups including alcohols, carboxyls, and amines. Fungi such as *Trichoderma* sp., *Aspergillus niger*, *Rhizopus arrhizus*, and white-rot fungi have demonstrated excellent dye adsorption through chemisorption and electrostatic interactions. Functional groups like amine, carboxyl, and hydroxyl are critical in binding dye molecules, often achieving removal efficiencies exceeding 90%, depending on the fungal species and dye type. Algal biomass also exhibits outstanding biosorption capacity due to cell wall components such as chitin, alginic acid, and polysaccharides rich in phosphate, sulfate, and amino groups. Pretreatment methods, including magnetic nanoparticle incorporation and citric acid functionalization, further enhance dye adsorption performance. Bacterial biomass, particularly *Pseudomonas* spp., effectively treats reactive dye wastewater, with adsorption influenced by operational parameters like temperature and pH. Yeast, a fungal microbe, is also notable for its high pollutant accumulation and rapid decolorization abilities. Collectively, these biomasses offer environmentally sustainable, economical, and efficient solutions for dye removal in wastewater treatment (Aragaw and Bogale, 2021).

1.5 Environmental Importance

Adsorption is a fundamental technology in environmental engineering due to its high efficiency, ease of operation, and capability to remove diverse contaminants from water and air. It plays a vital role in treating organic pollutants, heavy metals, dyes, and various industrial effluents. The effectiveness of adsorption is significantly influenced by operational parameters such as pH, contact time, and adsorbent dosage (Alaqarbeh, 2021). Sustainable water management is further promoted through the reuse and regeneration of adsorbents, reducing waste, and operational costs.

The design and optimization of fixed-bed adsorption units rely heavily on adsorption modeling—encompassing isotherms, kinetics, and thermodynamics—to enhance scalability and practical application (Busetty, 2019). The adsorption mechanism, primarily involving physical and chemical interactions between contaminants and adsorbent surfaces, also critically affects removal efficiency.

Key factors influencing adsorption performance include surface area, pore structure, the presence of functional groups, and surface charge.

Advances in material science and nanotechnology have paved the way for engineered adsorbents with improved selectivity, higher capacity, and better regeneration capabilities. Moreover, integrating adsorption with complementary treatment methods such as membrane filtration or advanced oxidation processes offers a synergistic approach to effectively treat complex wastewater streams. Future research should focus on expanding these integrated technologies while ensuring their economic viability and environmental safety (Rathi and Kumar, 2021).

1.5.1 Waste Valorization

Waste valorization is emerging as a key sustainable approach to producing valuable materials like structured porous carbons. Using biosorbents derived from agricultural waste offers an eco-friendly and cost-effective solution for water contamination by adsorption. These biosorbents align with green chemistry principles and address the global challenge of clean water scarcity. Recent advances focus on enhancing adsorptive capacity by integrating biosorbents with magnetic or polymeric materials and through chemical activation. These modified biosorbents effectively remove emerging contaminants like heavy metals, dyes, polycyclic aromatic hydrocarbons, and oils. Their demonstrated efficiency and reusability in treating real industrial wastewater underline their practical potential, supporting sustainable development goals (Chong et al., 2023). Figure 1.2 illustrates the concept of waste valorization via bioadsorption.

Food waste digestate (FWD), often overlooked due to its high ash content, is gaining recognition as a valuable feedstock for ash-biochar composites. These composites exhibit exceptional adsorption capacity for MB at 1123.5 mg/g. The alkaline

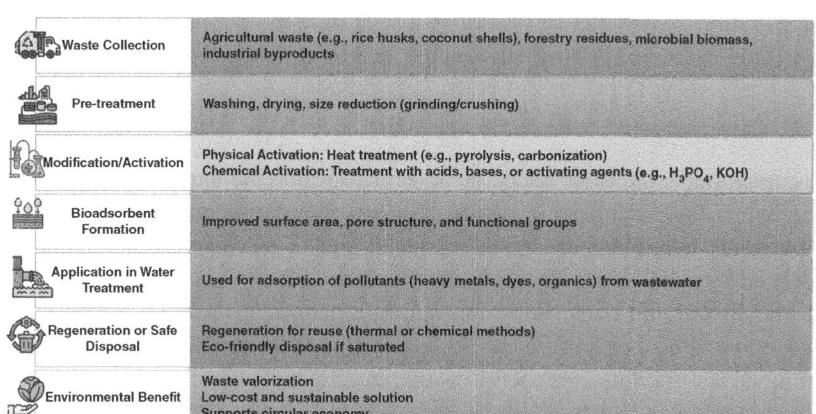

Figure 1.2 Waste valorization through bioadsorption.

environment and oxygen-rich interaction sites provided by the ash enhance adsorption, which is further improved by higher pyrolysis temperatures that increase surface area and π–π interactions. Thus, FWD emerges as a plentiful and sustainable raw material for wastewater treatment adsorbents (Yang et al., 2023).

Biomass-derived waste is a rich and economical source for producing ACs. Agricultural residues such as coconut shells, date seeds, olive stones, rice husks, cornstalks, and sugarcane bagasse have been extensively studied for AC synthesis due to their lignocellulosic content. These biobased ACs possess high porosity and surface areas (up to 2500 m^2/g), making them suitable for environmental applications like gas storage and pollutant adsorption. Advanced processing techniques, including chemical activation and flash air-carbonization, have further enhanced their properties and fostered circular economy practices (Kopac, 2023).

Traditional pyrolysis-activation methods often yield low-carbon porous materials with insufficient surface functional groups. Hydrothermal carbonization (180–250 °C) offers a superior alternative by converting wet biowaste into hydrochar with improved surface chemistry. For example, orange peel waste treated hydrothermally and activated with potassium oxalate ($K_2C_2O_4$) achieves a high surface area (~2130 m^2/g), substantial micropore volume (~1.12 cm^3/g), and significant CO_2 adsorption capacity (~6.67 mmol/g). $K_2C_2O_4$ provides safer handling and higher yields compared to conventional activators like KOH and K_2CO_3, enabling the production of heteroatom-doped carbons with tunable porosity for efficient CO_2 capture (Rehman et al., 2022).

Cigarette butts, one of the most common litter types, have been repurposed as precursors to synthesize ultra-high surface area porous carbons (up to 4300 m^2/g) via KOH activation. Optimizing synthesis parameters such as KOH-to-char ratio, activation temperature, heating rate, and time ensures reproducibility. These carbons function as superadsorbents due to their dual micro- and mesoporosity, demonstrating exceptional adsorption (>2.5 g/g) and rapid kinetics (85% adsorption within 15 minutes) (Dimitrakou et al., 2025).

1.5.2 Low-cost and Eco-friendly Approach

Dye-containing wastewater discharge poses significant environmental and health risks due to the carcinogenic nature of many dye compounds. Conventional treatment methods often fall short due to their high costs or inefficiencies. In this context, adsorption stands out as a promising technology because of its simplicity and low operational costs. Both natural and chemically modified agricultural wastes serve as cost-effective adsorbents and precursors for AC synthesis, highlighting an eco-friendly approach to waste recycling. However, scaling up these methods remains a challenge (Bushra et al., 2021).

Adsorption has also proven to be a highly practical and economical method for removing heavy metal ions from wastewater, even at trace concentrations.

The development of affordable and readily available adsorbents from industrial byproducts, animal residues, and agricultural wastes is crucial. These adsorbents inherently contain functional groups essential for metal ion binding, and chemical modifications often enhance their surface area and adsorption capacity compared to unmodified materials (Chakraborty et al., 2022).

Conventional adsorbents typically suffer from low surface areas and difficult recovery processes, issues that magnetic nanohybrids effectively address. Magnetizing biopolymer-based adsorbents such as chitosan improves pollutant interaction by enhancing dispersibility and structural stability. Green synthesis methods further reduce environmental hazards by employing eco-friendly precursors and processes, promoting sustainable and practical applications of these materials. To enable commercial-scale wastewater treatment, future research should focus on optimizing synthesis conditions, elucidating adsorption mechanisms, and scaling up production (Nasiri et al., 2024).

Eco-friendly and cost-effective adsorption techniques are increasingly vital for the removal of contaminants such as dyes from wastewater. While synthetic sorbents like zeolites, carbon-based materials, alumina, and nanoparticles are widely used, biosorption via biomass or agricultural waste is gaining popularity due to its affordability, sustainability, and biodegradability. Biomass materials rich in cellulose and lignin—such as apple stems, sawdust, and lemongrass leaves—exhibit innate adsorption capabilities, particularly for cationic dyes. Employing experimental design methods that consider parameter interactions enhances adsorption efficiency, positioning biosorption as a viable and environmentally friendly wastewater treatment strategy (Takabi et al., 2021). Figure 1.3 illustrates the production, application, regeneration, and disposal phases of bioadsorbents.

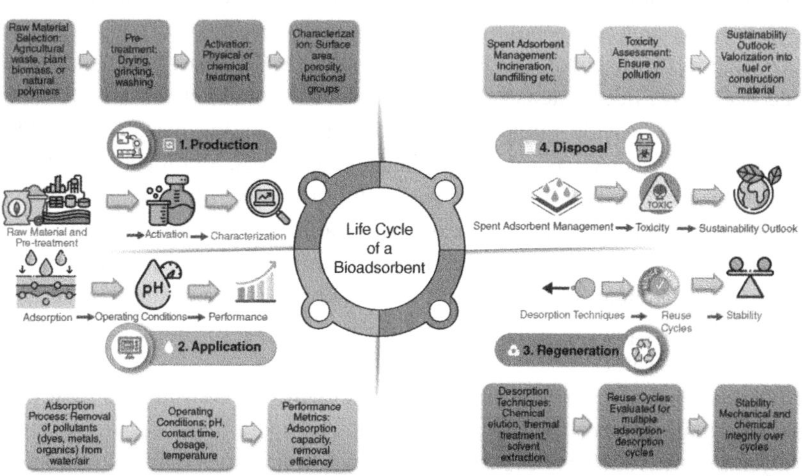

Figure 1.3 Life cycle of a bioadsorbent.

1.6 Scope and Applications

Globally, the development of engineered materials for effective water treatment remains predominantly at the bench-scale stage. Despite certain challenges limiting their scalability, adsorbent and photocatalytic materials have shown promising potential. In recent years, a variety of nanomaterials—including magnetic materials, AC, CNTs, graphene, graphene oxide, reduced graphene oxide, graphene quantum dots, carbon nanorods, carbon nano-onions, as well as zeolites, silica, and clay-based nanomaterials—have gained significant attention. These materials exhibit high specific surface areas and rapid kinetics, enabling effective photocatalytic degradation of pollutants even at very low concentrations, alongside efficient adsorption of organic dyes and heavy metal ions (Chenab et al., 2020).

1.6.1 Wastewater Treatment

Adsorption is a well-established, economical, and highly effective method for removing contaminants from wastewater, with removal efficiencies reaching up to 99.9%. This process relies on physico-chemical interactions, where solutes are bonded onto solid surfaces, facilitating mass transfer. Recognized by the US Environmental Protection Agency (USEPA) as one of the most reliable wastewater treatment techniques, adsorption commonly employs materials such as metal oxides, silica, alumina, clay, and especially AC due to its high porosity and large surface area. Recently, nanomaterials like TiO_2, Fe_3O_4, and ZnO have gained significant interest owing to their enhanced adsorption capacities, large surface areas, and recyclability, making them particularly suitable for advanced wastewater treatment even at low pollutant concentrations (Rashid et al., 2021).

According to De Gisi et al. (2016), agricultural byproducts such as banana and orange peels have been extensively studied as low-cost adsorbents due to their ability to remove a wide range of contaminants, including heavy metals like Cd(II) and Pb(II), as well as various acidic and basic dyes. Modification of these materials, for example treating orange peels with KCl or Mg^{2+} ions, improves their efficiency in removing metals like Cu(II), Zn(II), and Ni(II). Industrial byproducts such as red mud and blast furnace slag have also demonstrated promising removal capabilities for phenols, dyes, and heavy metals. Furthermore, adsorbents derived from sludge are frequently applied to capture antibiotics and pharmaceutical residues from wastewater.

Hydrogels represent a promising class of adsorbent materials due to their permeability, tunable properties, and environmental compatibility. They are typically employed in forms such as beads, films, and nanocomposites. The incorporation of functional particles within the hydrogel matrix enhances adsorption efficiency and improves mechanical strength, swelling behavior, pH responsiveness, and reusability. Nanocomposite hydrogels, which synergistically combine the properties of hydrogels and nanoparticles, exhibit superior performance in adsorbing dyes, heavy metals, and pharmaceuticals, making them highly suitable for sustainable wastewater treatment (Van Tran et al., 2018).

Water contamination by pollutants such as radionuclides, toxic metals, emerging contaminants, and dyes poses serious threats to ecosystems and human health. Adsorption has gained attention among water treatment methods because of its cost-effectiveness, rapid removal efficiency, and ability to eliminate contaminants even at trace levels. Unlike coagulation and flocculation, adsorption does not generate problematic secondary byproducts such as sewage sludge, which is a major environmental concern. Current research focuses on developing advanced adsorbent materials with high adsorption capacities and strong affinities for various pollutants, showing great promise for broader application in water and wastewater treatment (Tran, 2023).

1.6.2 Industrial Pollution Control

A major worldwide issue, water scarcity is made worse by growing water consumption in industry. Industrial processes discharge large volumes of wastewater that require effective treatment before disposal to prevent environmental harm. Adsorption has gained popularity as a remediation technique due to its simplicity, cost effectiveness, and efficiency. Recently, attention has shifted toward "green adsorbents"—low-cost materials derived from industrial and agricultural waste. These eco-friendly adsorbents efficiently remove a wide range of pollutants, promote waste valorization, and support the circular economy. By simultaneously reducing contaminants and recovering resources, they represent a sustainable option for treating industrial wastewater (Tony, 2022).

In 2010, global olive oil production reached 2.88 million metric tons, with 78.5% originating from European Union countries. The expansion of olive oil production has led to a significant increase in olive mill waste, creating serious environmental challenges due to inadequate waste management methods. Transforming these solid wastes into inexpensive adsorbents offers a promising approach to mitigating water pollution. Numerous studies have focused on the conversion, characterization, and adsorption capabilities of olive mill waste-derived adsorbents for removing aquatic contaminants. These valorized materials not only address water pollution but also facilitate sustainable waste utilization (Bhatnagar et al., 2014).

Waste generated from iron ore processing, primarily composed of quartz with minor amounts of feldspar, mica, and calcite, has shown potential as an antimicrobial adsorbent. Despite its rapid accumulation in tandem with increased iron production, only 20% of this waste is currently recycled. Iron ore waste has been explored as a precursor material and for environmental remediation, but its broader application remains limited. Therefore, developing innovative strategies for efficient reuse of iron ore waste is essential for pollution prevention and sustainable recycling (Fang et al., 2021).

Natural clay minerals and zeolites, belonging to the aluminosilicate family, possess unique structural features and surface active sites that vary according to their specific type and modification treatments, providing them with distinctive adsorption properties. This chapter explores how acidic and basic modifications influence the adsorption behavior of naturally occurring aluminosilicates in aqueous solutions. It highlights changes in surface and structural characteristics, the role of acid/base active sites as evidenced by catalytic conversion of 2-methylbut-3-yn-2-ol, and the adsorption of pollutants such as ammonium chloride, formaldehyde, and acetic acid. The main adsorption mechanisms include ion exchange, chemisorption, and hydrogen bonding (Novikova and Belchinskaya, 2016).

Heavy metals, such as Cd, Cu, Zn, and Pb, are toxic industrial pollutants harmful even at trace levels. Biochar, produced from diverse biomass feedstocks, has demonstrated high efficiency in adsorbing heavy metals through mechanisms like physical adsorption, precipitation, ion exchange, and electrostatic attraction. Research indicates that biochar can reduce Cr(VI) to the less toxic Cr(III) form and bind metals via functional groups forming stable complexes. The feedstock composition significantly influences biochar's adsorption performance due to variability in active sites and mineral content. Consequently, biochar is recognized as an economical and promising adsorbent for industrial wastewater treatment (Ambaye et al., 2021). Table 1.2 summarizes the application of bioadsorbents in industrial effluent treatment, detailing their sources, target pollutants, and removal performance.

1.7 Research Trends and Future Outlook

Adsorbents derived from carbon, biomass, and biowaste show great potential for removing heavy metals; however, their practical application faces significant challenges. Most research to date has focused on single-metal removal in controlled laboratory settings, while industrial wastewater is typically more complex. To scale up these technologies, improvements are needed in chemical usage, regeneration efficiency, and overall economic feasibility. Table 1.3 outlines the key challenges and emerging research opportunities in bioadsorption technologies. For example, cellulose hydrogels tend to lose adsorption capacity after regeneration,

Table 1.2 Industrial effluent treatment using bioadsorbents: sources, targets, and performance.

Type of industrial effluent	Bioadsorbent used	Source biomass	Target contaminants	Adsorption capacity (mg/g)	Treatment conditions (pH, Temp)	Removal efficiency (%)	Reference
Beverage industrial wastewater	Chitosan coated acid-treated coconut shell carbon	Coconut shell	Zn(II)	60.41	pH—6, 25 °C	90%	Amuda et al. (2007)
Beverage industrial wastewater	Chitosan coated coconut shell carbon	Coconut shell	Zn(II)	50.93	pH—6, 25 °C	58%	Amuda et al. (2007)
Beverage industrial wastewater	Acid-treated coconut shell carbon	Coconut shell	Zn(II)	45.14	pH—6, 25 °C	52%	Amuda et al. (2007)
Textile industrial wastewater	Flyash	Industry waste	Methylene blue	2.926	pH—7.5, 40 °C	85.7%	Rao and Rao (2006)
Textile industrial wastewater	Activated carbon	Industry waste	Methylene blue	13.35	pH—7.5, 50 °C	93%	Rao and Rao (2006)
Textile industrial wastewater	Flyash	Industry waste	Congo red dye	4.125	pH—7.5, 30 °C	96%	Rao and Rao (2006)
Textile industrial wastewater	Activated carbon	Industry waste	Congo red dye	15.80	pH—7.5, 30 °C	98.5%	Rao and Rao (2006)
Electroplating Industry	Dolochar	Industry waste	Cr(VI)	0.904	pH—2, 30 °C	95%	Tiadi et al. (2017)

(Continued)

Table 1.2 *(Continued)*

Type of industrial effluent	Bioadsorbent used	Source biomass	Target contaminants	Adsorption capacity (mg/g)	Treatment conditions (pH, Temp)	Removal efficiency (%)	Reference
Electroplating Industry	Activated carbon	Pine sawdust	Cr(VI)	2.14	pH—2, 40 °C	95%	Uysal and Ar (2007)
Electroplating Industry	Once-activated Moso bamboo carbon	Moso bamboo	Pb(II)	0.675	pH—6.07, 25 °C	99.87%	Lo et al. (2012)
Electroplating Industry	Once-activated Moso bamboo carbon	Moso bamboo	Cu(II)	0.401	pH—7.39, 25 °C	100%	Lo et al. (2012)
Electroplating Industry	Twice-activated Moso bamboo carbon	Moso bamboo	Cu(II)	0.638	pH—7.04, 25 °C	100%	Lo et al. (2012)
Electroplating Industry	Twice-activated Moso bamboo carbon	Moso bamboo	Pb(II)	0.400	pH—7.91, 25 °C	100%	Lo et al. (2012)
Electroplating Industry	Once-activated Ma bamboo carbon	Ma bamboo	Pb(II)	0.404	pH—10.02, 25 °C	100%	Lo et al. (2012)
Electroplating Industry	Once-activated Ma bamboo carbon	Ma bamboo	Cu(II)	0.634	pH—9.60, 25 °C	100%	Lo et al. (2012)
Electroplating Industry	Twice-activated Ma bamboo carbon	Ma bamboo	Pb(II)	0.397	pH—7.95, 25 °C	100%	Lo et al. (2012)
Electroplating Industry	Twice-activated Ma bamboo carbon	Ma bamboo	Cu(II)	0.398	pH—8.26, 25 °C	100%	Lo et al. (2012)

Table 1.3 Key challenges and emerging research opportunities in bioadsorption technologies.

S. No.	Current challenge	Research opportunity	Description	Reference
1	Limited real wastewater studies	Testing with actual industrial effluents	Most studies use synthetic wastewater with single contaminants	Foo and Hameed (2010)
2	Low regeneration efficiency	Development of stable, regenerable bioadsorbents	Many bioadsorbents lose efficiency after a few cycles	Babel and Kurniawan (2003)
3	Poor scalability	Design of pilot and full-scale systems	Scaling lab processes remains a major barrier to commercialization	Inyang et al. (2016)
4	High energy/chemical inputs	Eco-friendly synthesis and activation techniques	Some bioadsorbents require costly chemical or thermal treatments	Crini and Lichtfouse (2019)
5	Incomplete understanding of adsorption mechanisms	Advanced characterization and modeling tools	Detailed molecular-level mechanisms are often unclear	Mohan and Pittman (2006)
6	Lack of cost analysis	Comprehensive techno-economic evaluations	Few studies consider capital and operational costs	Sud et al. (2008)
7	Uncertainty of post-use adsorbent disposal	Development of safe disposal or reuse strategies	Spent adsorbents may contain toxic metals, posing disposal risks	Ahmad et al. (2014)
8	Variable performance in multi-metal systems	Investigation of competitive adsorption dynamics	Real wastewater contains multiple contaminants competing for adsorption sites	Wang and Chen (2009)

(Continued)

Table 1.3 *(Continued)*

S. No.	Current challenge	Research opportunity	Description	Reference
9	Insufficient long-term performance data	Conduct of long-term adsorption and regeneration studies	Most research reports only short-term adsorption behavior	Demirbas (2008)
10	Limited adsorbent selectivity	Functionalization to enhance selectivity toward specific heavy metals	Need to target specific metals like Pb^{2+}, Cd^{2+}, As^{3+}, etc., more efficiently	Guo and Wang (2021)
11	Environmental impact of modification methods	Greener surface modification methods using bio-based chemicals	Reducing use of hazardous chemicals during adsorbent preparation	Hubbe et al. (2011)
12	Poor performance in continuous systems	Transition from batch to column or continuous flow studies	Most bioadsorption research is in batch mode, not representative of industrial operations	Tan et al. (2015)

and graphene-based adsorbents are limited by the high cost of raw materials. Future research should prioritize large-scale, eco-friendly applications that minimize chemical inputs. Additionally, the toxicity and environmental impact of metal-laden adsorbents require further investigation.

To realize practical wastewater treatment solutions, future studies must address removal efficiency, operational costs, energy consumption, and adsorbent recyclability. Current adsorption research aims to develop high-performance, sustainable, and cost-effective adsorbents from industrial, agricultural, and natural waste sources. Advances in hybrid materials, surface modification, and nanotechnology have enhanced adsorption effectiveness and selectivity toward various contaminants. Future trends include real-time monitoring of adsorption processes, improved regeneration techniques for adsorbent reuse, and the integration of machine learning to optimize adsorbent design. Emphasizing the circular economy, waste-derived adsorbents contribute to waste-to-resource initiatives. Combining adsorption with other treatment methods is expected to yield more reliable and environmentally sustainable solutions for air and water purification.

1.8 Conclusion

This chapter provides a fundamental overview of adsorption and various types of adsorbents, with a particular focus on bioadsorbents derived from microbial, forestry, and agricultural biomass. It highlights the environmental benefits of bioadsorbents as cost-effective and sustainable alternatives to synthetic materials, underscoring their crucial roles in pollution prevention and waste valorization. The chapter explores a broad spectrum of applications, especially in wastewater treatment and industrial pollution control, while identifying current research gaps and emerging trends.

Carbon-based materials, agricultural residues, and biowaste-derived adsorbents show considerable potential for the efficient removal of heavy metals, offering both environmental sustainability and economic viability. However, challenges related to scalability, regeneration efficiency, cost-effectiveness, and toxicity assessment must be addressed to enable wider practical implementation. Future research should prioritize safe disposal of spent adsorbents, performance optimization in complex wastewater systems, and environmentally friendly modification methods. Overcoming these hurdles will significantly advance the role of bioadsorbents in sustainable wastewater treatment and environmental protection.

References

Aguilar-Rosero, J., Urbina-López, M. E., Rodríguez-González, B. E., León-Villegas, S. X., Luna-Cruz, I. E., & Cárdenas-Chávez, D. L. (2022). Development and characterization of bioadsorbents derived from different agricultural wastes for water reclamation: a review. Applied Sciences, 12(5), 2740.

Ahmad, M., Rajapaksha, A. U., Lim, J. E., Zhang, M., Bolan, N., Mohan, D., Vithanage, M. S., Lee, S. S., & Ok, Y. S. (2014). Biochar as a sorbent for contaminant management in soil and water: a review. Chemosphere, 99, 19–33.

Ajmal, M., Khan, A. H., Ahmad, S., & Ahmad, A. (2000). Role of sawdust in the removal of copper (II) from industrial wastes. Water Research, 34(17), 4261–4272.

Akindolie, M. S., & Choi, H. J. (2023). Acid modification of lignocellulosic derived material for dye and heavy metals removal: a review. Environmental Engineering Research, 28(2), 210574.

Alaqarbeh, M. (2021). Adsorption phenomena: definition, mechanisms, and adsorption types: short review. RHAZES: Green and Applied Chemistry, 13, 43–51.

Ambaye, T. G., Vaccari, M., van Hullebusch, E. D., Amrane, A., & Rtimi, S. (2021). Mechanisms and adsorption capacities of biochar for the removal of organic and inorganic pollutants from industrial wastewater. International Journal of Environmental Science and Technology, 18(10), 3273–3294.

Ameri, A., Tamjidi, S., Dehghankhalili, F., Farhadi, A., & Saati, M. A. (2020). Application of algae as low cost and effective bio-adsorbent for removal of heavy metals from wastewater: a review study. Environmental Technology Reviews, 9(1), 85–110.

Amuda, O. S., Giwa, A., & Bello, I. A. (2007). Removal of heavy metal from industrial wastewater using modified activated coconut shell carbon. Biochemical Engineering Journal, 36(2), 174–181.

Annadurai, G., Juang, R. S., & Lee, D. J. (2002). Use of cellulose-based wastes for adsorption of dyes from aqueous solutions. Journal of Hazardous Materials, 92(3), 263–274.

Aragaw, T. A., & Bogale, F. M. (2021). Biomass-based adsorbents for removal of dyes from wastewater: a review. Frontiers in Environmental Science, 9, 764958.

Awad, A. M., Jalab, R., Benamor, A., Nasser, M. S., Ba-Abbad, M. M., El-Naas, M., & Mohammad, A. W. (2020). Adsorption of organic pollutants by nanomaterial-based adsorbents: an overview. Journal of Molecular Liquids, 301, 112335.

Babel, S., & Kurniawan, T. A. (2003). Low-cost adsorbents for heavy metals uptake from contaminated water: a review. Journal of Hazardous Materials, 97(1–3), 219–243.

Bhatnagar, A., Kaczala, F., Hogland, W., Marques, M., Paraskeva, C. A., Papadakis, V. G., & Sillanpää, M. (2014). Valorization of solid waste products from olive oil industry as potential adsorbents for water pollution control—a review. Environmental Science and Pollution Research, 21, 268–298.

Brown, P. A., Gill, S. A., & Allen, S. J. (2000). Metal removal from wastewater using peat. Water Research, 34(16), 3907–3916.

Busetty, S. (2019). Environmental treatment technologies: adsorption. In: Hussain, C. (eds) Handbook of Environmental Materials Management (pp. 1367–1397). Cham: Springer.

Bushra, R., Mohamad, S., Alias, Y., Jin, Y., & Ahmad, M. (2021). Current approaches and methodologies to explore the perceptive adsorption mechanism of dyes on low-cost agricultural waste: a review. Microporous and Mesoporous Materials, 319, 111040.

Cela-Dablanca, R., Barreiro, A., López, L. R., Santás-Miguel, V., Arias-Estévez, M., Núñez-Delgado, A., Álvarez-Rodríguez, E., & Fernández-Sanjurjo, M. J. (2022). Potential of low-cost bio-adsorbents to retain amoxicillin in contaminated water. Environmental Research, 213, 113621.

Chakraborty, R., Asthana, A., Singh, A. K., Jain, B., & Susan, A. B. H. (2022). Adsorption of heavy metal ions by various low-cost adsorbents: a review. International Journal of Environmental Analytical Chemistry, 102(2), 342–379.

Chenab, K. K., Sohrabi, B., Jafari, A., & Ramakrishna, S. (2020). Water treatment: functional nanomaterials and applications from adsorption to photodegradation. Materials Today Chemistry, 16, 100262.

Chong, Z. T., Soh, L. S., & Yong, W. F. (2023). Valorization of agriculture wastes as biosorbents for adsorption of emerging pollutants: modification, remediation and industry application. Results in Engineering, 17, 100960.

Crini, G., & Lichtfouse, E. (2019). Advantages and disadvantages of techniques used for wastewater treatment. Environmental Chemistry Letters, 17, 145–155.

Crini, G., Lichtfouse, E., Wilson, L. D., & Morin-Crini, N. (2019). Conventional and non-conventional adsorbents for wastewater treatment. Environmental Chemistry Letters, 17, 195–213.

Dąbrowski, A. (2001). Adsorption—from theory to practice. Advances in Colloid and Interface Science, 93(1–3), 135–224.

De Gisi, S., Lofrano, G., Grassi, M., & Notarnicola, M. (2016). Characteristics and adsorption capacities of low-cost sorbents for wastewater treatment: a review. Sustainable Materials and Technologies, 9, 10–40.

Demirbas, A. (2008). Heavy metal adsorption onto agro-based waste materials: a review. Journal of Hazardous Materials, 157(2–3), 220–229.

Dimitrakou, A., Koutselas, I. B., Karakassides, M. A., Salmas, C. E., & Dimos, K. (2025). Waste valorization towards hierarchical porous carbon with ultra-high surface area and dye adsorption efficiency. International Journal of Environmental Science and Technology, 22(6), 4027–4042.

Fakhri, V., Jafari, A., Vahed, F. L., Su, C. H., & Pirouzfar, V. (2023). Polysaccharides as eco-friendly bio-adsorbents for wastewater remediation: current state and future perspective. Journal of Water Process Engineering, 54, 103980.

Fang, N., He, Q., Sheng, L., Xi, Y., Zhang, L., Liu, H., & Cheng, H. (2021). Toward broader applications of iron ore waste in pollution control: adsorption of norfloxacin. Journal of Hazardous Materials, 418, 126273.

Foo, K. Y., & Hameed, B. H. (2009). A short review of activated carbon derived from lignocellulosic biomass for water treatment. Bioresource Technology, 99(4), 1117–1128.

Foo, K. Y., & Hameed, B. H. (2010). An overview of dye removal via activated carbon adsorption process. Desalination and Water Treatment, 19(1–3), 255–274.

Gadd, G. M. (2009). Biosorption: critical review of scientific rationale, environmental importance and significance for pollution treatment. Journal of Chemical Technology & Biotechnology: International Research in Process, Environmental & Clean Technology, 84(1), 13–28.

Guo, X., & Wang, J. (2021). Projecting the sorption capacity of heavy metal ions onto microplastics in global aquatic environments using artificial neural networks. Journal of Hazardous Materials, 402, 123709.

Hajjizadeh, M., Ganjidoust, H., & Farsad, F. (2020). Review the types of adsorbents in water and wastewater treatment. Journal of Environmental Science Studies, 5(4), 3173–3182.

Hamad, H. N., & Idrus, S. (2022). Recent developments in the application of bio-waste-derived adsorbents for the removal of methylene blue from wastewater: a review. Polymers, 14(4), 783.

Harshala, K., & Wagh, N. D. (2022). Use of agricultural waste-based biosorbents for the removal of heavy metals from aqueous solution: a review. Nature Environment & Pollution Technology, 21(3), 1003–1014.

Hossain, M. M., Islam, T., Jalil, M. A., Rakibuzzaman, S. M., Surid, S. M., Zabed, M. R. I., Talukder, A., & Hossain, S. (2024). Advancements of eco-friendly natural antimicrobial agents and their transformative role in sustainable textiles. SPE Polymers, 5(3), 241–276.

Hubbe, M. A., Hasan, S. H., & Ducoste, J. J. (2011). Cellulosic substrates for removal of pollutants from aqueous systems: a review. 1. Metals. Bio Resources, 6(2), 2161–2287.

Hussain, M. K., Khatoon, S., Nizami, G., Fatma, U. K., Ali, M., Singh, B., Quraishi, A., Ail Assiri, M., Ahamad, S., & Saquib, M. (2024). Unleashing the power of bio-adsorbents: efficient heavy metal removal for sustainable water purification. Journal of Water Process Engineering, 64, 105705.

Inyang, M. I., Gao, B., Yao, Y., Xue, Y., Zimmerman, A., Mosa, A., Pullammanappallil, P., Ok, Y. S., & Cao, X. (2016). A review of biochar as a low-cost adsorbent for aqueous heavy metal removal. Critical Reviews in Environmental Science and Technology, 46(4), 406–433.

Kali, A., Amar, A., Loulidi, I., Jabri, M., Hadey, C., Lgaz, H., Alrashdi, A., & Boukhlifi, F. (2024). Characterization and adsorption capacity of four low-cost adsorbents based on coconut, almond, walnut, and peanut shells for copper removal. Biomass Conversion and Biorefinery, 14(3), 3655–3666.

Keshvardoostchokami, M., Braghiroli, F. L., Neculita, C. M., & Koubaa, A. (2023). Advances in modified wood-based adsorbents for contaminant removal: valorization methods, modification mechanisms, and environmental applications. Current Forestry Reports, 9(6), 444–460.

Kopac, T. (2023). Current overview of the valorization of bio-wastes for adsorbed natural gas applications. Carbon Letters, 33(6), 1519–1547.

Kumar, P. S., & Balasubramani, S. R. (2025). Bio-adsorbents from biomass for environmental remediation. In Biomass for Environmental Remediation (pp. 41–54). Elsevier.

Kyzas, G. Z., & Kostoglou, M. (2014). Green adsorbents for wastewaters: a critical review. Materials, 7(1), 333–364.

Lin, S. H., & Juang, R. S. (2009). Adsorption of phenol and its derivatives from water using synthetic resins and low-cost natural adsorbents: a review. Journal of Environmental Management, 90(3), 1336–1349.

Liu, L., Gao, Z. Y., Su, X. P., Chen, X., Jiang, L., & Yao, J. M. (2015). Adsorption removal of dyes from single and binary solutions using a cellulose-based bioadsorbent. ACS Sustainable Chemistry & Engineering, 3(3), 432–442.

Lo, S. F., Wang, S. Y., Tsai, M. J., & Lin, L. D. (2012). Adsorption capacity and removal efficiency of heavy metal ions by Moso and Ma bamboo activated carbons. Chemical Engineering Research and Design, 90(9), 1397–1406.

Malkoc, E., & Nuhoglu, Y. (2005). Investigations of nickel (II) removal from aqueous solutions using tea factory waste. Journal of Hazardous Materials, 127(1–3), 120–128.

Masoumi, H. (2024). Types of polymeric adsorbents. In: Polymeric Adsorbents (pp. 1–46). Elsevier.

Mathew, B. B., Jaishankar, M., Biju, V. G., & Beeregowda, K. N. (2016). Role of bioadsorbents in reducing toxic metals. Journal of Toxicology, 2016(1), 4369604.

Mathew, S., Soans, J. C., Rachitha, R., Shilpalekha, M. S., Gowda, S. G. S., Juvvi, P., & Chakka, A. K. (2023). Green technology approach for heavy metal adsorption by agricultural and food industry solid wastes as bio-adsorbents: a review. Journal of Food Science and Technology, 60(7), 1923–1932.

Meftah, S., Meftah, K., Drissi, M., Radah, I., Malous, K., Amahrous, A., Chahid, A., Tamri, T., Rayyad, A., Darkaoui, B., Hanine, S., El-Hassan, O., & Bouyazza, L. (2025). Heavy metal polluted water: effects and sustainable treatment solutions using bio-adsorbents aligned with the SDGs. Discover Sustainability, 6(1), 137.

Meftah, S., Meftah, K., Lamkhanter, H., Taib, M., Amahrous, A., Assadi, F., Oukani, E. H., & Lahboub, B. (2024). The main types of derivatives of plant matter and agricultural waste used as bio-adsorbents for the removal of heavy metals and dyes: a review. Egyptian Journal of Chemistry, 67(9), 185–207.

Menéndez-Díaz, J. A., & Martín-Gullón, I. (2006). Types of carbon adsorbents and their production. In: Interface Science and Technology (Vol. 7, pp. 1–47). Elsevier.

Mhemeed, A. H. (2018). A general overview on the adsorption. Indian Journal of Natural Sciences, 9(51), 16127–16131.

Míguez-González, A., Cela-Dablanca, R., Barreiro, A., Rodríguez-López, L., Rodríguez-Seijo, A., Arias-Estévez, M., Núñez-Delgado, A., J. Fernández-Sanjurjo, M., Castillo-Ramos, V., & Álvarez-Rodríguez, E. (2023). Adsorption of antibiotics on bio-adsorbents derived from the forestry and agro-food industries. Environmental Research, 233, 116360.

Mohan, D., & Pittman Jr, C. U. (2006). Activated carbons and low cost adsorbents for remediation of tri-and hexavalent chromium from water. Journal of Hazardous Materials, 137(2), 762–811.

Moradi, O., & Sharma, G. (2021). Emerging novel polymeric adsorbents for removing dyes from wastewater: a comprehensive review and comparison with other adsorbents. Environmental Research, 201, 111534.

Mubarak, A. A., Ilyas, R. A., Nordin, A. H., Ngadi, N., & Alkbir, M. F. M. (2024). Recent developments in sugarcane bagasse fibre-based adsorbent and their potential industrial applications: a review. International Journal of Biological Macromolecules, 277, 134165.

Nasiri, A., Golestani, N., Rajabi, S., & Hashemi, M. (2024). Facile and green synthesis of recyclable, environmentally friendly, chemically stable, and cost-effective magnetic nanohybrid adsorbent for tetracycline adsorption. Heliyon, 10(2), e24179.

Novikova, L., & Belchinskaya, L. (2016). Adsorption of industrial pollutants by natural and modified aluminosilicates. In: Clays, Clay Minerals and Ceramic Materials Based on Clay Minerals (pp. 89–128). IntechOpen.

Núñez-Delgado, A., Álvarez-Rodríguez, E., Fernández-Sanjurjo, M. J., Arias-Estévez, M., Nóvoa-Muñoz, J. C., Fernández-Calviño, D., & Conde-Cid, M. (2020). By-Products from forest activities as low-cost sorbents for bioremediation of effluents and other polluted media. In: Combined Application of Physico-Chemical & Microbiological Processes for Industrial Effluent Treatment Plant (pp. 1–14). Springer.

Nyika, J., & Dinka, M. O. (2023). The potential of reusing fruit bio-adsorbents for water purification: a minireview. Materials Today: Proceedings.

Pourhakkak, P., Taghizadeh, A., Taghizadeh, M., Ghaedi, M., & Haghdoust, S. (2021). Fundamentals of adsorption technology. In: Interface Science and Technology (Vol. 33, pp. 1–70). Elsevier.

Quintero-Jaramillo, J. A., Carrero-Mantilla, J. I., & Sanabria-González, N. R. (2021). A review of caffeine adsorption studies onto various types of adsorbents. The Scientific World Journal, 2021(1), 9998924.

Raji, Z., Karim, A., Karam, A., & Khalloufi, S. (2023, September). Adsorption of heavy metals: mechanisms, kinetics, and applications of various adsorbents in wastewater remediation—a review. Waste, 1(3), 775–805.

Rao, V. B., & Rao, S. R. M. (2006). Adsorption studies on treatment of textile dyeing industrial effluent by flyash. Chemical Engineering Journal, 116(1), 77–84.

Rashid, R., Shafiq, I., Akhter, P., Iqbal, M. J., & Hussain, M. (2021). A state-of-the-art review on wastewater treatment techniques: the effectiveness of adsorption method. Environmental Science and Pollution Research, 28, 9050–9066.

Rathi, B. S., & Kumar, P. S. (2021). Application of adsorption process for effective removal of emerging contaminants from water and wastewater. Environmental Pollution, 280, 116995.

Rehman, A., Nazir, G., Rhee, K. Y., & Park, S. J. (2022). Valorization of orange peel waste to tunable heteroatom-doped hydrochar-derived microporous carbons for selective CO2 adsorption and separation. Science of the Total Environment, 849, 157805.

Rodríguez-López, L., Santás-Miguel, V., Cela-Dablanca, R., Pérez-Rodríguez, P., Núnez-Delgado, A., Álvarez-Rodríguez, E., Rodríguez-Seijo, A., & Arias-Estévez, M. (2023). Valorization of forest by-products as bio-adsorbents for emerging contaminants. Journal of Environmental Chemical Engineering, 11(6), 111437.

Romera, E., González, F., Ballester, A., Blázquez, M. L., & Muñoz, J. A. (2007). Comparative study of biosorption of heavy metals using different types of algae. Bioresource Technology, 98(17), 3344–3353.

Solangi, N. H., Kumar, J., Mazari, S. A., Ahmed, S., Fatima, N., & Mubarak, N. M. (2021). Development of fruit waste derived bio-adsorbents for wastewater treatment: a review. Journal of Hazardous Materials, 416, 125848.

Sud, D., Mahajan, G., & Kaur, M. P. (2008). Agricultural waste material as potential adsorbent for sequestering heavy metal ions from aqueous solutions–a review. Bioresource Technology, 99(14), 6017–6027.

Takabi, A. S., Shirani, M., & Semnani, A. (2021). Apple stem as a high performance cellulose based biosorbent for low cost and eco-friendly adsorption of crystal violet from aqueous solutions using experimental design: mechanism, kinetic and thermodynamics. Environmental Technology & Innovation, 24, 101947.

Tan, I. A. W., Ahmad, A. L., & Hameed, B. H. (2008). Adsorption of basic dye on high-surface-area activated carbon prepared from coconut husk: equilibrium, kinetic, and thermodynamic studies. Journal of Hazardous Materials, 154(1–3), 337–346.

Tan, X., Liu, Y., Zeng, G., Wang, X., Hu, X., Gu, Y., & Yang, Z. (2015). Application of biochar for the removal of pollutants from aqueous solutions. *Chemosphere*, 125, 70–85.

Tiadi, N., Mohanty, M., Mohanty, C. R., & Panda, H. P. (2017). Studies on adsorption behavior of an industrial waste for removal of chromium from aqueous solution. South African Journal of Chemical Engineering, 23(1), 132–138.

Tony, M. A. (2022). Low-cost adsorbents for environmental pollution control: a concise systematic review from the prospective of principles, mechanism and their applications. Journal of Dispersion Science and Technology, 43(11), 1612–1633.

Tran, H. N. (2023). Adsorption technology for water and wastewater treatments. Water, 15(15), 2857.

Turner, J. S., Ebner, A. D., & Ritter, J. A. (2024). Five definitions of adsorption and their relevance to the formulation of dynamic mass balances in gas adsorption columns. Adsorption, 30(8), 2267–2281.

Uchegbulam, I., Momoh, E. O., & Agan, S. A. (2022). Potentials of palm kernel shell derivatives: a critical review on waste recovery for environmental sustainability. Cleaner Materials, 6, 100154.

Uysal, M., & Ar, I. (2007). Removal of Cr (VI) from industrial wastewaters by adsorption: Part I: Determination of optimum conditions. Journal of Hazardous Materials, 149(2), 482–491.

Van Tran, V., Park, D., & Lee, Y. C. (2018). Hydrogel applications for adsorption of contaminants in water and wastewater treatment. Environmental Science and Pollution Research, 25, 24569–24599.

Volesky, B. (2001). Detoxification of metal-bearing effluents: biosorption for the next century. Hydrometallurgy, 59(2–3), 203–216.

Wang, J., & Chen, C. (2009). Biosorbents for heavy metals removal and their future. Biotechnology Advances, 27(2), 195–226.

Wang, J., & Guo, X. (2023). Adsorption kinetics and isotherm models of heavy metals by various adsorbents: an overview. Critical Reviews in Environmental Science and Technology, 53(21), 1837–1865.

Wu, J., Sun, X., Wu, J., & Yu, X. (2025). Eggshell-enhanced biochar with in-situ formed $CaO/Ca(OH)_2$ for efficient removal of Pb^{2+} and Cd^{2+} from wastewater: performance and mechanistic insights. Separation and Purification Technology, 354, 129352.

Xu, S., Xiao, H., Jiang, X., Liu, L., Cao, M., & Wang, Z. (2020). Exploring of toxic Pb (II) removal by low-cost bio-adsorbent of camphor leaf forestry waste after camphor oil extraction. Environmental Science and Pollution Research, 27, 43625–43637.

Yang, C., Wu, H., Cai, M., Li, Y., Guo, C., Han, Y., Zhang, Y., & Song, B. (2023). Valorization of food waste digestate to ash and biochar composites for high performance adsorption of methylene blue. Journal of Cleaner Production, 397, 136612.

Zahoor, I. (2024). Biomass-based biosorbents potential for wastewater treatment. Canadian Journal of Scientific Research, 1(1), 1–20.

2

Biomass-derived Adsorbents

2.1 Introduction

Petroleum pollutants generated by petrochemical operations have had a significant negative impact on the environment and human health in recent decades. These contaminants pose serious threats to wastewater systems, necessitating the development of effective remediation strategies. Among these, adsorbents have garnered attention as a practical solution for removing organic pollutants. Biomass-derived adsorbents, in particular, are showing considerable promise due to their low cost and environmental sustainability. Research indicates that these materials can significantly reduce treatment expenses while offering reliable pollutant removal from aquatic environments, often matching the performance of more expensive techniques such as ion exchange (Vahabisani and An, 2021).

Currently, a wide variety of adsorbents derived from industrial and agricultural waste—including bacteria, seaweeds, crab shells, wool, rice straw, coconut husks, and spent tea leaves—are employed for pollutant removal. These low-cost sorbents, either naturally occurring or generated as byproducts from other industries, offer both technologically and economically viable alternatives for effluent treatment through sorption or biosorption processes. Adsorbents are typically categorized as conventional or nonconventional based on their origin and properties. The development of diverse adsorbent materials reflects their growing importance in cost-effective wastewater treatment technologies (Hashmi et al., 2024).

Biomass is predominantly composed of organic matter, with a smaller fraction consisting of inorganic components such as minerals, which vary depending on the source. In plant-based biomass, cellulose, hemicellulose, and lignin are the major structural constituents of the cell wall. Characterization of biomass often involves proximate and ultimate analyses to determine moisture content, volatile matter, ash content, and elemental composition. Additionally, ash samples are molded and subjected to high temperatures in either reducing or oxidizing atmospheres to evaluate ash melting behavior—an essential property for gasification and combustion processes (de Jong, 2014).

Thorough characterization is crucial to understanding the adsorption mechanisms of biomass-derived adsorbents. Two key factors influencing adsorption efficiency are the surface functional groups and the porous structure. Textural properties such as surface area, pore size distribution, and pore volume are typically assessed using physisorption techniques, including nitrogen (N_2) adsorption–desorption at −196 °C. For ultra microporous materials, carbon dioxide (CO_2) adsorption at 0 °C is preferred, as it more effectively penetrates smaller pores. Furthermore, surface chemistry is characterized using techniques like Fourier transform infrared (FTIR) spectroscopy, which identifies functional groups, and X-ray photoelectron spectroscopy (XPS), which provides insights into elemental composition, chemical states, and heteroatom distribution on the adsorbent surface (Bedia et al., 2018).

Physical, chemical, and biological pretreatment techniques each have inherent limitations, including restricted applicability, high energy or chemical requirements, and relatively low efficiency. To address these challenges, combined pretreatment technologies—integrating two or more methods—have gained attention due to their enhanced conversion efficiency and synergistic effects. These approaches disrupt the complex lignocellulosic matrix, thereby improving cellulose accessibility and facilitating enzymatic hydrolysis and fermentation. Moreover, they offer greater flexibility for treating diverse biomass types, such as wood residues and agricultural byproducts. Systematic research into combined pretreatment strategies is therefore essential for the effective utilization of biomass (Chen et al., 2024).

In the field of environmental remediation, activated carbon derived from biomass waste has demonstrated significant potential as an adsorbent, particularly in addressing pollution resulting from large-scale contaminant discharge. Several factors influence its performance, including the type of biomass precursor, production method, activation technique (chemical or physical), and processing conditions. However, comprehensive evaluations of these parameters remain limited. For large-scale applications, it is crucial to optimize key characteristics such as yield, surface area, total pore volume, and pore size distribution. Further research is needed to refine production techniques and deepen the understanding of adsorbent–adsorbate interactions, thereby enabling the development of more sustainable and efficient activated carbon materials (Gayathiri et al., 2022).

This chapter investigates the potential of lignocellulosic biomass, algal biomass, and fruit/vegetable peels as raw materials for the production of adsorbents. It highlights their structural features, functional groups, and chemical composition, along with the importance of pretreatment methods such as grinding, washing, and acid or alkali treatments. In addition to assessing the advantages and limitations of using raw biomass, the chapter provides a detailed analysis of how biomass origin influences adsorption efficiency and mechanisms. Unlike previous studies, this work emphasizes the potential for reusability and presents a comparative evaluation of different biomass types. Ultimately, the chapter offers new insights into the optimal utilization of biomass-based adsorbents for environmentally sustainable remediation.

2.2 Types of Biomass Used

Bio-based carbon materials—including biomass, biochar, and residual wastes—are abundant, cost-effective, and sustainable resources that are gaining significant attention for applications in water treatment, gas capture, and carbon sequestration. Globally, nearly 140 billion tons of biomass are generated each year, comprising wood waste, agricultural residues, and other byproducts. Wood wastes are particularly suitable for carbon capture and biogas upgrading due to their high carbon content and low levels of ash and sulfur (Singh et al., 2022). Similarly, hard-shelled biomass such as peanut, coconut, and almond shells serve as valuable feedstocks, owing to their inherently porous structures that are well-suited for activated carbon production.

In addition, the generation of municipal solid waste is projected to rise from 2.01 billion metric tons in 2018 to 3.4 billion metric tons by 2050, representing a growing environmental concern (Karimi et al., 2022). Biomass feedstocks such as wood debris, tree trimmings, bagasse, rice husks, sunflower seed husks, and rapeseed pomace—primarily originating from the food and forestry industries—are renewable and readily available. Despite this, these materials are frequently underutilized, with large quantities discarded annually, contributing to environmental degradation.

Of the estimated 7–10 billion tons of agricultural waste produced globally each year, only 30–40% is effectively managed. Converting these wastes into activated carbon or biochar presents a sustainable waste valorization strategy. The diverse composition of these biomass materials directly influences their surface characteristics and adsorption performance (Tomczyk et al., 2020). Their reuse thus represents an eco-friendly approach to waste management and resource recovery.

2.2.1 Lignocellulosic Biomass

Lignocellulosic materials are nonedible plant residues obtained from forests, agricultural lands, and savannas. These include agricultural byproducts such as corn stover, sugarcane bagasse, wheat straw, and rice husks; dedicated energy crops like switchgrass, timothy grass, poplar, and willow; wood residues such as pinewood and spruce; as well as sawmill byproducts, paper mill waste, and municipal paper refuse. Lignocellulosic feedstocks are abundant, cost-effective, and hold considerable potential for supporting sustainable biofuel production. With advancements in conversion technologies, it is projected that by 2050, energy crops, forest biomass, and agricultural residues could contribute 250–500 exajoules (EJ) of energy annually (Nanda et al., 2015).

Lignocellulosic biomass is the world's most abundant renewable resource and plays a pivotal role in biofuel production by offering a sustainable alternative to fossil fuels and contributing to climate change mitigation. To optimize biomass conversion processes, it is essential to understand the material's physicochemical properties, including particle size, bulk density, flowability, moisture sorption behavior, proximate and elemental composition, energy content, and chemical makeup. When combined with advanced analytical characterization techniques, this comprehensive understanding enables the efficient design and operation of biofuel production facilities (Cai et al., 2017).

Lignocellulosic biomass is primarily composed of three major polymers: cellulose, hemicellulose, and lignin. The relative proportions of these components vary based on the biomass type, species, and source, resulting in a complex heterogeneous matrix. These compositional differences significantly influence the energy content and conversion efficiency of the biomass. Since the ease of biomass deconstruction and accessibility of fermentable sugars depend heavily on this composition, a detailed understanding is critical for improving pretreatment strategies and enhancing biofuel production efficiency (Bajpai, 2016).

The primary herbaceous energy crops in urban areas include switchgrass, Miscanthus, canary grass, giant reed, alfalfa, and Napier grass. These species are considered important sources of lignocellulosic biomass. Compared to food crops, these energy crops offer energetic, economic, and environmental benefits for ethanol production, helping to address the "food versus fuel" debate. Woody biomass is primarily derived from fast-growing forest trees with short rotation cycles and includes bark, timber residues, forestry byproducts (such as sawdust and wood chips), and natural forest litter. Nonwoody biomass, obtained from agricultural residues such as straw and crop stubble, along with processing wastes like sugarcane bagasse and cotton gin trash, is both abundant and economical. It is generally characterized by a lower lignin content and lower energy density (Yousuf et al., 2020).

Globally, large quantities of agricultural lignocellulosic waste are generated. According to the Food and Agriculture Organization (FAO), annual global production includes 2,900 million tons of grains, 160 million tons of pulses, 14 million tons of oilseeds, and 540 million tons from plantations. Additionally, of the 600 million tons of biomass generated annually from palm oil production, nearly 90% is discarded as waste. The cellulose content in residues such as corncobs, rice straw, wheat straw, and palm waste varies significantly. With an estimated global production of 1.5 trillion tons annually, cellulose biomass represents a highly sustainable source of bioenergy. However, despite extensive research, a fully efficient and economically viable method for converting lignocellulosic biomass into bioenergy has yet to be achieved (Kumar et al., 2008).

2.2.2 Algal Biomass

Dead algal biomass is considered a valuable natural sorbent for industrial wastewater treatment due to its strong metal-binding capabilities (Vilar et al., 2007). The negatively charged functional groups present in algal cell walls and exopolymers attract cationic metal ions, facilitating biosorption. Factors such as pH, metal ion concentration, biomass type, temperature, and contact time significantly influence biosorption efficiency. While batch experiments help optimize operating conditions, continuous-flow systems—such as packed bed columns—are more practical and cost-effective for large-scale applications. Solar-driven cultivation systems can rapidly generate algal biomass, including cyanobacteria, making it a sustainable and renewable resource.

Despite the advantages of algal biofuels, high processing costs and energy demands hinder their commercial viability. Approximately 70% of algal biomass remains as defatted algal biomass (DAB) after lipid extraction for biofuel. When utilized effectively, DAB can improve the overall economics of algae-based bioenergy systems (Park et al., 2012). DAB can be anaerobically digested to produce methane or used as animal feed. Alternatively, it can serve as a low-cost adsorbent for the removal of dyes and heavy metals from wastewater—often without requiring additional pretreatment (Rashid et al., 2013). While coal-based activated carbon remains widely used, it is expensive and nonrenewable, increasing the demand for renewable biomass-based alternatives (Tseng et al., 2006). Functional groups such as hydroxyl and carboxyl in microalgal cell walls exhibit strong binding affinities for cationic dyes. Although macroalgal biomass has been extensively studied for adsorption, research on the use of DAB as an adsorbent remains limited (Chandra et al., 2015).

Algae are crucial components of aquatic ecosystems—they form the base of food webs, generate oxygen, absorb CO_2, and provide habitat for numerous species. They also play a role in the uptake of organic pollutants from seawater. Historically, algae have been used as food, fodder, fertilizer, and sources of bioactive compounds with antibacterial, antifungal, antiviral, and anticancer properties. More recently, algae have been explored for biofuel production, with studies showing that even small additions of algae to cattle feed can reduce methane emissions by over 95%. Additionally, phycocolloids such as agar and carrageenan (from red algae) and alginate (from brown algae) are widely used in the food, pharmaceutical, and cosmetic industries due to their gelling and stabilizing properties. Marine algae are promising biosorbents capable of converting waste into valuable resources by removing contaminants and heavy metals. With their high productivity and natural abundance, algal biomass presents vast potential, and emerging technologies offer opportunities to transform coastal waste into significant environmental and economic benefits (Fernández et al., 2023).

2.2.3 Fruit and Vegetable Peels

Fruit and vegetable peel waste constitutes ~25–30% of all processed products and is generated extensively by households and the food industry. These peels are rich in bioactive compounds such as carotenoids, polyphenols, vitamins, and enzymes. Their potential spans a wide array of industrial applications, including probiotics, edible films, and bio-based products. The sustainable utilization of this waste can enhance its economic value while mitigating environmental impact (Kumar et al., 2020). Owing to their lignocellulosic composition, fruit and vegetable peels—commonly considered organic waste—hold significant potential as low-cost adsorbents. A study using SEM, FTIR, and TGA/DTG analyzed the surface area, porosity, and water absorption capacity of commonly discarded peels, including those of pomegranate, pineapple, watermelon, garlic, green peas, and pigeon peas. All samples exhibited acidic surface pH (ranging from 3 to 6) and thermal stability below 150 °C. Watermelon peel showed the highest water absorption capacity (11.5 mL/g). Their irregular, porous surfaces and favorable physicochemical properties suggest strong potential for use in environmental applications, particularly wastewater treatment (Pathak et al., 2016).

Further investigations evaluated the biosorption capabilities of potato, cucumber, and banana peels for removing both anionic (Orange G—OG) and cationic (Methylene Blue—MB) dyes from wastewater. The biosorbents were characterized using ATR-FTIR, SEM, XRD, and nitrogen sorption analysis to better understand their surface and structural features. ATR-FTIR analysis helped identify the functional groups involved in the adsorption process. Among the tested materials, banana peel demonstrated the highest adsorption capacity for MB (211.9 mg/g at pH 6), followed by cucumber and potato peels. For OG, cucumber peel exhibited the maximum adsorption capacity (40.5 mg/g at pH 2). MB adsorption by banana and potato peels followed the Langmuir isotherm, while cucumber peel fit the Freundlich model at low concentrations and the Langmuir model at higher concentrations. OG adsorption for all three peels conformed to the Langmuir model (Stavrinou et al., 2018).

Romero-Cano et al. (2017) investigated the enhancement of Cu(II) adsorption from aqueous solutions using biosorbents derived from agro-industrial wastes, specifically orange and pineapple peels. The sorption efficiency of orange peel-based biosorbents was significantly improved through a physicochemical modification process involving Instant Controlled Pressure Drop (DIC) treatment followed by citric acid functionalization. The modified orange peels exhibited a maximum adsorption capacity of 163 mg/g, which was approximately 4.19 times higher than that of the untreated peels. FTIR analysis confirmed the presence of hydroxyl (—OH) and carboxyl (—COOH) functional groups as key contributors to metal binding. The DIC treatment enhanced porosity and exposed additional active sites, thereby improving adsorption efficiency. This study highlights the potential of fruit peels, when properly modified, to serve as effective and selective biosorbents for water purification applications.

Similarly, Pathak et al. (2015) provided a comprehensive evaluation of fruit peel waste as a cost-effective biosorbent for wastewater treatment. Among various fruit peels, orange and banana peels were the most extensively studied, particularly for the removal of MB dye and Pb^{2+} ions. Most studies reported that the adsorption behavior followed pseudo-second-order kinetics and conformed well to both Langmuir and Freundlich isotherms. However, despite promising laboratory-scale results, limited research has been conducted on column operations or pilot-scale applications, and the reusability of these biosorbents remains a challenge. Although fruit peel waste is inexpensive and widely available, its potential for removing organic and gaseous pollutants remains underexplored—highlighting a key area for future research.

2.3 Composition and Characteristics

The chemical composition of biomass varies widely depending on its origin, making it a complex and heterogeneous resource. The primary constituents of plant biomass include cellulose, hemicelluloses, lignin, starch, and proteins, with trees predominantly composed of cellulose, hemicelluloses, and lignin. In a comprehensive review analyzing the chemical composition of 86 different biomass types, Vassilev et al. (2010) examined proximate, ultimate, and ash properties.

Their findings revealed significant variability in biomass composition, particularly influenced by the presence and type of inorganic materials, which impacted moisture and ash content. When recalculated on a dry or dry ash-free basis, these differences became less pronounced, showing narrower ranges in chemical characteristics. Common elements found in biomass, ranked by decreasing abundance, include carbon (C), oxygen (O), hydrogen (H), nitrogen (N), calcium (Ca), potassium (K), silicon (Si), magnesium (Mg), aluminum (Al), sulfur (S), iron (Fe), phosphorus (P), chlorine (Cl), sodium (Na), manganese (Mn), and titanium (Ti). Compared to coal, natural biomass generally contains lower amounts of carbon, ash, silicon, and certain metals, but higher concentrations of manganese, potassium, phosphorus, and chlorine. The study also identified five key elemental correlations—C–H, N–S–Cl, Si–Al–Fe–Na–Ti, Ca–Mg–Mn, and K–P–S–Cl—which may be useful for biomass classification and quality prediction. However, to fully understand these chemical patterns and their underlying mechanisms, further research into the mineral phases of biomass is necessary (Vassilev et al., 2010).

2.3.1 Cellulose, Hemicellulose, Lignin

The three primary components of lignocellulosic biomass—cellulose, hemicellulose, and lignin—each exhibit distinct pyrolysis behaviors. Cellulose, characterized by its regular and crystalline structure, decomposes below 300 °C into "active cellulose," a key intermediate in generating light volatiles such as levoglucosan. This decomposition follows the Bradbury-Shafizadeh (B–S) model, which involves depolymerization and dehydration stages. Levoglucosan is stable in the gas phase but undergoes secondary reactions in the liquid phase, forming coke and heavier compounds (Zhong et al., 2022). Chemically, cellulose is a highly polymerized polysaccharide made of D-glucose units linked by β-glucosidic bonds ($C_6H_{12}O_6$).

Hemicellulose, a complex and branched polymer consisting of various five- and six-carbon sugar monomers including xylose, arabinose, mannose, galactose, and glucose, decomposes less predictably. Studies using xylan as a model compound reveal two degradation stages: primary volatilization occurring between 210 and 340 °C, and secondary reactions from 340–500 °C. Hemicellulose contributes significantly to the formation of solid residues such as coke and heavy oils and exhibits lower thermal stability than cellulose.

Lignin begins to degrade gradually around 200 °C, producing a complex array of aromatic compounds. Its pyrolysis products are influenced by the types of chemical bonds and aromatic substituents present. Lignin tends to generate a high proportion of heavy components and coke, forming high molecular weight oligomers (200–1300 Da) that are often undetectable by conventional GC-MS analysis (Zhong et al., 2022). Structurally, lignin is a three-dimensional polymer composed of phenylpropane units that act as a natural binder, gluing cellulose and hemicellulose fibers together.

Typically, plant biomass consists of approximately 25% lignin and 75% carbohydrates (cellulose and hemicellulose), though these proportions vary among different biomass types such as straw, softwoods, and hardwoods. These polymers constitute the most abundant carbon sources on Earth (Parmar, 2017). Pyrolysis of biomass produces biochar, a porous, carbon-rich material with diverse applications in environmental remediation, agriculture, and energy sectors. The properties of biochar depend on both the pyrolysis temperature and the composition of the biomass. Comparative studies indicate that lignin-derived biochar is more stable and carbon-rich than that derived from cellulose or hemicellulose. Increasing pyrolysis temperature enhances carbon content while reducing hydrogen and oxygen levels, accompanied by structural changes such as the formation of graphite-like domains and the loss of cellulose crystallinity. These findings suggest that the pseudo-components of biomass can be used to predict the characteristics of biochar produced from real biomass samples (Ma et al., 2019). Table 2.1 presents the cellulose, hemicellulose, and lignin content in various biomass sources.

2.4 Pre-treatment Techniques

Pre-treatment of biomass plays a crucial role in enhancing its digestibility by reducing the content of lignin and amorphous cellulose, thereby increasing accessibility for subsequent processing (Agbor et al., 2011). The main pre-treatment methods can be classified into four categories: chemical, biological, physicochemical, and physical approaches. Optimizing these techniques and their operating parameters is essential to tailor pre-treatment strategies to specific biomass types, ultimately improving the efficiency of downstream conversion processes (Baksi et al., 2023). Figure 2.1 illustrates the various pre-treatment processes involved in biomass preparation.

2.4.1 Washing, Drying, Grinding

To prepare biomass for use as an adsorbent, essential steps include washing, drying, and grinding. Washing removes dirt and surface impurities, ensuring the material is cleaner and more effective. Drying, typically performed in an oven at around 60 °C for 24 hours, reduces moisture content, thereby enhancing thermal stability and preventing microbiological degradation. Grinding reduces particle size, which increases the surface area and, consequently, improves adsorption efficiency. For instance, Al-Mashhadani et al. (2023) prepared pod shells by

Table 2.1 Cellulose, hemicellulose, and lignin content in different biomass sources.

Biomass type	Cellulose (%)	Hemicellulose (%)	Lignin (%)	Other components	Reference
Sawdust	35–60	15–35	15–30	Extractives	Kazmi et al. (2019)
Citrus peel	12.7–13.6	5.3–6.1	2	Essential oils, pectin	Ververis et al. (2007)
Coconut husk	54–65	4–5	30–42	Lignin-rich	Wang and Sarkar (2018)
Sugarcane bagasse	37.2	22.1	26.5	Extractives, ash	Imman et al. (2021)
Rice husk	35	25	20	Ash, extractives	Ma'ruf et al. (2017)
Pine wood	46.7	18.4	20.5	Extractives	Anoop et al. (2014)
Sugar beet pulp	20–25	25–36	1–2	Pectin	Pawłowska et al. (2023)
Banana peel	7.6–9.6	6.4–9.4	6–12	Pectin, ash	Mishra et al. (2022)
Peanut shell	17–32	10–21	27–40	Extractives	Tomé et al. (2023)
Coffee husk	39.2 ± 0.2	26.2 ± 0.5	26.2 ± 0.5	Polyphenols	Van Nguyen et al. (2023)
Sugar palm leaf	45.31 ± 3.20	$23.21 \pm 3.73\%$	27.23 ± 4.23	Extractives	Hakim et al. (2024)
Sunflower hull	30	38.5	26.4	Extractives	De'Nobili et al. (2021)
Corn stover	37.72	20.62	30.50	Extractives	Mensah et al. (2021)
Pine bark	18.3	21.9	40.7	Extractives, ash	Díez et al. (2020)
Spruce bark	13.9	29.7	45.1	Extractives, ash	Díez et al. (2020)
Poplar	21.7	42.7	26.9	Extractives, ash	Díez et al. (2020)
Willow	22.6	44.3	25.1	Extractives, ash	Díez et al. (2020)
Wheat straw	23.8	37.5	20.5	Extractives, ash	Díez et al. (2020)

Figure 2.1 Pre-treatment processes for biomass preparation.

washing, drying in a Memmert UNB300 oven, and grinding to particle sizes of less than 300 μm and between 600 and 1000 μm to optimize their performance in water treatment applications.

In another study, the fungus *Bjerkandera adusta* was cultivated under static conditions on malt extract agar at pH 5.6 and 25 °C for 7 days. After sporulation, the fungal biomass was deactivated by autoclaving at 121 °C and 103.42 kPa for 45 minutes. The biomass was then washed, dried for 36 hours at 70 ± 1 °C, crushed, and sieved to particle sizes of 0.2–0.6 mm as part of pretreatment procedure A. In pretreatment method B, 10 g of the autoclaved biomass was treated with 0.1 M HCl for 1 hour at 250 rpm and room temperature, followed by washing and grinding into powder for subsequent use (Astuti and Muda, 2020).

2.4.2 Acid/Alkali Pre-treatment

The impact of pretreatment on cork biomass for the biosorption of heavy metals—specifically Cu, Zn, and Ni—was thoroughly investigated. The importance of carboxylic groups in metal binding was confirmed using the Fischer reaction. A 0.5 M NaCl pretreatment enhanced copper sorption by 30%, while treatment with 0.5 M $CaCl_2$ showed no significant effect. Alkaline heating further improved copper adsorption, whereas boiling in water or washing with 0.1 M NaCl or NaOH had minimal influence. Notably, treatment with 7% active chlorine for 30 minutes, using oxidizing agents such as NaClO and $NaIO_3$, increased sorption capacity by 70–80%. Furthermore, steam activation of cork carbon at 700–800 °C resulted in a sorption capacity that was 4–6 times greater (Chubar et al., 2004).

Alkaline pretreatment was also employed to enhance the heavy metal adsorption capacity of chestnut (Castanea sativa) shells. A 4% NaOH solution was applied for 4 hours at 25 °C, maximizing cadmium (Cd^{2+}) adsorption. The removal of Cu^{2+}, Pb^{2+}, and Zn^{2+} ions was also evaluated, with cadmium showing the highest adsorption capacity (9.9 mg/g), following the affinity order $Cd^{2+} > Cu^{2+} \approx Zn^{2+} > Pb^{2+}$. Adsorption data fit well to the Sips and Freundlich isotherm models and followed pseudo-second-order kinetics. Chemical and structural modifications were confirmed through FTIR, EDX, and SEM analyses. When carboxyl and hydroxyl groups were blocked, metal uptake significantly decreased, indicating their crucial role in biosorption (Vázquez et al., 2012).

To study acid and alkali pretreatment effects on various minerals—including bentonite, biotite, illite, kaolin, vermiculite, and zeolite—samples were treated with HCl or NaOH (0.1–5.0 M) at room temperature for 2 weeks. Acid treatment caused greater dissolution of aluminum compared to silicon, while alkali treatment favored silicon dissolution. XRD analysis revealed significant crystal structure disruption in acid-treated samples, whereas alkali-treated samples exhibited peak sharpening. Kaolin and illite were the most resistant to acid degradation. Overall, surface areas increased following treatment, especially with stronger chemical concentrations. Nitrogen adsorption surface areas increased more rapidly compared to water adsorption. However, differences in adsorption energies and site distributions suggested mineral-specific responses to treatment (Jozefaciuk and Bowanko, 2002).

Pretreatment of waste biomass is a crucial step in advancing the waste-to-wealth concept, enabling the production of valuable products such as biofuels, biochemicals, and biopolymers. Conventional pretreatment techniques include biological, physicochemical, and mechanical methods. However, advanced pretreatment approaches—such as hydrothermal processing, ionic liquid treatment, and supercritical fluid pretreatment—have emerged to improve biomass accessibility to enzymes, catalysts, and microorganisms. These novel techniques reduce processing time and energy consumption while enhancing product yield and overall cost-effectiveness (Roy and Chakraborty, 2024). Despite these advantages, contemporary pretreatment technologies often face limitations, including high costs, operational complexity, and considerable energy demands. Nonetheless, with further development and integration into biorefinery systems, these methods hold significant potential for more sustainable and efficient biomass conversion processes.

To offer a cost-effective alternative to commercial activated carbon, biomass adsorbents can be surface-modified to improve their capacity for removing organic pollutants from wastewater. Adegoke et al. (2023) reviewed various biomass modification techniques for the adsorption of dyes, pesticides, and pharmaceuticals. Among the materials studied, acid-activated *Posidonia oceanica* exhibited the highest adsorption capacity for Rhodamine B dye at 2681.9 mg/g, while HNO_3-modified rice husk showed the lowest adsorption capacity of 0.35 mg/g for the pesticide methyl parathion.

2.5 Influence of Biomass Source on Adsorption

The adsorption efficiency of biomass-derived adsorbents is heavily influenced by their intrinsic properties such as surface area, porosity, and functional groups, which vary significantly depending on the biomass source. Functional groups present in biomass components like cellulose, hemicellulose, lignin, and chitosan provide active sites that facilitate the binding of dyes and heavy metals. The adsorptive properties can be further enhanced through surface treatments or by creating hybrid materials.

Wu et al. (2013) demonstrated this effect by using sewage sludge and maize cob as precursors to produce carbon-based adsorbents via pyrolysis. The addition of maize cob substantially increased the surface area from 287 to 591 m^2/g and microporosity from 5 to 48%, which led to improved adsorption capacity for sulfur dioxide. This finding underscores the importance of selecting and modifying biomass sources to tailor adsorbent performance.

Activated biomass wastes such as groundnut shells, yam peels, and cassava peels have shown promising results for the adsorption of lead (II) ions from wastewater. Among these, cassava peels exhibited the highest adsorption capacity of 50.1 mg/g and a removal efficiency of 96.83% at equilibrium within 80 minutes. SEM analysis revealed enhanced microporosity following carbonization and chemical treatment. While cassava peels maintained consistent adsorption across different temperatures, yam and groundnut peels performed optimally at 60°C. The high removal efficiency at pH 7.6 was attributed to the predominantly negative charge on the adsorbent surfaces, favoring lead ion uptake (Thompson et al., 2020).

Chatterjee et al. (2020) investigated how biomass origin affects biochar properties and CO_2 adsorption after various treatments. Biochars from eight biomass types—including crop residues, herbaceous plants, and agro-industrial residues—were subjected to sono-chemical activation, amination, and sonic treatment. Results showed that sono-amination notably increased adsorption capacity by 2–2.5 times compared to chemical or physical activation alone. Among the biochars, agro-industrial and herbaceous sources performed best, with miscanthus showing a 7.5-fold increase in nitrogen content. Regeneration studies confirmed that ultrasono-aminated biochars retained 68–76% of their initial CO_2 adsorption capacity after 15 cycles, demonstrating good durability for repeated use.

2.5.1 Adsorption Efficiency

Adsorption efficiency refers to the ability of a material to selectively and effectively capture pollutants such as CO_2 from gaseous or aqueous streams. The efficiency of biomass-derived activated carbons (ACs) is influenced by several factors, including the type of feedstock, activation method, surface chemistry, and pore structure. For instance, narrow microporosity combined with a large surface area promotes strong physical adsorption at low temperatures and high CO_2 partial pressures. However, chemical functionalities—such as incorporated $Ca(OH)_2$, MgO, or amine groups like polyethyleneimine (PEI)—are crucial for enhancing chemisorption and selectivity, particularly under simulated post-combustion conditions. PEI-impregnated ACs have demonstrated up to fourfold higher CO_2 uptake compared to unmodified ACs, along with improved regeneration performance and reduced energy consumption relative to conventional monoethanolamine solvents (Salituro, 2016).

A notable example of biomass-based AC is derived from Brazilian babassu mesocarp, a widely available biomass resource. Thermal pretreatments such as hydrothermal carbonization and pyrolysis, followed by chemical activation with KOH, yield ACs possessing high surface area and microporous/mesoporous structures. These ACs exhibit strong adsorption for cationic dyes due to their negatively charged oxygen-containing surface groups. The adsorption of MB and crystal violet dyes is rapid and highly efficient, with equilibrium capacities of 388 mg/g and 220 mg/g, respectively, after 6 hours. This adsorption is primarily driven by Coulombic interactions between the dyes and surface groups. The work of da Silva Lima et al. (2023) highlights babassu-derived AC as a sustainable and renewable adsorbent for effective wastewater treatment.

Beyond ACs, biological materials such as granular and powdered anaerobic sludge have shown promising metal removal capabilities, offering sustainable and cost-effective alternatives. Aguiar et al. (2022) reported maximum removal efficiencies of 85.0% for copper and 92.8% for zinc using powdered and granular anaerobic biomass, respectively, with zinc removal efficiencies of 62.7 and 72.9%. Adsorption kinetics followed a pseudo-second-order model, indicating chemisorption as the predominant mechanism. These findings suggest that sludge-derived biosorbents not only reduce residual sludge volume but also effectively lower metal concentrations in industrial effluents, providing a practical solution for low-cost wastewater treatment.

Adsorption effectiveness refers to an adsorbent's capacity to remove contaminants from aqueous solutions, influenced by parameters such as pH, adsorbent dosage, contact time, and temperature. In a study using activated carbon derived from *Borassus aethiopum* flower biomass, the highest adsorption of Malachite Green dye occurred at pH 6–8. The adsorption process adhered to the Langmuir isotherm, indicating monolayer adsorption on a homogeneous surface. Kinetic modeling showed that the pseudo-second-order model best described the adsorption rate, suggesting chemisorption as the rate-limiting step. Both film diffusion and intraparticle diffusion contributed to the adsorption mechanism, highlighting the adsorbent's high efficiency and potential for practical wastewater treatment applications (Nethaji et al., 2010).

The adsorption capacity of biological adsorbents such as fungal biomass depends on dye structure, concentration, and the presence of competing dyes. Lyophilized granular biomass of *Cunninghamella elegans* exhibited remarkable adsorption capacities for Acid Blue 62, Acid Red 266, and Acid Yellow 49, ranging from 300 to 600 mg dye/g dry mass. However, in multi-dye systems, total adsorption efficiency decreased due to competition for adsorption sites. Kinetic studies indicated rapid adsorption (within 1–10 minutes), following pseudo-first-order or pseudo-second-order models depending on saturation. The biomass showed a selective affinity for Acid Red 266, attributed to electrostatic interactions between the dye's $-CF_3$ group and the biomass's $-NH_x$ sites (Russo et al., 2010).

Chemical treatment was used to enhance Cr(VI) adsorption on teff straw, an underutilized Ethiopian agricultural byproduct. Treatments with H_3PO_4 and KOH were optimized using Box–Behnken design and response surface methodology, achieving maximum Cr(VI) removal of 95.2% (1.5 M KOH, 105 °C, 3.5 hours) and 92.5% (2 M H_3PO_4, 110 °C, 4 hours). Adsorption experiments under optimized conditions (25 °C, 75 mg/L initial Cr(VI), 4 hours contact time, 2 g biosorbent dosage) fit best with the pseudo-second-order kinetic model and the Langmuir isotherm. Thermodynamic analysis confirmed that the Cr(VI) adsorption was spontaneous and endothermic (Ayele et al., 2022).

Seven Pseudomonas strains isolated from wastewater were evaluated for alginate production under different conditions, with optimal synthesis at 37°C and mildly acidic pH. *Pseudomonas alcaligenes* YLS18 produced the highest alginate yield (29.4 mg/g). Alginate was combined with bacterial biomass to form alginate-biomass hydrogel beads, which were tested for Pb^{2+} and Cd^{2+} adsorption. *Pseudomonas nitroreducens* YLB32 biomass showed superior biosorption capacities of 110.7 mg/g for Pb^{2+} and 54.3 mg/g for Cd^{2+} at pH 5. The hydrogel beads further enhanced adsorption capacities to 184.0 mg/g (Pb^{2+}) and 92.4 mg/g (Cd^{2+}) and demonstrated excellent reusability. Adsorption mechanisms involved ion trapping, electrostatic interactions, complexation, cation exchange, and precipitation. Figure 2.2 illustrates the adsorption mechanism and interaction between biomass and pollutants. This study supports the potential of alginate recovery from activated sludge and its application in heavy metal wastewater treatment (Zhang et al., 2022). Table 2.2 summarizes the adsorption performance of various biomass sources for heavy metal and dye removal.

2.5.2 Reusability and Regeneration

While adsorption is an effective method for removing contaminants, challenges related to the regeneration and recycling of adsorbents limit its broader practical

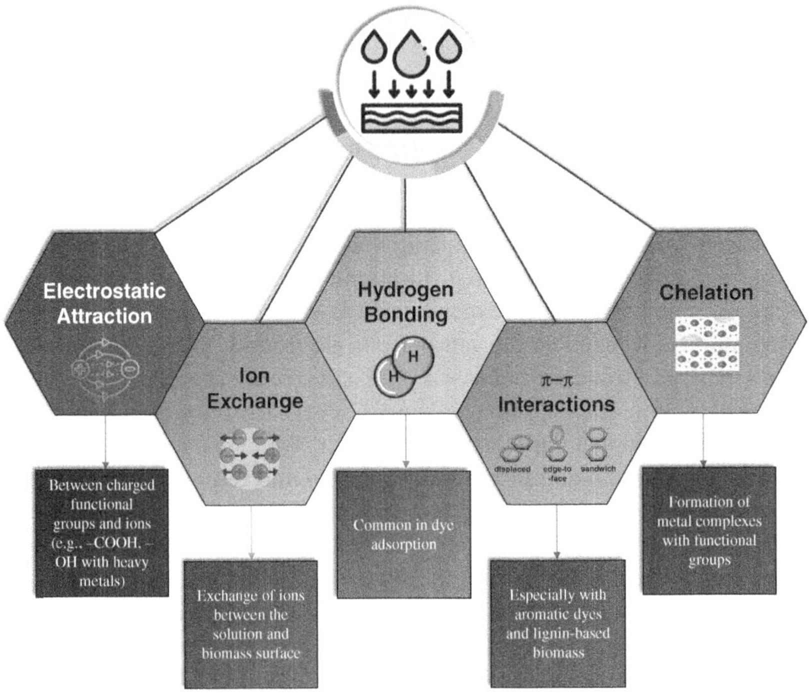

Figure 2.2 Adsorption mechanism showing interaction between biomass and pollutants.

application. Without appropriate regeneration techniques, contaminants trapped in spent adsorbents risk leaching back into the environment during disposal or storage. Effective regeneration not only ensures adsorbent reusability and long-term performance but also significantly impacts the economic viability of the adsorption process. The high cost of some adsorbents further raises concerns about their practical use unless efficient regeneration methods allow multiple reuse cycles. Therefore, developing robust regeneration protocols is critical to optimize adsorption benefits while minimizing environmental risks (Gkika et al., 2022). Figure 2.3 illustrates the main regeneration techniques applied to biomass adsorbents.

Urbanization and industrialization have increased chemical pollution in aquatic systems, particularly from dye-containing effluents. Although adsorption efficiently removes dyes, it generates dye-saturated adsorbents that can cause secondary pollution if not properly managed. Regeneration of adsorbents plays a vital role in dye recovery and adsorbent reuse. Various chemical agents have been employed for desorption, including alcohols, solvents, salts, bases, acids,

Table 2.2 Adsorption performance of various biomass sources for heavy metal and dye removal.

Biomass source	Pollutant adsorbed	Maximum adsorption capacity (mg/g)	Adsorption conditions	Reference
Banana peel	Methylene blue	91.84	pH 7, 30 °C, 60 min	Annadurai et al. (2002)
Tamarind pulp	Pb^{2+}	1.772	pH 5, 35 °C, 4 hours	Fatma et al. (2024)
Tamarind pulp	Cu^{2+}	2.281	pH 5, 35 °C, 4 hours	Fatma et al. (2024)
Tamarind pulp	Zn^{2+}	0.0660	pH 5, 35 °C, 4 hours	Fatma et al. (2024)
Tamarind pulp	Ni^{2+}	3.797	pH 5, 35 °C, 4 hours	Fatma et al. (2024)
Palm kernel fiber	Methylene blue	671.78	pH 7.1, 24 °C, 60 minutes	Ofomaja (2007)
Coconut shell	Cr^{6+}	15.47	pH 5.2, 25 °C, 180 minutes	Babel and Kurniawan (2004)
Coconut shell	Cr^{6+}	25.8	pH 2, 30 °C, 140 minutes	Chandana et al. (2020)
Sorghum vulgare	Cu (II)	161	pH 7.2, 30 °C, 60 minutes	Jimoh et al. (2024)
Sorghum vulgare	Zn (II)	140.85	pH 8.5, 30 °C, 60 minutes	Jimoh et al. (2024)
Sorghum vulgare	Cd (II)	204.08	pH 7, 30 °C, 45 minutes	Jimoh et al. (2024)
Sawdust	Malachite green oxalate	74.5	pH 7, 26 °C, 120 minutes	Garg et al. (2003)
Paper sludge	Cu (II)	4.742	pH 7, 25 °C, 90 minutes	Samešová et al. (2025)
Paper sludge	Zn (II)	2.472	pH 7, 25 °C, 90 minutes	Samešová et al. (2025)
Paper sludge	Cd (II)	2.685	pH 7, 25 °C, 90 minutes	Samešová et al. (2025)
Paper sludge	Mn (II)	3.413	pH 7, 25 °C, 90 minutes	Samešová et al. (2025)

(Continued)

Table 2.2 *(Continued)*

Biomass source	Pollutant adsorbed	Maximum adsorption capacity (mg/g)	Adsorption conditions	Reference
Pineapple peel	Cu (II)	27.68	pH 5.4, 30 °C, 30 minutes	Hu et al. (2011)
Pineapple peel	Cd (II)	34.18	pH 7.5, 30 °C, 30 minutes	Hu et al. (2011)
Pineapple peel	Pb (II)	70.29	pH 5.6, 30 °C, 30 minutes	Hu et al. (2011)
Palm ash	Acid green 25	123.4	pH 7.0, 30 °C, 120 minutes	Hameed et al. (2007)
Olive stone	Cd (II)	1.65	pH 6.15, 30 °C, 60 minutes	Kula et al. (2008)
Coffee grounds	Cd (II)	6.72	pH 7.8, 30 °C, 24 hours	Kaikake et al. (2007)
Chitosan	Cu (II)	318	pH 6, 25 °C, 24 hours	Kyzas et al. (2009)
Chitosan	Cr (VI)	935	pH 4, 25 °C, 24 hours	Kyzas et al. (2009)
Beech wood	Pb (II)	8.2–9.0	pH 5, 57 °C, 4 hours	Demirbas (2004)
Beech wood	Cd (II)	6.7–7.5	pH 5, 17 °C, 10 hours	Demirbas (2004)
Poplar wood	Pb (II)	8.2–9.0	pH 5, 57 °C, 4 hours	Demirbas (2004)
Poplar wood	Cd (II)	6.7–7.5	pH 5, 17 °C, 10 hours	Demirbas (2004)

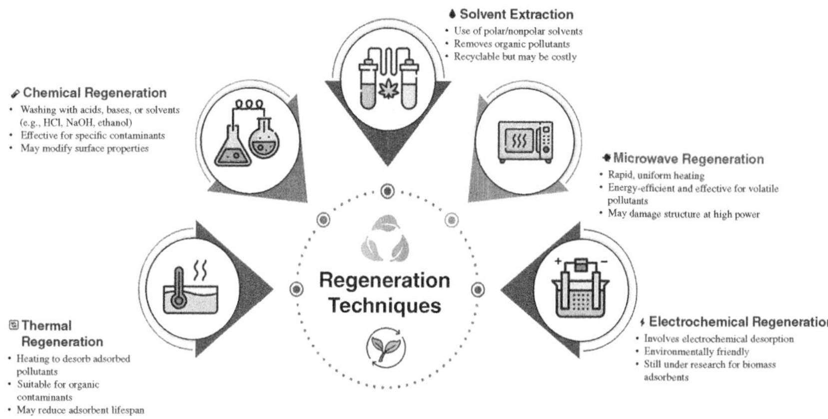

Figure 2.3 Regeneration techniques of biomass adsorbents.

and chelators. Among adsorbent sources, agricultural solid wastes are most frequently regenerated (39.56%), followed by sea biomasses (22.31%), natural materials (19.42%), and industrial wastes (18.71%). These approaches show promising regeneration efficiency and reusability (El Messaoudi et al., 2024). However, selecting optimal regeneration techniques and desorption agents remains challenging, limiting wider adoption.

Sustainable wastewater treatment requires careful consideration of adsorption, desorption, regeneration, and reusability processes, ensuring adsorbent structural integrity is maintained. Although many studies focus primarily on adsorption performance, fewer thoroughly analyze desorption behavior, regeneration efficiency, or metal recovery factors. Comprehensive evaluations and future-focused recommendations are essential to improve regeneration methods and promote sustainable heavy metal removal from wastewater (Bayuo et al., 2024). Table 2.3 summarizes regeneration efficiencies and cycle stability data for various biomass-derived adsorbents.

Several physical and chemical regeneration methods are employed to restore spent adsorbents, including chemical, thermal, microwave-assisted, electrochemical, and ultrasonic techniques. Reported desorption efficiencies vary by method: 66.61% for ultrasonic, 92.6% for thermal, an impressive 284% for microwave-assisted, 150% for electrochemical, and 99.5% for chemical regeneration. Typically, adsorbents undergo between one and ten regeneration cycles. However, chemical regeneration is the most energy-intensive, consuming approximately 6.6 kWh/kg of adsorbent. These findings highlight the importance of balancing regeneration efficiency against energy consumption to enable sustainable reuse of adsorbents in wastewater treatment applications (Sithole, 2024).

Table 2.3 Regeneration efficiency and cycle stability of biomass-derived adsorbents.

Biomass type	Pollutant adsorbed	No. of cycles	Retention efficiency (%)	Desorption method	Reference
Rhizopus nigricans	Cr (VI)	25	75–78	0.01 N Na_2CO_3	Bai and Abraham (2003)
Sargassum muticum	Cd (II)	11	90	0.1 M HNO_3	Lodeiro et al. (2006)
Coconut shell	Peach red	3	60	Acetone	Lu et al. (2011)
Coconut shell	Mustard yellow dyes	3	40	Isopropanol	Lu et al. (2011)
Coconut shell	Mustard yellow dyes	3	60	Acetone	Lu et al. (2011)
Coconut shell	Peach red	3	40	Isopropanol	Lu et al. (2011)
Commercial AC	p-Nitrophenol	5	70	Thermal regeneration	Ledesma et al. (2014)
Almond shell	Methylene blue	2	80.96	Water	Sana and Jalila (2017)
Sheep manure waste	Methylene blue	2	88.9	Water	Sana and Jalila (2017)
Sawdust	Methylene blue	2	35.64	Water	Sana and Jalila (2017)
Orange peel biochar	Cu (II)	5	68	0.8 M H_2SO_4	Sireesha et al. (2022)
Orange peel biochar	Ni (II)	5	65	0.8 M H_2SO_4	Sireesha et al. (2022)
Waste tea	Methylene blue	6	98.90	Cold atmospheric plasma	Zhou et al. (2016)
Pinus kesiya	Alizarin Red S	4	61.43	0.1 M NaOH	Bhomick et al. (2018)
Jatoba fruits	Paracetamol	5	31.92	Thermal regeneration	Spessato et al. (2019)

2.6 Conclusion

This chapter demonstrated the promising potential of various biomass types as sustainable adsorbents through an extensive review of lignocellulosic materials, algae, and fruit and vegetable peels. Their primary composition—cellulose, hemicellulose, lignin, and diverse surface functional groups—was thoroughly analyzed to illustrate how these inherent characteristics influence adsorption capacity. Pretreatment techniques such as washing, drying, grinding, and acid/alkali treatments were shown to significantly enhance adsorption efficiency by modifying surface area and increasing the number of active sites. It became evident that structural variations and biomass origin play a crucial role in determining adsorption performance. Although raw biomass is a cost-effective and eco-friendly adsorbent, it has limitations including relatively low adsorption capacity and challenges related to regeneration and reusability. These findings underscore the importance of developing effective regeneration strategies and optimizing pretreatment methods to improve the performance and sustainability of biomass-based adsorbents in wastewater treatment applications.

References

Adegoke, K. A., Akinnawo, S. O., Adebusuyi, T. A., Ajala, O. A., Adegoke, R. O., Maxakato, N. W., & Bello, O. S. (2023). Modified biomass adsorbents for removal of organic pollutants: a review of batch and optimization studies. International Journal of Environmental Science and Technology, 20(10), 11615–11644.

Agbor, V. B., Cicek, N., Sparling, R., Berlin, A., & Levin, D. B. (2011). Biomass pretreatment: fundamentals toward application. Biotechnology Advances, 29(6), 675–685.

Aguiar, A. B. S., Costa, J. M., Santos, G. E., Sancinetti, G. P., & Rodriguez, R. P. (2022). Removal of metals by biomass derived adsorbent in its granular and powdered forms: adsorption capacity and kinetics analysis. Sustainable Chemistry, 3(4), 535–550.

Al-Mashhadani, M. K. H., Hadi, S. M., Abed, K. M., & Hassan, H. A. (2023). The thermal pre-processing technique of the bio-waste for contaminated water treatment: histological and experimental study. Iranian Journal of Chemistry & Chemical Engineering, 42(1), 349–360.

Annadurai, G., Juang, R. S., & Lee, D. J. (2002). Use of cellulose-based wastes for adsorption of dyes from aqueous solutions. Journal of Hazardous Materials, 92(3), 263–274.

Anoop, E. V., Ajayghosh, V., Nijil, J. M., & Jijeesh, C. M. (2014). Evaluation of pulp wood quality of selected tropical pines raised in the high ranges of Idukki District, Kerala. Journal of Tropical Agriculture, 52(1), 59–66.

Astuti, A. D., & Muda, K. (2020). Effect of pretreatment of biosorbent in biosorption: a comparative study. Journal of Environmental Treatment Techniques, 8(3), 894–899.

Ayele, A. L., Tizazu, B. Z., & Wassie, A. B. (2022). Chemical modification of *Teff straw* biomass for adsorptive removal of Cr (VI) from aqueous solution: characterization, optimization, kinetics, and thermodynamic aspects. Adsorption Science & Technology, 2022, 5820207.

Babel, S., & Kurniawan, T. A. (2004). Cr (VI) removal from synthetic wastewater using coconut shell charcoal and commercial activated carbon modified with oxidizing agents and/or chitosan. Chemosphere, 54(7), 951–967.

Bai, R. S., & Abraham, T. E. (2003). Studies on chromium (VI) adsorption–desorption using immobilized fungal biomass. Bioresource Technology, 87(1), 17–26.

Bajpai, P. (2016). Structure of lignocellulosic biomass. In: Pretreatment of Lignocellulosic Biomass for Biofuel Production (pp. 7–12). Singapore: Springer.

Baksi, S., Saha, D., Saha, S., Sarkar, U., Basu, D., & Kuniyal, J. C. (2023). Pre-treatment of lignocellulosic biomass: review of various physico-chemical and biological methods influencing the extent of biomass depolymerization. International Journal of Environmental Science and Technology, 20(12), 13895–13922.

Bayuo, J., Rwiza, M. J., Choi, J. W., Mtei, K. M., Hosseini-Bandegharaei, A., & Sillanpää, M. (2024). Adsorption and desorption processes of toxic heavy metals, regeneration and reusability of spent adsorbents: economic and environmental sustainability approach. Advances in Colloid and Interface Science, 329, 103196.

Bedia, J., Peñas-Garzón, M., Gómez-Avilés, A., Rodriguez, J. J., & Belver, C. (2018). A review on the synthesis and characterization of biomass-derived carbons for adsorption of emerging contaminants from water. C: Journal of Carbon Research, 4(4), 63.

Bhomick, P. C., Supong, A., Baruah, M., Pongener, C., & Sinha, D. (2018). Pine Cone biomass as an efficient precursor for the synthesis of activated biocarbon for adsorption of anionic dye from aqueous solution: isotherm, kinetic, thermodynamic and regeneration studies. Sustainable Chemistry and Pharmacy, 10, 41–49.

Cai, J., He, Y., Yu, X., Banks, S. W., Yang, Y., Zhang, X., Yu, Y., Liu, R., & Bridgwater, A. V. (2017). Review of physicochemical properties and analytical characterization of lignocellulosic biomass. Renewable and Sustainable Energy Reviews, 76, 309–322.

Chandana, L., Krushnamurty, K., Suryakala, D., & Subrahmanyam, C. (2020). Low-cost adsorbent derived from the coconut shell for the removal of hexavalent chromium from aqueous medium. Materials Today: Proceedings, 26, 44–51.

Chandra, T. S., Mudliar, S. N., Vidyashankar, S., Mukherji, S., Sarada, R., Krishnamurthi, K., & Chauhan, V. S. (2015). Defatted algal biomass as a non-conventional low-cost adsorbent: surface characterization and methylene blue adsorption characteristics. Bioresource Technology, 184, 395–404.

Chatterjee, R., Sajjadi, B., Chen, W. Y., Mattern, D. L., Hammer, N., Raman, V., & Dorris, A. (2020). Impact of biomass sources on acoustic-based chemical functionalization of bio-chars for improved CO_2 adsorption. Energy & Fuels, 34(7), 8608–8627.

Chen, J., Ma, X., Liang, M., Guo, Z., Cai, Y., Zhu, C., Wang, Z., Wang, S., Xu, J.-L., & Ying, H. (2024). Physical–chemical–biological pretreatment for biomass degradation and industrial applications: a review. Waste 2(4), 451–473.

Chubar, N., Carvalho, J. R., & Correia, M. J. N. (2004). Heavy metals biosorption on cork biomass: effect of the pre-treatment. Colloids and Surfaces A: Physicochemical and Engineering Aspects, 238(1–3), 51–58.

da Silva Lima, P. N., Ghosh, A., Nascimento, A., Vieira, L. H. S., Costa, R. S., Ferreira, O. P., Abreu, G.J.P., Fujisawa, K., Silva-Filho, E.C., de Souza Filho, A.G., Lobo, A.O., Terrones, M., & Viana, B. C. (2023). Advanced sustainable carbon material from babassu biomass and its adsorption performance. Journal of Physics and Chemistry of Solids, 176, 111188.

de Jong, W. (2014). Biomass composition, properties, and characterization. In: Biomass as a Sustainable Energy Source for the Future: Fundamentals of Conversion Processes (pp. 36–68). Wiley.

De'Nobili, M. D., Bernhardt, D. C., Basanta, M. F., & Rojas, A. M. (2021). Sunflower (*Helianthus annuus* L.) seed hull waste: composition, antioxidant activity, and filler performance in pectin-based film composites. Frontiers in Nutrition, 8, 777214.

Demirbas, A. (2004). Adsorption of lead and cadmium ions in aqueous solutions onto modified lignin from alkali glycerol delignication. Journal of Hazardous Materials, 109(1–3), 221–226.

Díez, D., Urueña, A., Piñero, R., Barrio, A., & Tamminen, T. (2020). Determination of hemicellulose, cellulose, and lignin content in different types of biomasses by thermogravimetric analysis and pseudo component kinetic model (TGA-PKM method). Processes, 8(9), 1048.

El Messaoudi, N., El Khomri, M., El Mouden, A., Bouich, A., Jada, A., Lacherai, A., Iqbal, H. M. N., Mulla, S. I., Kumar, V., & Américo-Pinheiro, J. H. P. (2024). Regeneration and reusability of non-conventional low-cost adsorbents to remove dyes from wastewaters in multiple consecutive adsorption–desorption cycles: a review. Biomass Conversion and Biorefinery, 14(11), 11739–11756.

Fatma, U. K., Nizami, G., Ahamad, S., Saquib, M., & Hussain, M. K. (2024). Activated Green Tamarind Pulp (AGTP) as an efficient adsorbent for removal of Pb^{2+}, Cu^{2+}, Zn^{2+} & Ni^{2+} from contaminated water. Journal of Water Process Engineering, 59, 105048.

Fernández, L. A. G., Ramos, V. C., Polo, M. S., & Castillo, N. A. M. (2023). Fundamentals in applications of algae biomass: a review. Journal of Environmental Management, 338, 117830.

Garg, V. K., Gupta, R., Yadav, A. B., & Kumar, R. (2003). Dye removal from aqueous solution by adsorption on treated sawdust. Bioresource Technology, 89(2), 121–124.

Gayathiri, M., Pulingam, T., Lee, K. T., & Sudesh, K. (2022). Activated carbon from biomass waste precursors: factors affecting production and adsorption mechanism. Chemosphere, 294, 133764.

Gkika, D. A., Mitropoulos, A. C., & Kyzas, G. Z. (2022). Why reuse spent adsorbents? The latest challenges and limitations. Science of the Total Environment, 822, 153612.

Hakim, L., Iswanto, A. H., Herawati, E., Batubara, R., Lubis, Y. S., & Aini, E. N. (2024). Characterization of Indonesian sugar palm bunch (*Arenga longipes* Mogea) properties for various utilization purposes. Forests, 15(2), 239.

Hameed, B. H., Ahmad, A. A., & Aziz, N. (2007). Isotherms, kinetics and thermodynamics of acid dye adsorption on activated palm ash. Chemical Engineering Journal, 133(1–3), 195–203.

Hashmi, Z., Jatoi, A. S., Nadeem, S., Anjum, A., Imam, S. M., & Jangda, H. (2024). Comparative analysis of conventional to biomass-derived adsorbent for wastewater treatment: a review. Biomass Conversion and Biorefinery, 14(1), 45–76.

Hu, X., Zhao, M., Song, G., & Huang, H. (2011). Modification of pineapple peel fibre with succinic anhydride for Cu^{2+}, Cd^{2+} and Pb^{2+} removal from aqueous solutions. Environmental Technology, 32(7), 739–746.

Imman, S., Khongchamnan, P., Wanmolee, W., Laosiripojana, N., Kreetachat, T., Sakulthaew, C., Chokejaroenrat, C., & Suriyachai, N. (2021). Fractionation and characterization of lignin from sugarcane bagasse using a sulfuric acid catalyzed solvothermal process. RSC Advances, 11(43), 26773–26784.

Jimoh, I., Sumaila, A., Asipita, P. O., & Musa, L. A. (2024). The removal of Cu(II), Zn(II), and Cd(II) ions from aqueous solutions through adsorption of *Sorghum vulgare*: a comparative analysis using batch adsorption techniques. New Healthcare Advancements and Explorations, 1(2), 1–10.

Jozefaciuk, G., & Bowanko, G. (2002). Effect of acid and alkali treatments on surface areas and adsorption energies of selected minerals. Clays and Clay Minerals, 50(6), 771–783.

Kaikake, K., Hoaki, K., Sunada, H., Dhakal, R. P., & Baba, Y. (2007). Removal characteristics of metal ions using degreased coffee beans: adsorption equilibrium of cadmium (II). Bioresource Technology, 98(15), 2787–2791.

Karimi, M., Shirzad, M., Silva, J. A., & Rodrigues, A. E. (2022). Biomass/Biochar carbon materials for CO_2 capture and sequestration by cyclic adsorption processes: a review and prospects for future directions. Journal of CO2 Utilization, 57, 101890.

Kazmi, M. Z. H., Karmakar, A., Michaelis, V. K., & Williams, F. J. (2019). Separation of cellulose/hemicellulose from lignin in white pine sawdust using boron trihalide reagents. Tetrahedron, 75(11), 1465–1470.

Kula, I., Uğurlu, M., Karaoğlu, H., & Celik, A. (2008). Adsorption of Cd (II) ions from aqueous solutions using activated carbon prepared from olive stone by $ZnCl_2$ activation. Bioresource Technology, 99(3), 492–501.

Kumar, H., Bhardwaj, K., Sharma, R., Nepovimova, E., Kuča, K., Dhanjal, D. S., Verma, R., Bhardwaj, P., Sharma, S., & Kumar, D. (2020). Fruit and vegetable peels: utilization of high value horticultural waste in novel industrial applications. Molecules, 25(12), 2812.

Kumar, R., Singh, S., & Singh, O. V. (2008). Bioconversion of lignocellulosic biomass: biochemical and molecular perspectives. Journal of Industrial Microbiology and Biotechnology, 35(5), 377–391.

Kyzas, G. Z., Kostoglou, M., & Lazaridis, N. K. (2009). Copper and chromium (VI) removal by chitosan derivatives—equilibrium and kinetic studies. Chemical Engineering Journal, 152(2–3), 440–448.

Ledesma, B., Román, S., Álvarez-Murillo, A., Sabio, E., & González, J. F. (2014). Cyclic adsorption/thermal regeneration of activated carbons. Journal of Analytical and Applied Pyrolysis, 106, 112–117.

Lodeiro, P., Herrero, R., & De Vicente, M. S. (2006). Batch desorption studies and multiple sorption–regeneration cycles in a fixed-bed column for Cd (II) elimination by protonated *Sargassum muticum*. Journal of Hazardous Materials, 137(3), 1649–1655.

Lu, P. J., Lin, H. C., Yu, W. T., & Chern, J. M. (2011). Chemical regeneration of activated carbon used for dye adsorption. Journal of the Taiwan Institute of Chemical Engineers, 42(2), 305–311.

Ma, Z., Yang, Y., Wu, Y., Xu, J., Peng, H., Liu, X., Zhang, W., & Wang, S. (2019). In-depth comparison of the physicochemical characteristics of bio-char derived from biomass pseudo components: hemicellulose, cellulose, and lignin. Journal of Analytical and Applied Pyrolysis, 140, 195–204.

Ma'ruf, A., Pramudono, B., & Aryanti, N. (2017, March). Lignin isolation process from rice husk by alkaline hydrogen peroxide: lignin and silica extracted. In: AIP Conference Proceedings (Vol. 1823, No. 1). AIP Publishing.

Mensah, M. B., Jumpah, H., Boadi, N. O., & Awudza, J. A. (2021). Assessment of quantities and composition of corn stover in Ghana and their conversion into bioethanol. Scientific African, 12, e00731.

Mishra, S., Prabhakar, B., Kharkar, P. S., & Pethe, A. M. (2022). Banana peel waste: an emerging cellulosic material to extract nanocrystalline cellulose. ACS Omega, 8(1), 1140–1145.

Nanda, S., Azargohar, R., Dalai, A. K., & Kozinski, J. A. (2015). An assessment on the sustainability of lignocellulosic biomass for biorefining. Renewable and Sustainable Energy Reviews, 50, 925–941.

Nethaji, S., Sivasamy, A., Thennarasu, G., & Saravanan, S. (2010). Adsorption of Malachite Green dye onto activated carbon derived from *Borassus aethiopum* flower biomass. Journal of Hazardous Materials, 181(1–3), 271–280.

Ofomaja, A. E. (2007). Kinetics and mechanism of methylene blue sorption onto palm kernel fibre. Process Biochemistry, 42(1), 16–24.
Park, J. B. K., Craggs, R. J., & Shilton, A. N. (2012). Life cycle assessment of biodiesel production from microalgae grown on wastewater treatment high rate ponds. Bioresource Technology, 124, 230–238.
Parmar, K. (2017). Biomass-an overview on composition characteristics and properties. IRA-International Journal of Applied Sciences. (ISSN 2455-4499), 7(1), 42.
Pathak, P. D., Mandavgane, S. A., & Kulkarni, B. D. (2015). Fruit peel waste as a novel low-cost bio adsorbent. Reviews in Chemical Engineering, 31(4), 361–381.
Pathak, P. D., Mandavgane, S. A., & Kulkarni, B. D. (2016). Characterizing fruit and vegetable peels as bio adsorbents. Current Science, 110(11), 2114–2123.
Pawłowska, M., Zdeb, M., & Nieścioruk, M. (2023). Efficiency of chemical pretreatment of sugar beet pulp biomass intended to energy production via biological processes. Energies, 16(2), 574.
Rashid, N. R. N. A., Noor, Z. Z., & Hassan, M. A. (2013). Potential of microalgae for wastewater treatment. Desalination and Water Treatment, 51(1–3), 475–481.
Romero-Cano, L. A., García-Rosero, H., Gonzalez-Gutierrez, L. V., Baldenegro-Pérez, L. A., & Carrasco-Marín, F. (2017). Functionalized adsorbents prepared from fruit peels: equilibrium, kinetic and thermodynamic studies for copper adsorption in aqueous solution. Journal of Cleaner Production, 162, 195–204.
Roy, S., & Chakraborty, S. (2024). Emerging technologies for waste biomass pretreatment: pros and cons. In: Processing of Biomass Waste (pp. 41–54). Elsevier.
Russo, M. E., Di Natale, F., Prigione, V., Tigini, V., Marzocchella, A., & Varese, G. C. (2010). Adsorption of acid dyes on fungal biomass: equilibrium and kinetics characterization. Chemical Engineering Journal, 162(2), 537–545.
Salituro, A. (2016). CO_2 capture through sorption onto activated carbons derived from biomass. Doctoral dissertation. University of Leeds.
Samešová, D., Pochyba, A., Ďuricová, A., Poništ, J., Prepilková, V. Š., Schwarz, M., Veverková, D., Salva, J., & Schmidtová, J. (2025). Comparative adsorption of Cu (II), Zn (II), Cd (II), and Mn (II) from aquatic solution and neutral mine drainage using paper sludge. Water, 17(10), 1471.
Sana, D., & Jalila, S. (2017). A comparative study of adsorption and regeneration with different agricultural wastes as adsorbents for the removal of methylene blue from aqueous solution. Chinese Journal of Chemical Engineering, 25(9), 1282–1287.
Singh, H., Tomar, S., Qureshi, K. A., Jaremko, M., & Rai, P. K. (2022). Recent advances in biomass pretreatment technologies for biohydrogen production. Energies, 15(3), 999.
Sireesha, S., Upadhyay, U., & Sreedhar, I. (2022). Comparative studies of heavy metal removal from aqueous solution using novel biomass and biochar-based adsorbents: characterization, process optimization, and regeneration. Biomass Conversion and Biorefinery, 15, 1–13.

Sithole, T. (2024). A review on regeneration of adsorbent and recovery of metals: adsorbent disposal and regeneration mechanism. South African Journal of Chemical Engineering, 50, 39–50.

Spessato, L., Bedin, K. C., Cazetta, A. L., Souza, I. P., Duarte, V. A., Crespo, L. H., Silva, M.C., Pontes, R.M., & Almeida, V. C. (2019). KOH-super activated carbon from biomass waste: insights into the paracetamol adsorption mechanism and thermal regeneration cycles. Journal of Hazardous Materials, 371, 499–505.

Stavrinou, A., Aggelopoulos, C. A., & Tsakiroglou, C. D. (2018). Exploring the adsorption mechanisms of cationic and anionic dyes onto agricultural waste peels of banana, cucumber and potato: adsorption kinetics and equilibrium isotherms as a tool. Journal of Environmental Chemical Engineering, 6(6), 6958–6970.

Thompson, C. O., Ndukwe, A. O., & Asadu, C. O. (2020). Application of activated biomass waste as an adsorbent for the removal of lead (II) ion from wastewater. Emerging Contaminants, 6, 259–267.

Tomczyk, A., Sokołowska, Z., & Boguta, P. (2020). Biomass type effect on biochar surface characteristic and adsorption capacity relative to silver and copper. Fuel, 278, 118168.

Tomé, A. G., Ribeiro, E. A., Lima, M., Brocenschi, R. F., & Ribeiro, L. N. (2023). Biorefinery of peanut shell agro industrial lignocellulosic waste for synthesis of a natural coagulant applied in the treatment of dairy wastewater. Journal of Environmental Chemical Engineering, 11(6), 111535.

Tseng, R. L., Wu, F. C., & Juang, R. S. (2006). Activated carbon prepared from bagasse by zinc chloride activation for adsorption of dyes from aqueous solutions. Journal of Hazardous Materials, 131(1–3), 51–59.

Vahabisani, A., & An, C. (2021). Use of biomass-derived adsorbents for the removal of petroleum pollutants from water: a mini-review. Environmental Systems Research, 10(1), 25.

Van Nguyen, D., Duong, C. T. T., Vu, C. N. M., Nguyen, H. M., Pham, T. T., Tran-Thuy, T. M., & Nguyen, L. Q. (2023). Data on chemical composition of coffee husks and lignin microparticles as their extracted product. Data in Brief, 51, 109781.

Vassilev, S. V., Baxter, D., Andersen, L. K., & Vassileva, C. G. (2010). An overview of the chemical composition of biomass. Fuel, 89(5), 913–933.

Vázquez, G., Mosquera, O., Freire, M. S., Antorrena, G., & González-Álvarez, J. (2012). Alkaline pre-treatment of waste chestnut shell from a food industry to enhance cadmium, copper, lead and zinc ions removal. Chemical Engineering Journal, 184, 147–155.

Ververis, C., Georghiou, K., Danielidis, D., Hatzinikolaou, D. G., Santas, P., Santas, R., & Corleti, V. (2007). Cellulose, hemicelluloses, lignin and ash content of some organic materials and their suitability for use as paper pulp supplements. Bioresource Technology, 98(2), 296–301.

Vilar, V. J., Botelho, C. M., & Boaventura, R. A. (2007). Methylene blue adsorption by algal biomass based materials: biosorbents characterization and process behaviour. Journal of hazardous materials, 147(1–2), 120–132.

Wang, Q., & Sarkar, J. (2018). Pyrolysis behaviors of waste coconut shell and husk biomasses. In: Towards Energy Sustainability (pp. 111–120). WIT Press.

Wu, C., Song, M., Jin, B., Wu, Y., & Huang, Y. (2013). Effect of biomass addition on the surface and adsorption characterization of carbon-based adsorbents from sewage sludge. Journal of Environmental Sciences, 25(2), 405–412.

Yousuf, A., Pirozzi, D., & Sannino, F. (2020). Fundamentals of lignocellulosic biomass. In: Lignocellulosic Biomass to Liquid Biofuels (pp. 1–15). Academic Press.

Zhang, P., Yuan, L., Zeng, J., Zou, K., Liu, B., Qing, T., & Feng, B. (2022). Alginate production of Pseudomonas strains and its application in preparation of alginate-biomass hydrogel for heavy metal adsorption. International Journal of Biological Macromolecules, 222, 1511–1521.

Zhong, D., Zeng, K., Li, J., Qiu, Y., Flamant, G., Nzihou, A., Vladimirovich, V. S., Yang, H., & Chen, H. (2022). Characteristics and evolution of heavy components in bio-oil from the pyrolysis of cellulose, hemicellulose and lignin. Renewable and Sustainable Energy Reviews, 157, 111989.

Zhou, R., Zhou, R., Zhang, X., Tu, S., Yin, Y., Yang, S., & Ye, L. (2016). An efficient bio-adsorbent for the removal of dye: adsorption studies and cold atmospheric plasma regeneration. Journal of the Taiwan Institute of Chemical Engineers, 68, 372–378.

3

Synthesis Techniques of Novel Bioadsorbents

3.1 Introduction

Access to potable water is increasingly threatened by global pollution and the rising costs associated with conventional water treatment technologies. Among the available methods, adsorption is recognized as a cost-effective and efficient technique for removing a wide range of water contaminants. However, conventional adsorbents derived from fossil fuels are often nonrenewable and may contribute to secondary pollution. In contrast, bioadsorbents made from renewable biomass sources—such as wood, algae, bacteria, and agricultural waste—offer a more environmentally friendly and sustainable alternative. These materials not only utilize waste that would otherwise harm the environment but also support the principles of a circular economy. Bioadsorbents have shown significant potential in removing pollutants such as heavy metals, dyes, and pharmaceuticals (Osman et al., 2023).

Thermochemical biomass conversion yields biochar (BC), a carbon-rich solid characterized by high surface area, porosity, and cation exchange capacity—properties that make it a popular choice for wastewater treatment. Hydrothermal carbonization (HTC) and pyrolysis are two widely used production methods. Pyrolysis, conducted at 300–800 °C under a nitrogen atmosphere, tends to produce more charcoal when performed at slower rates. HTC, on the other hand, operates at lower temperatures (180–280 °C) and pressures (2–10 MPa), resulting in higher solid yields. BC is stable, nontoxic, and efficient as a bioadsorbent, with its adsorption capacity strongly influenced by its structure and synthesis conditions (Jian et al., 2018).

Mechanical grinding—particularly ball milling (BM)—induces significant surface, structural, and morphological changes in minerals, thereby enhancing their adsorption capacities. Studies on minerals such as calcite, palygorskite, pyrophyllite, montmorillonite, and kaolinite have shown that dry grinding leads to deformation, exfoliation, increased surface area, and improved cation exchange capacity. These modifications significantly enhance metal ion adsorption. Structural changes such as distortion, fragmentation, and particle size reduction increase the availability of active sites for adsorption. Additionally, grinding can promote amorphization and increase the material's chemical reactivity. Despite extensive research on the effects of grinding on mineral structure, relatively few studies have focused specifically on the impact of BM-induced particle size changes on the adsorption of heavy metals from aqueous solutions (Maleki and Karimi-Jashni, 2017).

Improved adsorption performance can be achieved through chemical surface modification, which alters the surface properties of adsorbents without changing their bulk structure Loganathan et al. (2013). These modifications enhance characteristics such as porosity, surface charge, and the availability of functional groups by promoting chemical interactions between the adsorbent surface and modifying agents. Table 3.1 presents a comparative analysis of bioadsorbents, detailing raw materials, functional groups, and applications.

To enhance binding affinity toward various contaminants, acidic, alkaline, and neutral chemicals are commonly used to tailor surface functionalities (Senturk et al., 2009). As a result, chemically modified adsorbents are widely applied in environmental remediation, particularly for the effective removal of organic pollutants and heavy metals.

A promising strategy for enhancing the properties and applications of polysaccharides is through chemical or enzymatic modification (Karaki et al., 2016). Enzymatic approaches, in particular, are gaining traction due to their high specificity, lower energy requirements, and minimal environmental impact compared to conventional chemical treatments. Enzymes such as laccases, peroxidases, lipases, tyrosinases, and transglutaminases are frequently employed to induce targeted structural transformations. These biocatalysts facilitate various reactions, resulting in functionalized derivatives with improved physicochemical properties. These enhancements—such as better solubility, stability, and reactivity—make polysaccharide-based materials suitable for advanced applications in environmental remediation, packaging, and biomedicine.

Advanced synthesis methods, including pyrolytic, microwave-assisted, and thermal processes, are used to produce adsorbents from chemical precursors. These techniques have gained popularity in recent research due to their efficiency, reproducibility, cost-effectiveness, and environmental friendliness compared to traditional approaches (Usman and Cheng, 2024).

Starch, a biodegradable polymer rich in hydroxyl groups, has limited adsorption efficiency due to its low porosity. However, its properties can be significantly enhanced through physical or chemical modifications, such as nanoclay

Table 3.1 Comparative analysis of bioadsorbents: raw materials, functional groups, and applications.

Bioadsorbent type	Raw source	Main functional groups	Typical pollutants adsorbed	Advantages	Reference
Activated carbon	Coconut shell, wood, bamboo	—OH, —COOH, —C=O	Heavy metals, dyes, organics	High surface area, strong adsorption	Babel and Kurniawan (2003)
Chitosan	Crustacean shells (shrimp, crab)	$-NH_2$, —OH	Heavy metals, dyes	Biodegradable, good metal affinity	Rinaudo (2006)
Cellulose	Agricultural waste, wood pulp	—OH, —COOH	Dyes, metals	Abundant, renewable, modifiable	Aslam et al. (2023)
Biochar	Agricultural residues, sawdust	—OH, —COOH, —C=O	Metals, dyes, pharmaceuticals	Cost-effective, porous, high stability	Mohan et al. (2014)
Alginate	Brown algae	$-COO^-$, —OH	Heavy metals, cationic dyes	Gel-forming, eco-friendly	Wang and Chen (2009)
Peat	Decomposed plant matter	—OH, —COOH, phenolic groups	Heavy metals, organics	High cation exchange capacity	Ho and McKay (2000)
Lignin	Wood, paper industry waste	—OH, —COOH, aromatic groups	Dyes, organics	Aromatic backbone, abundant by-product	Cazier et al. (2024)
Sawdust	Timber industry waste	—OH, —COOH	Heavy metals, dyes	Easily available, low cost	Malik (2004)

(Continued)

Table 3.1 (*Continued*)

Bioadsorbent type	Raw source	Main functional groups	Typical pollutants adsorbed	Advantages	Reference
Rice husk	Rice milling waste	—OH, —COOH, —Si—OH	Metals, dyes	Silica-rich, porous structure	Kurniawan et al. (2006)
Orange peel	Citrus fruit processing waste	—OH, —COOH, pectin	Metal ions, dyes	Biodegradable, pectin-rich	Annadurai et al. (2002)
Banana peel	Food waste	—OH, —COOH, $—NH_2$	Methylene blue, Pb^{2+}, Cd^{2+}	Cost-free, effective in aqueous media	Annadurai et al. (2002)
Moringa seed cake	Oil extraction residue	—OH, $—NH_2$	Turbidity, metals, organics	Natural coagulant and bioadsorbent	Gassenschmidt et al. (1995)

reinforcement. Starch-based bio-hybrid composites show great promise in environmental pollutant removal, as they improve adsorption performance while retaining the natural advantages of starch (Nambiar et al., 2023).

This chapter provides a comprehensive and critical assessment of various synthesis methodologies for novel bioadsorbents, encompassing physical, chemical, biological, and advanced techniques. It also delves into the development of new composite and hybrid materials, including magnetic biosorbents, while addressing challenges associated with scalable production. By offering a synthesis-to-application perspective, this study bridges the gap between laboratory research and industrial-scale implementation, thereby contributing to the advancement of sustainable bioadsorbent technologies for environmental remediation.

3.2 Physical Modification

Physical activation and chemical activation are the two primary techniques used for the preparation of activated carbons (ACs). Physical activation is a two-step process that begins with the carbonization of the raw material, followed by the activation of the resulting char at temperatures ranging from 600 to 1000°C. This activation takes place either in an oxidizing atmosphere (such as carbon dioxide or oxygen) or in an inert environment (such as argon or nitrogen). Figure 3.1 illustrates the physical modification techniques.

Studies have demonstrated that activation using agents such as steam, CO_2, or a steam–CO_2 mixture results in ACs with a high Brunauer–Emmett–Teller (BET) surface area (Oumam et al., 2020). Table 3.2 presents various physical modification techniques and their corresponding influence on adsorption performance.

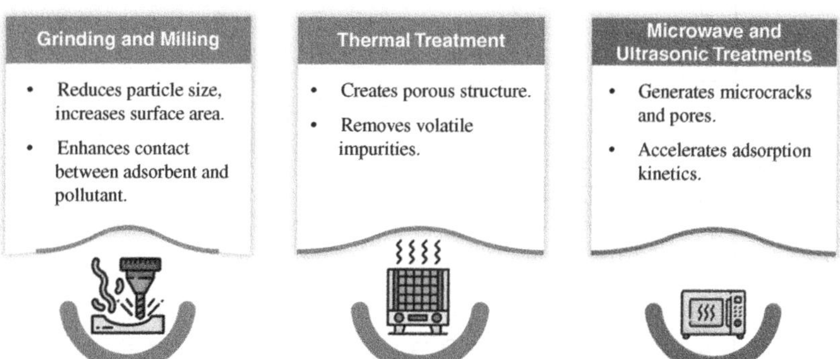

Figure 3.1 Physical modification techniques.

Table 3.2 Physical modification techniques and their influence on adsorption performance.

Modification technique	Description	Effect on adsorbent properties	Typical bioadsorbents	Improved pollutant adsorption	Reference
Grinding/milling	Reduces particle size to increase surface area	Higher surface area, faster kinetics	Sawdust, rice husk, biochar	Heavy metals, dyes	Foo and Hameed (2012)
Heat treatment (carbonization)	Heating biomass in absence of air to form biochar	Increased porosity and stability	Coconut shell, wood, bamboo	Organic pollutants, metals	Mohan et al. (2014)
Thermal activation	Further heating to activate surface functional groups	Enhanced pore volume and reactivity	Activated carbon, peat	Dyes, phenols	Babel and Kurniawan (2003)
Magnetic impregnation (Fe_3O_4)	Addition of magnetic particles for easier recovery	Enables magnetic separation	Chitosan, alginate, cellulose	Heavy metals, dyes	Al-Bagawi (2024)
Steam activation	High-temperature steam used to activate biochar	Increased mesoporosity, surface chemistry	Agricultural waste biochar	Pharmaceuticals, VOCs	Tan et al. (2015)
Freeze-drying	Removes moisture without damaging structure	Preserves porosity, low density	Cellulose, chitosan beads	Dyes, lead ions	Rinaudo (2006)
Sonication (ultrasound)	Uses ultrasound waves to disperse and activate materials	Breaks agglomerates, increases surface sites	Magnetic nanoparticles, nanocellulose	Dyes, Cr(VI)	Yang et al. (2024)
Plasma treatment	Exposes surface to ionized gas to introduce new functionalities	Surface activation, better wettability	Cellulose, lignin	Heavy metals, dyes	Vajpayee et al. (2021)
Drying (sun/oven)	Simple moisture removal method	Reduced weight, easier storage	Fruit peels, leaves	Organic dyes	Annadurai et al. (2002)

3.2.1 Heating and Carbonization

Heat treatment is a widely used technique for enhancing the adsorption properties of various adsorbents. Its primary effect is to increase the surface area and porosity of the material, thereby providing more active sites for contaminant adsorption. Studies have shown that thermally AC derived from biological materials exhibits improved adsorption of a wide range of dissolved organic compounds (Yin et al., 2007; Gupta et al., 2011). Similarly, heat activation of inorganic adsorbents such as red mud and alumina has been found to enhance their uptake of heavy metals and phosphates (Bhatnagar et al., 2011).

The physicochemical properties of layered double hydroxides (LDHs) are also significantly altered by controlled heating. Heating up to 400 °C removes weakly bound interlayer water and physically adsorbed water, resulting in increased porosity and surface area. However, according to Islam and Patel (2009, 2011), Islam et al. (2010), the nitrate adsorption capacity of LDHs improved only up to 200 °C; beyond this temperature, structural deformation led to a decline in performance.

Additionally, Socías-Viciana et al. (2008) reported that heating LDHs to 500 and 650 °C produced amorphous Mg–Al mixed oxides with enhanced surface area and a greater number of active sites for nitrate adsorption. Interestingly, nitrate adsorption capacity increased from 62 to 105 mg/g with temperature due to the decomposition of carbonate ions and their subsequent replacement by nitrate ions. However, heating to 850 °C resulted in the formation of spinel-like crystalline phases, which reduced the surface area. While samples treated at 500 and 650 °C notably regained their layered structure upon rehydration, the sample heated to 850 °C only partially restored its structure, leading to diminished adsorption performance.

Although heat treatment is effective, its feasibility depends on balancing the energy costs with the achieved adsorption improvements. In some cases—such as heat treatment of $ZnCl_2$-AC—adsorption capacity increased by more than 100%, making the process highly attractive. However, in other cases, the energy input may outweigh the benefits, especially when the raw materials already exhibit strong adsorption capabilities without thermal modification.

3.2.2 Ball Milling and Activation

BM has garnered attention as a cost-effective, simple, and eco-friendly technique for enhancing the physicochemical properties of AC. The kinetic energy transferred by the moving balls breaks down larger particles, generating highly reactive surfaces and increasing the number of available adsorption sites. Among various lignocellulosic feedstocks, mangosteen peel (MP)—an agricultural waste rich in polysaccharides and phenolic compounds—has shown particular promise as a precursor for AC, especially in the adsorption of methylene blue due to its superior adsorption capacity. Studies have demonstrated that BM significantly improves adsorption efficiency and increases internal surface area for pollutants such as phenol and Ni(II). While many studies have employed prolonged milling durations, recent findings suggest that short-duration BM can also substantially enhance adsorption performance while reducing operational costs (Nasrullah et al., 2021).

In addition to particle size reduction, wet BM alters the surface chemistry of carbon materials—an essential factor in adsorption performance. Surface functional group modifications during milling can improve the material's selectivity and binding affinity for specific pollutants. For instance, oxidation reactions during milling may increase metal ion adsorption while reducing the adsorption of certain organic compounds. The enhancement of surface area and porosity further contributes to elevated adsorption capacities. As noted by Partlan et al. (2016), BM significantly boosts the performance of AC for environmental applications, particularly in water and wastewater treatment systems targeting specific contaminants.

BM enhances adsorption by increasing surface area, reducing particle size, and introducing functional groups that facilitate stronger interactions with pollutants through mechanisms such as ion exchange, electrostatic attraction, and hydrophobic interactions. While adsorption mechanisms and regeneration strategies have been extensively studied, further research is needed to evaluate performance under real-world environmental conditions, scale-up feasibility, and ecological safety. A comprehensive life cycle analysis (LCA) is essential for ensuring the sustainable development and practical deployment of ball-milled adsorbents (Gao et al., 2024).

Moreover, BM is an efficient and eco-friendly method for producing nanostructured materials with improved surface uniformity and properties. BM pretreatment has been shown to significantly enhance the physicochemical characteristics of BC, boosting its ability to adsorb both heavy metals and organic pollutants. However, the nanoscale size and high surface activity of BC particles present challenges for recovery from treated solutions, necessitating the development of innovative separation techniques for large-scale applications (Li et al., 2020).

3.3 Chemical Modification

ACs exhibit neutral, basic, or acidic surface chemistries, depending on their surface functional groups. Figure 3.2 illustrates the various chemical modification techniques used to tailor these surface characteristics. By introducing or

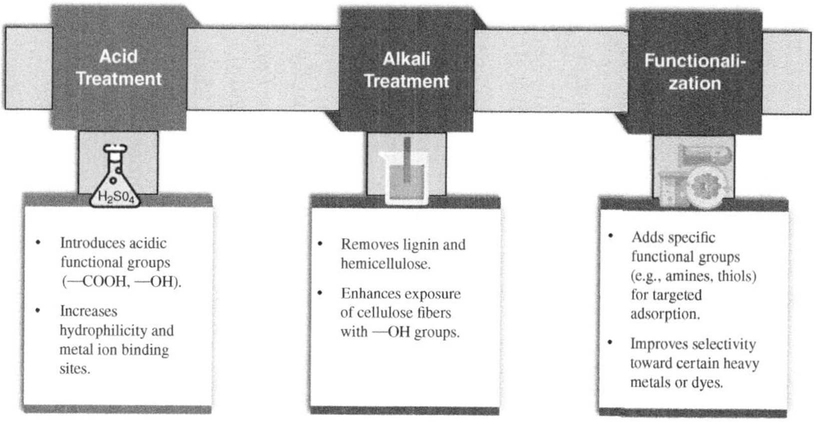

Figure 3.2 Chemical modification techniques.

altering specific functional groups, the surface chemistry of ACs can be precisely tuned to enhance adsorption capacity and selectivity toward different pollutants. These chemical modifications enable the development of adsorbents specifically designed to target certain contaminants, thereby improving their performance in environmental remediation applications (Rehman et al., 2019). Table 3.3 presents a comparative overview of surface functionalization methods for bioadsorbents, including treatments using acids, alkalis, and chemical reagents.

3.3.1 Acid and Alkali Treatment

Oxidation is a commonly employed technique to enhance the acidic nature and hydrophilicity of AC surfaces by reducing mineral content. Acids such as phosphoric, sulfuric, and nitric acids are frequently used for this purpose. Through oxidation, oxygen-containing functional groups—such as carboxylic, lactonic, and phenolic hydroxyl groups—are introduced onto the carbon surface. The nature and extent of these acidic groups vary depending on whether liquid-phase or gas-phase oxidants (e.g., CO_2, O_2, steam, or air) are used, each requiring different energy inputs. These surface functionalities enhance heavy metal adsorption by forming metal complexes at negatively charged sites, and they also facilitate the removal of organic pollutants from water (Rehman et al., 2019).

Structural changes induced by acid treatments often lead to increased porosity and surface roughness, making adsorption sites more accessible to contaminants. The improved surface chemistry supports stronger interactions with pollutants, including complexation and ion exchange. Furthermore, the type

Table 3.3 Surface functionalization of bioadsorbents through acid, alkali, and reagent treatments.

Bioadsorbent	Chemical treatment	Functional groups introduced	Purpose of modification	Reference
Orange peel	HCl treatment	—COOH, —OH	Increase active sites for metal binding	Annadurai et al. (2002)
Sawdust	NaOH treatment	—OH, $—COO^-$	Enhance porosity and deprotonate surface	Malik (2004)
Banana peel	H_2SO_4 activation	—COOH, —OH	Improve surface acidity and adsorption sites	Annadurai et al. (2002)
Rice husk	H_3PO_4 activation	$—PO_4^{3-}$, —OH	Develop mesoporous structure	Kurniawan et al. (2006)
Chitosan	Glutaraldehyde crosslinking	$—NH_2$, Schiff base (—C=N—)	Enhance mechanical strength and selectivity	Rinaudo (2006)
Biochar	HNO_3 oxidation	—COOH, —OH, $—NO_2$	Increase acidic functional groups	Mohan et al. (2014)
Coconut shell activated carbon	KOH activation	—OH, —C=O	Develop micro/mesopores and basic sites	Babel and Kurniawan (2003)
Cellulose	Citric acid functionalization	—COOH, ester groups	Improve affinity for cationic pollutants	Aslam et al. (2023); He et al. (2018)
Alginate beads	$CaCl_2$ gelation + functionalization	$—COO^-$, carboxylates	Stabilize and provide ion exchange capacity	Wang and Chen (2009)
Lignin	H_2SO_4 sulfonation	$—SO_3H$, —OH	Improve water solubility and adsorption sites	Cazier et al. (2024); Gao et al. (2019)
Moringa seed cake	NaOH followed by citric acid	—OH, —COOH	Activate binding sites for turbidity/ions	Gassenschmidt et al. (1995)
Peat	HNO_3 and H_2O_2 oxidation	—OH, —COOH, —C=O	Increase cation exchange and oxidative groups	Ho and McKay (2000)

and concentration of acid used can be optimized to tailor surface properties for specific applications. However, it is critical to control treatment conditions to prevent damage to the adsorbent structure. According to Abegunde et al. (2020), acid modification remains a versatile and effective method for improving the efficiency of adsorbents in environmental remediation.

Base treatment, particularly using ammonia at high temperatures (400–900 °C), introduces nitrogen-containing functional groups—such as amide, imide, lactam, pyrrolic, and pyridinic groups—thereby enhancing the basicity of the AC surface (Bumajdad et al., 2023). These modifications improve the adsorption of acidic species through interactions like hydrogen bonding, covalent bonding, and dipole–dipole interactions. Studies have shown that ammonia treatment at 650–700 °C significantly boosts adsorption performance, including a fourfold increase in perchlorate removal (He et al., 2025). The porous structure of the adsorbent is largely preserved during this treatment, while nitrogen doping is facilitated by free radicals generated during ammonia exposure. These radicals react with surface carboxylic acids to form amides and nitriles, further enhancing adsorption properties (Rehman et al., 2019).

Alkali modification using agents like NaOH or KOH is another effective method to alter surface functional groups and improve adsorption performance, particularly for nonpolar pollutants. This treatment introduces positive surface charges, which facilitate the adsorption of negatively charged species (Cashin et al., 2018). Additionally, alkali treatments reduce oxygen-containing groups and increase both surface area and pore volume, thereby improving the uptake of gases such as methane and organic dyes. For instance, alkali-treated AC demonstrates significantly higher dye adsorption capacity compared to untreated material. Similarly, alkaline modification of biomass enhances the removal of heavy metals from aqueous environments (Sultana et al., 2022). Among alkali treatments, KOH activation is particularly effective, producing adsorbents with superior surface area and porosity, which translate into exceptional adsorption capacities for specific dyes (Abegunde et al., 2020). Overall, alkali treatment is a valuable strategy for improving the surface characteristics and performance of adsorbents.

3.3.2 Surface Functionalization

Surface functionalization significantly enhances the adsorption capacity and selectivity of biomass-derived adsorbents. The introduction of functional groups onto the adsorbent surface increases the number of active sites and facilitates targeted adsorption through mechanisms such as electrostatic interactions. The incorporation of charged moieties enables the selective attraction of oppositely charged pollutants, thereby improving removal efficiency. Several factors influence adsorption performance, including surface chemistry, pore structure, surface area, and the physico-chemical characteristics of both the adsorbent and adsorbate. Common surface modification techniques—such as amine or carboxylic group functionalization and acid/base activation—enhance these interactions and improve pollutant uptake (Karim et al., 2024).

Carbon-based adsorbents, such as AC, inherently possess hydrophobic and nonpolar surfaces that adsorb molecules primarily through van der Waals forces and dipole–dipole interactions, resulting in nonselective adsorption. However, surface modification—through the incorporation of functional groups that facilitate hydrogen bonding, ionic interactions, or covalent bonding—substantially improves adsorption selectivity. In liquid-phase adsorption, chemical surface characteristics play a critical role, while in gas-phase adsorption, physical properties such as surface area and pore volume are more influential. Oxidation techniques, typically applied at defect sites and structural edges, introduce oxygen-containing groups such as carboxyl, phenolic, ketone, carbonyl, and ether groups. These groups enhance surface acidity and increase affinity for metal ions and organic molecules, thus improving aqueous-phase pollutant removal (Kwak et al., 2022). For optimal adsorbent performance, both chemical and physical properties must be considered during surface modification.

In the case of inorganic adsorbents, surface characteristics profoundly affect adsorption capacity and stability, particularly due to interactions with hydroxyl (—OH) groups and saturated oxygens that often result in negatively charged surfaces. These characteristics can cause instability and reduce adsorption efficiency. Functionalization strategies, such as amine grafting using compounds like 3-aminopropyltriethoxysilane, are employed to mitigate these issues by reducing hydroxyl content and improving both adsorption performance and material durability. Compared to unmodified variants, amine-functionalized inorganic adsorbents exhibit superior reusability and enhanced uptake of hazardous compounds. Despite their relatively lower adsorption capacity, the customizable surface charge of inorganic materials allows for tailored interactions via silane coupling, physical adsorption, or electrostatic interactions, improving water treatment efficacy (Bouazizi et al., 2022).

Naturally occurring hydroxyl groups on adsorbent surfaces provide reactive sites essential for enhanced pollutant adsorption. Techniques such as acid/base activation are employed to increase hydroxyl density, thus improving adsorption effectiveness. However, processes like high-temperature hydrothermal calcination can reduce hydroxyl content, making alternative methods—such as solvent extraction or microwave digestion—more suitable for preserving these groups. Additionally, carboxylic acid functionalization significantly boosts adsorption due to the electron-rich nature of carboxyl groups, which contribute to metal ion complexation and chelation. For instance, silica functionalized with sinapinaldehyde has shown selective Pb^{2+} adsorption, though to a lesser extent than silica modified with EDTA. A range of functional groups—including hydrazide and aminopropylcarboxylic acids—have been explored to optimize mesoporous silica for heavy metal adsorption via covalent bonding or electrostatic attraction (Cashin et al., 2018).

3.4 Biological Modification

Biological modification significantly enhances the adsorption capacity and selectivity of bioadsorbents by altering their surface chemistry, morphology, or functionality through the use of microorganisms, enzymes, or genetic tools. These modifications introduce new capabilities such as biodegradation and biomineralization, while also improving the adsorbents' stability, reusability, and overall adsorption efficiency. Key strategies include genetic programming, enzymatic augmentation, and microbiological alteration. Beyond increasing pollutant affinity, these biological modifications promote sustainable and environmentally friendly water treatment approaches. Recent studies highlight promising advances in the use of biologically modified adsorbents for the removal of dyes and heavy metals (Andriambahiny et al., 2025). An overview of these techniques is illustrated in Figure 3.3.

A common biological modification approach involves combining microbial degradation with adsorption by immobilizing microorganisms within the porous structure of AC. Under optimal environmental conditions, these microbes proliferate and form biofilms on the carbon surface. The biofilms enhance the adsorption process by degrading complex organic pollutants into simpler compounds such as carbon dioxide and biomass, thereby extending the lifespan and improving the performance of the carbon adsorbent. Studies have shown that using polyvinyl alcohol (PVA) as an entrapment matrix improves mass transfer

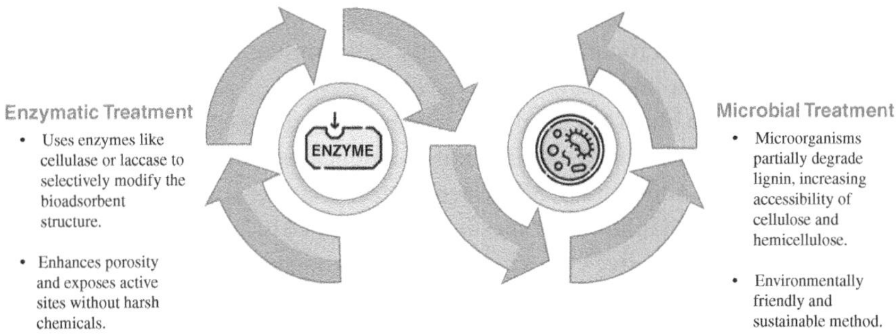

Figure 3.3 Biological modification techniques.

efficiency within these biofilm systems. Moreover, AC acts as an effective microbial carrier, significantly increasing the removal efficiency of contaminants like p-phenol (Jha et al., 2021).

3.5 Advanced Synthesis Techniques

Advanced production techniques have significantly enhanced AC's capacity for contaminant removal. Various modifications aimed at improving surface area, porosity, and functional groups have been explored, including chemical activation, composite development, and hybrid physical-chemical methods. For instance, chemically modified magnetic AC has shown high adsorption capacity, while KOH-activated AC has proven effective in dye removal. Composite materials such as AC-polyaniline exhibit excellent initial removal efficiencies but often suffer from reduced performance after regeneration. Other promising approaches include HTC combined with microwave-assisted activation, particularly effective for dye removal within specific pH ranges. Despite these advances, few studies comprehensively compare chemical versus physical modifications, and many lack consistent evaluations of adsorbent regeneration performance. There remains a critical need for standardized testing protocols and assessments of adsorbent efficacy against multiple contaminants to ensure practical applicability (Khan et al., 2024).

3.5.1 Microwave-assisted Synthesis

Compared to traditional methods, microwave-assisted synthesis significantly enhances adsorbent properties by providing rapid and uniform heating through ionic conduction and dipole rotation. Bayode (2016) demonstrated improved dye removal efficiency using microwave-activated hybrid clay adsorbents. Lim and Foo (2022) successfully synthesized a multifunctional carbonaceous adsorbent (SBAC) from soybean industrial bio-residue via one-step microwave irradiation. The SBAC exhibited a microporous–mesoporous structure with a high pore volume (0.94 cm^3/g) and an impressive surface area of 1696 m^2/g. Its adsorption capacities for methylene blue (434.57 mg/g), acetaminophen (393.31 mg/g), and 2,4-D (253.17 mg/g) followed the Langmuir and Freundlich isotherms, indicating excellent performance. Additionally, SBAC demonstrated strong antibacterial and antifungal activities, with inhibition zones ranging from 14.0 to 28.0 mm. Regeneration tests showed that after five cycles of microwave irradiation, the adsorbent maintained effective performance (69.42–79.31%). This study highlights a sustainable, low-energy "waste-to-wealth" approach for converting agro-waste into high-performance, multipurpose environmental adsorbents.

Nilavu et al. (2023) used a microwave-assisted green synthesis method to prepare a phosphorus-doped carbon-based adsorbent from Mangifera indica as a natural precursor. This material possessed a high surface area (390.17 m^2/g) and a mesoporous structure, enabling adsorption of 64.1 mg/g of dibenzothiophene (DBT) from model fuel. Comprehensive characterization techniques—including FTIR, Raman, XRD, XRF, TGA, XPS, and SEM-EDS—confirmed structural integrity before and after adsorption. Adsorption studies revealed an endothermic and spontaneous process governed by pseudo-first-order kinetics. The adsorbent efficiently removed 66% of sulfur from commercial diesel using n-pentane as a solvent, retaining high regeneration efficiency over six cycles. This work demonstrates microwave-assisted synthesis as an environmentally friendly and effective route for deep fuel desulfurization.

Ojoro (2015) reported the microwave synthesis of a ZnO-activated carbon (Ac-ZnO) nanocomposite from leftover palm seeds. This rapid and cost-effective method yielded a material capable of both adsorbing and photocatalytically degrading phenol in water. Successful synthesis was validated via FTIR, SEM, and point of zero charge measurements. Under optimal conditions—0.2 g of adsorbent and 400 mg/L phenol at pH 5.0 in sunlight—the material exhibited a maximum phenol adsorption capacity of 39.62 mg/g. The adsorption followed pseudo-first-order kinetics. Moreover, photocatalytic degradation was enhanced by sunlight and hydroxyl radicals generated from H_2O_2, confirming that coupling adsorption with photocatalysis effectively removed phenol from water.

AC produced from pomegranate peel waste (PPW) using home microwave-assisted irradiation has proven effective for the removal of Congo Red (CR) dye from water. Batch adsorption studies investigated the effects of varying temperature, adsorbent dosage, initial dye concentration, and contact time. The material's physicochemical properties were characterized by XRD, FTIR, BET surface area analysis, zeta potential measurements, and SEM imaging. Adsorption data fitted best to the Langmuir isotherm model, revealing a maximum adsorption capacity of 26.93 mg/g at 40 °C. Thermodynamic analysis indicated that the adsorption was both endothermic and spontaneous, while kinetic data corresponded well with the Elovich and pseudo-second-order models. Molecular dynamics simulations and density functional theory calculations further validated the experimental findings, confirming that PPW-derived AC is an economical and eco-friendly adsorbent for dye removal (Benahdach et al., 2025).

In another innovative approach, microwave-assisted surface imprinting was used to synthesize a novel sorbent based on surface molecularly imprinted polymers (MIPs) for selective adsorption and extraction of 4-nitrophenol (4-NP) from aqueous solutions. The $Fe_3O_4@SiO_2$@PNP-SMIP polymer was prepared using γ-ureidopropyltrimethoxysilane as the functional monomer supported on $Fe_3O_4@SiO_2$ nanoparticles. This imprinted polymer exhibited superior adsorption capacity (134.23 mg/g) compared to its non-imprinted counterpart. Characterization techniques such as SEM, FTIR, TEM, and BET confirmed the polymer's structure and morphology. Adsorption followed the Langmuir isotherm and pseudo-second-order kinetic models. Demonstrating 99% extraction efficiency at neutral pH and high selectivity even in the presence of structurally similar compounds, this MIP shows strong potential for solid-phase extraction applications targeting 4-NP removal from wastewater (Liang et al., 2020).

3.5.2 Hydrothermal Carbonization

HTC is a thermochemical process that converts wet biomass into valuable carbonaceous materials such as AC and high-energy density solid fuels. These products find applications in fuel production, soil amendment, and pollution remediation. One significant advantage of HTC is its ability to process wet biomass directly, eliminating the need for energy-intensive pre-treatment. For example, Ugbebor (2019) demonstrated that producing 100 kg of AC from sewage sludge via HTC followed by steam activation requires 2120 kg of sludge, 1059.22 kg of water, and 2520 MJ of energy. The process showed economic feasibility with a projected production cost of €91, a BET surface area of 226 m^2/g, and a 9.66% return rate.

Anaerobically digested effluent (ADE), a byproduct of wastewater treatment rich in water, carbon compounds, microbes, and nutrients, poses disposal challenges due to the limitations of conventional methods like landfilling, composting, and incineration. HTC provides a sustainable alternative by converting ADE into hydrochar, which can serve as an effective adsorbent for dye removal. Sivaprasad (2021) produced hydrochar by processing ADE at 200°C for 1 hour and enhanced its adsorption performance for methylene blue dye by modifying it with maleic anhydride. The modified hydrochar exhibited a maximum adsorption capacity of 107.6 mg/g and demonstrated excellent regeneration potential over multiple cycles.

In the realm of biosorbents, *Mucor rouxii* (MR) has garnered attention for heavy metal removal, particularly chromium (Cr VI) from wastewater. Immobilization of MR in silica-doped calcium alginate beads significantly improves its adsorption capacity, reaching up to 132 mg/g (Ho, 2020). Further enhancement is achieved through HTC combined with nitrogen doping using polyethyleneimine (PEI), elevating adsorption capacities above 400 mg/g. The removal mechanism involves electrostatic adsorption and the reduction of toxic Cr(VI) to the less harmful Cr(III), facilitating detoxification. Magnetic functionalization with Fe_3O_4 nanoparticles allows for facile adsorbent recovery, underscoring its potential for sustainable water treatment applications.

Optimization of HTC parameters using a response surface methodology (RSM) was employed for rice husk hydrochar to maximize adsorption capacity (q_e) and mass yield (MY). While MY remained relatively low (~40%), temperature was the dominant factor enhancing methylene blue (MB) adsorption capacity. Multi-response optimization successfully increased MY to above 60%. Subsequent alkali activation improved the pseudo-second-order rate constant (k_2), likely due to increased surface area, but did not significantly affect q_e. Characterization by SEM, FTIR, XRD, and BET revealed that surface functional groups and porosity critically influenced MB adsorption, indicating potential for commercial hydrochar production (Naranjo et al., 2023).

3.6 Composite and Hybrid Bioadsorbents

Heavy metals in wastewater present significant long-term environmental risks due to their toxicity and nonbiodegradability. Conventional removal methods—such as membrane filtration, ion exchange, and chemical precipitation—often suffer from drawbacks like high operational costs, sludge generation, and poor efficiency at low metal concentrations. As a promising alternative, hybrid bioadsorbents have gained attention for their affordability, reusability, and eco-friendly nature. Key factors influencing their adsorption performance include pH, temperature, contact time, and adsorbent dosage. Studies have identified ion exchange and surface complexation as the principal mechanisms for metal uptake, emphasizing their potential for scalable wastewater remediation (Abu Taleb et al., 2022).

The industrial revolution has improved living standards but has simultaneously exacerbated environmental pollution, especially from textile and allied industries that discharge persistent dyes. Traditional wastewater treatments are often ineffective against such dyes, which has led to increased research into nanohybrid materials for remediation. Nanocomposites such as graphene-metal oxide hybrids, carbon nanotubes, and nanorods have been engineered for their high adsorption capacity, cost efficiency, and reusability. For example, chitosan-magnetic iron oxide and carrageenan-silica nanohybrids have demonstrated impressive dye removal efficiencies, with some achieving up to 95.4%. These unique structural and functional attributes position nanohybrid adsorbents as strong candidates for sustainable water purification (Mahalakshmi and Growther, 2023).

Among these, the biohybrid nanocomposite Chit-PVA/MgO/Fe_3O_4 shows effective adsorption of remazol brilliant blue R (RBBR) dye. Characterization techniques including BET surface area analysis, FTIR, SEM-EDX, and XRD confirmed its porous and functionalized nature. Utilizing RSM to optimize parameters such as dosage, pH, temperature, and contact time, the system followed pseudo-second-order kinetics and fitted well with the Langmuir isotherm model, yielding a maximum adsorption capacity of 163.7 mg/g. The adsorption mechanism likely involves a combination of electrostatic attractions, hydrogen bonding, and n–π interactions, underscoring the nanocomposite's potential for efficient dye removal in wastewater treatment applications (Jawad et al., 2023).

3.7 Magnetic Biosorbents

Magnetic biosorbents, produced by impregnating biosorbents with metal or metal oxide nanoparticles, have emerged as highly effective materials for wastewater remediation. Their appeal lies in their large surface area, tunable active sites, easy magnetic separation, and excellent reusability. Key factors influencing magnetic biosorbents performance include the biomass source, nanoparticle characteristics, synthesis or modification methods, and processing conditions. Magnetic biosorbents utilize multiple adsorption mechanisms such as ion exchange and electrostatic interactions for ionic contaminants, and hydrophobic and hydrogen bonding for organic pollutants. Compared to conventional adsorbents like AC and modified BC, magnetic biosorbents typically exhibit superior adsorption efficiency and regeneration capacity, making them cost-effective for removing both heavy metals and organic pollutants (Hassan et al., 2020).

Magnetic biosorbents commonly integrate biomass-derived BC with metal or metal oxide nanoparticles. BC, often produced via pyrolysis of plant residues, gains adsorption capabilities through chemical functionalities such as hydroxyl (—OH), carboxyl (—COOH), and sulfhydryl (—SH) groups, which act as active binding sites for pollutants and provide anchorage points for metal nanoparticles. Typical magnetic components include Fe_3O_4, $FeCl_3$, and zero-valent iron (ZVI). While nanoparticles offer high surface area, their tendency to aggregate can reduce effectiveness; the synergy between carbonaceous biosorbents and magnetic nanoparticles enhances nanoparticle dispersion, stability, and ease of magnetic separation. This synergy makes magnetic biosorbents efficient, eco-friendly, and reusable adsorbents for contaminant removal in wastewater treatment (Tee et al., 2022).

A specific example is a novel magnetic biosorbent fabricated by immobilizing Bacillus subtilis onto nano-sized magnetic silica. Characterized by FTIR, TGA, VSM, and SEM, magnetic biosorbents demonstrated effective removal of Methylene Blue dye at pH 6.8 and 30 °C, reaching adsorption equilibrium within 3 hours. The biosorption capacity was measured at 59 ± 0.6 mg/g. Equilibrium data fit the Freundlich isotherm better than the Langmuir model, suggesting heterogeneous adsorption sites, while kinetics followed a pseudo-second-order model, indicating chemisorption as the rate-limiting step. Importantly, magnetic biosorbents retained high adsorption performance after five reuse cycles, highlighting its sustainability and potential for practical wastewater dye removal (Tural et al., 2017).

Two magnetic biosorbents synthesized from epichlorohydrin-cross-linked Saccharomyces cerevisiae cell walls demonstrated significantly enhanced metal ion adsorption capacities for Cu^{2+}, Cd^{2+}, and Ag^{+} compared to their non-magnetic cell wall counterparts, confirming the beneficial effect of magnetic modification on biosorption performance (Patzak et al., 1997).

In another approach, marine-derived fungus *Penicillium janthinellum* was used to generate self-immobilized mycelial pellets embedded with Fe_3O_4 nanoparticles during cultivation. Characterization via microscopy, EDX, FTIR, and TGA revealed these magnetic pellets achieved a maximum CR dye adsorption capacity of 102.4 mg/g. Furthermore, magnetic induction heating enhanced adsorption kinetics and energy efficiency relative to conventional heating methods. The magnetic nature allowed easy separation of pellets from water, underscoring their practical potential for efficient wastewater treatment (Zhang et al., 2016).

Magnetic biosorbents have also been explored for pharmaceutical pollutant removal. Zheng et al. (2020) developed polymer brush-modified core-shell magnetic composites (PMCCs) via radical polymerization, which introduced abundant active sites and elevated ciprofloxacin (CIP) adsorption capacity to an exceptional 527.93 mg/g—surpassing many existing adsorbents. Adsorption conformed to the Langmuir isotherm and pseudo-second-order kinetics, indicating monolayer chemisorption. Thermodynamic analysis showed the process to be spontaneous and exothermic. Notably, PMCCs retained over 80% adsorption efficiency after five reuse cycles, highlighting their excellent recyclability and making them promising candidates for sustainable removal of pharmaceutical contaminants in water.

3.8 Process Scale-up, Cost Factors, and Its Challenges

Recent improvements in adsorption-based water treatment have emphasized both economic feasibility and ethical responsibility. Green synthesis technologies—including sol-gel, hydrothermal, and bio-fabrication methods—enable the production of low-cost, high-performance adsorbents while minimizing environmental impact. Ethical concerns regarding waste generation have driven the use of biodegradable materials and circular economy solutions such as regeneration and reuse. Surface functionalization and composite synthesis enhance adsorbent lifetime and pollutant selectivity. The integration of technologies such as membrane filtration and photocatalysis has led to hybrid systems that improve treatment efficiency. Life cycle assessment (LCA) and cost-benefit analysis (CBA) are two evaluation methodologies that help decision-makers quantify sustainability measures. Additionally, surrogate-based optimization facilitates the translation of lab-scale concepts into scalable solutions (Satyam and Patra, 2024).

The overall economics of adsorption treatment facilities are heavily influenced by operating costs, regeneration expenses, and adsorbent prices, with the latter accounting for approximately 70% of total expenditure. Thus, producing low-cost adsorbents from biomass and waste materials is critical for long-term water treatment sustainability (Dotto and McKay, 2020). Although industry has improved effluent standards, further pollution reduction remains necessary. Most academic research focuses on maximum adsorption capacity derived from equilibrium isotherms at high adsorbate concentrations; however, pollutant levels in industrial wastewater are often much lower, and optimal adsorbent capacity must align with actual operating conditions.

Notably, fixed-bed adsorption units, which are widely used in industry, behave differently than the batch studies frequently reported in the literature. While kinetic and diffusion models are mentioned in around half of the research, less than 10% investigate fixed-bed columns, highlighting a gap in scale-up relevance. Alternative reactor designs—such as fluidized beds, spouted beds, and moving beds—deserve further exploration. Furthermore, accurate isotherm and diffusion models for multicomponent systems, as well as advanced hybrid modeling techniques, are required.

Sustainable regeneration methods are crucial for adsorbent reuse and adsorbate recovery, as they reduce waste and operational costs while enhancing process viability in environmental applications. Variability in biomass source and pretreatment significantly impacts adsorption capacity and selectivity. Biosorption modeling is challenging due to the diversity of surfaces and multimetal interactions, complicating accurate prediction. Nanomaterials raise concerns about environmental stability and toxicity. Surface modification techniques, while effective, are often expensive and technically complex, limiting scalability. Additionally, regeneration and reusability of plant-based adsorbents require further investigation, particularly under real-world conditions. The adsorption mechanisms for complex ions remain poorly understood, indicating a notable gap in current research. Addressing these challenges is critical to developing remediation systems that are efficient, sustainable, and economically viable (Raji et al., 2023).

3.9 Conclusion

This chapter reviewed various modification strategies employed to enhance the performance of bio-based adsorbents for wastewater treatment. Physical modifications such as heating, carbonization, BM, and activation increase surface area and porosity. Chemical treatments—including acid or alkali modification and surface functionalization methods like amination and esterification—alter functional groups to enable selective adsorption. Although still emerging, biological modifications offer environmentally friendly options for improving adsorbent efficacy. Advanced synthesis techniques, such as microwave-assisted synthesis and HTC, facilitate the production of structurally modified adsorbents with improved properties. Composite and hybrid bioadsorbents, particularly magnetic variants, demonstrate superior separation efficiency and recyclability. However, challenges remain in scaling these processes, ensuring cost-effectiveness, and addressing regeneration and environmental safety concerns. Overcoming these challenges is essential for practical, real-world applications. The future of sustainable water remediation lies in developing scalable, affordable, and ecologically safe adsorbents through strategic surface modifications that support long-term reuse and align with green chemistry principles.

References

Abegunde, S. M., Idowu, K. S., Adejuwon, O. M., & Adeyemi-Adejolu, T. (2020). A review on the influence of chemical modification on the performance of adsorbents. Resources, Environment and Sustainability, 1, 100001.

Abu Taleb, M., Halawani, R., Neamtallah, A., Kumar, R., & Barakat, M. (2022). Hybrid bioadsorbents for heavy metal decontamination from wastewater: a review. International Journal of Materials Technology and Innovation, 2(1), 5–19.

Al-Bagawi, A. H. (2024). Adsorption of Pb^{2+} ions from wastewater via microwave-assisted hydrothermal preparation of chitosan-modified magnetic nanoparticles. Biomass Conversion and Biorefinery, 14(23), 30149–30159.

Andriambahiny, R. N. A., Das, J., Roy, B., Lodh, B., & Pal, B. (2025). A review on the recent advancement of acid modified bio-adsorbents for the removal of methyl orange dye from wastewater treatment. Discover Chemistry, 2(1), 92.

Annadurai, G., Juang, R. S., & Lee, D. J. (2002). Use of cellulose-based wastes for adsorption of dyes from aqueous solutions. Journal of Hazardous Materials, 92(3), 263–274.

Aslam, A. A., Hassan, S. U., Saeed, M. H., Kokab, O., Ali, Z., Nazir, M. S., Siddiqi, W., & Aslam, A. A. (2023). Cellulose-based adsorbent materials for water remediation: harnessing their potential in heavy metals and dyes removal. Journal of Cleaner Production, 421, 138555.

Babel, S., & Kurniawan, T. A. (2003). Low-cost adsorbents for heavy metals uptake from contaminated water: a review. Journal of Hazardous Materials, 97(1–3), 219–243.

Bayode, A. A. (2016). Removal of cationic and anionic dyes from aqueous solution by microwave-assisted hybrid clay material. MSc Project, 3.

Benahdach, K., Ali, Y. A. E. H., Azzouz, A., González, M. D. M. C., Chraka, A., El Mail, R., Laghdach, A. E., Wahby, A., & Arhoun, B. (2025). Synthesis of activated carbon by microwave-assisted sulfuric acid activation for the adsorption of Congo Red in aqueous medium: kinetic, isotherm, thermodynamic, molecular dynamic, and functional theory studies. Journal of Molecular Liquids, 420, 126821.

Bhatnagar, A., Sillanpää, M., & Witek-Krowiak, A. (2011). Agricultural waste peels as versatile biomass for water purification–a review. Chemical Engineering Journal, 168(2), 493–504.

Bouazizi, N., Vieillard, J., Samir, B., & Le Derf, F. (2022). Advances in amine-surface functionalization of inorganic adsorbents for water treatment and antimicrobial activities: a review. Polymers, 14(3), 378.

Bumajdad, A., Khan, M. J. H., & Lukaszewicz, J. P. (2023). Nitrogen-enriched activated carbon derived from plant biomasses: a review on reaction mechanism and applications in wastewater treatment. Frontiers in Materials, 10, 1218028.

Cashin, V. B., Eldridge, D. S., Yu, A., & Zhao, D. (2018). Surface functionalization and manipulation of mesoporous silica adsorbents for improved removal of pollutants: a review. Environmental Science: Water Research & Technology, 4(2), 110–128.

Cazier, E. A., Pham, T. N., Cossus, L., Abla, M., Ilc, T., & Lawrence, P. (2024). Exploring industrial lignocellulosic waste: sources, types, and potential as high-value molecules. Waste Management, 188, 11–38.

Dotto, G. L., & McKay, G. (2020). Current scenario and challenges in adsorption for water treatment. Journal of Environmental Chemical Engineering, 8(4), 103988.

Foo, K. Y., & Hameed, B. H. (2012). Preparation, characterization and evaluation of adsorptive properties of orange peel based activated carbon via microwave induced K_2CO_3 activation. Bioresource Technology, 104, 679–686.

Gao, P., Fan, X., Sun, D., Zeng, G., Wang, Q., & Wang, Q. (2024). Recent advances in ball-milled materials and their applications for adsorptive removal of aqueous pollutants. Water, 16(12), 1639.

Gao, W., Inwood, J. P., & Fatehi, P. (2019). Sulfonation of hydroxy methylated lignin and its application. Journal of Bioresources and Bioproducts, 4(2), 80–88.

Gassenschmidt, U., Jany, K. D., Tauscher, B., & Niebergall, H. (1995). Isolation and characterization of a flocculating protein from *Moringa oleifera* Lam. Biochimica et Biophysica Acta (BBA) - General Subjects, 1243(3), 477–481.

Gupta, V. K., Carrott, P. J. M., Ribeiro Carrott, M. M. L., & Suhas. (2011). Low-cost adsorbents: growing approach to wastewater treatment—a review. Journal of Hazardous Materials, 190(1–3), 659–668.

Hassan, M., Naidu, R., Du, J., Liu, Y., & Qi, F. (2020). Critical review of magnetic biosorbents: their preparation, application, and regeneration for wastewater treatment. Science of the Total Environment, 702, 134893.

He, X., Luzi, F., Yang, W., Xiao, Z., Torre, L., Xie, Y., & Puglia, D. (2018). Citric acid as green modifier for tuned hydrophilicity of surface modified cellulose and lignin nanoparticles. ACS Sustainable Chemistry & Engineering, 6(8), 9966–9978.

He, Y., Guan, B., Zhuang, Z., Chen, J., Zhu, L., Ma, Z., Hu, M., Zhu, C., Zhao, S., Shu, K., Dang, H., Zhu, T., & Huang, Z. (2025). Advances in ammonia (NH3) adsorption and storage: materials, mechanisms, and applications. Adsorption, 31(2), 48.

Ho, Y. S., & McKay, G. (2000). The kinetics of sorption of divalent metal ions onto sphagnum moss peat. Water Research, 34(3), 735–742.

Islam, M., & Patel, R. (2009). Nitrate sorption by thermally activated Mg/Al chloride hydrotalcite-like compound. Journal of Hazardous Materials, 169(1–3), 524–531.

Islam, M., & Patel, R. (2011). Physicochemical characterization and adsorption behavior of Ca/Al chloride hydrotalcite-like compound towards removal of nitrate. Journal of Hazardous Materials, 190(1–3), 659–668.

Islam, M., Mishra, P. C., & Patel, R. (2010). Physicochemical characterization of hydroxyapatite and its application towards removal of nitrate from water. Journal of Environmental Management, 91(9), 1883–1891.

Jawad, A. H., Hameed, B. H., & Abdulhameed, A. S. (2023). Synthesis of biohybrid magnetic chitosan-polyvinyl alcohol/MgO nanocomposite blend for remazol brilliant blue R dye adsorption: solo and collective parametric optimization. Polymer Bulletin, 80(5), 4927–4947.

Jha, M. K., Joshi, S., Sharma, R. K., Kim, A. A., Pant, B., Park, M., & Pant, H. R. (2021). Surface modified activated carbons: sustainable bio-based materials for environmental remediation. Nanomaterials, 11(11), 3140.

Jian, X., Zhuang, X., Li, B., Xu, X., Wei, Z., Song, Y., & Jiang, E. (2018). Comparison of characterization and adsorption of bio-chars produced from hydrothermal carbonization and pyrolysis. Environmental Technology & Innovation, 10, 27–35.

Karaki, N., Aljawish, A., Humeau, C., Muniglia, L., & Jasniewski, J. (2016). Enzymatic modification of polysaccharides: mechanisms, properties, and potential applications: a review. Enzyme and Microbial Technology, 90, 1–18.

Karim, A. R., Danish, M., Alam, M. G., Majeed, S., & Alanazi, A. M. (2024). A review of pre-and post-surface-modified neem (Azadirachta indica) biomass adsorbent: surface functionalization mechanism and application. Chemosphere, 351, 141180.

Khan, I. A., Khan, A. U., Deen, K. M., Asselin, E., Sadiq, R., Yasir, M., & Ahmad, N. M. (2024). Ultrasonic assisted removal of methyl orange and bovine serum albumin from wastewater using modified activated carbons: RSM optimization and reusability. Materials Research Express, 11(9), 095505.

Kurniawan, T. A., Chan, G. Y. S., Lo, W. H., & Babel, S. (2006). Physico–chemical treatment techniques for wastewater laden with heavy metals. Chemical Engineering Journal, 118(1–2), 83–98.

Kwak, C. H., Lim, C., Kim, S., & Lee, Y. S. (2022). Surface modification of carbon materials and its application as adsorbents. Journal of Industrial and Engineering Chemistry, 116, 21–31.

Li, Y., Zimmerman, A. R., He, F., Chen, J., Han, L., Chen, H., Hu, X., & Gao, B. (2020). Solvent-free synthesis of magnetic biochar and activated carbon through ball-mill extrusion with Fe_3O_4 nanoparticles for enhancing adsorption of methylene blue. Science of the Total Environment, 722, 137972.

Liang, W., Lu, Y., Li, N., Li, H., & Zhu, F. (2020). Microwave-assisted synthesis of magnetic surface molecular imprinted polymer for adsorption and solid phase extraction of 4-nitrophenol in wastewater. Microchemical Journal, 159, 105316.

Lim, K. Y., & Foo, K. Y. (2022). One-step synthesis of carbonaceous adsorbent from soybean bio-residue by microwave heating: adsorptive, antimicrobial and antifungal behavior. Environmental Research, 204, 112044.

Loganathan, P., Vigneswaran, S., & Kandasamy, J. (2013). Enhanced removal of nitrate from water using surface modification of adsorbents–a review. Journal of Environmental Management, 131, 363–374.

Mahalakshmi, V., & Growther, L. (2023). Survey of nanohybrid materials in textile dyes removal. In: Nanohybrid Materials for Treatment of Textiles Dyes (pp. 19–40). Singapore: Springer Nature Singapore.

Maleki, S., & Karimi-Jashni, A. (2017). Effect of ball milling process on the structure of local clay and its adsorption performance for Ni (II) removal. Applied Clay Science, 137, 213–224.

Malik, P. K. (2004). Dye removal from wastewater using activated carbon developed from sawdust: adsorption equilibrium and kinetics. Journal of Hazardous Materials, 113(1–3), 81–88.

Mohan, D., Sarswat, A., Ok, Y. S., & Pittman Jr., C. U. (2014). Organic and inorganic contaminants removal from water with biochar, a renewable, low cost and sustainable adsorbent – a critical review. Bioresource Technology, 160, 191–202.

Nambiar, A. P., Pillai, R., Sanyal, M., Vadikkeettil, Y., & Shrivastav, P. S. (2023). A starch based sustainable bio-hybrid composite for surface assimilation of methylene blue: preparation, characterization, and adsorption study. Environmental Science: Advances, 2(6), 861–876.

Naranjo, J., Juiña, E., Loyo, C., Romero, M., Vizuete, K., Debut, A., Ponce, S., & Murillo, H. A. (2023). Preparation of adsorbent materials from rice husk via hydrothermal carbonization: optimization of operating conditions and alkali activation. Resources, 12(12), 145.

Nasrullah, A., Khan, A. S., Bhat, A. H., Din, I. U., Inayat, A., Muhammad, N., Bakhsh, E. M., & Khan, S. B. (2021). Effect of short time ball milling on physicochemical and adsorption performance of activated carbon prepared from mangosteen peel waste. Renewable Energy, 168, 723–733.

Nilavu, M. C., Arunraj, B., Aggarwal, H., & Rajesh, N. (2023). Expeditious microwave assisted green synthesis of phosphorous doped carbon obtained from Mangifera Indica for adsorptive desulfurization of fuel. Fuel, 345, 128172.

Ojoro, Z. E. (2015). Microwave assisted facile synthesis of zinc-oxide-activated carbon nanocomposite for photo-fenton degradation of phenol. Master's thesis. Eastern Mediterranean University (EMU)-Doğu Akdeniz Üniversitesi (DAÜ).

Osman, A. I., El-Monaem, E. M. A., Elgarahy, A. M., Aniagor, C. O., Hosny, M., Farghali, M., Rashed, E., Ejimofor, M. I., López-Maldonado, E. A., Ihara, I., Yap, P.-S., Rooney, D. W., & Eltaweil, A. S. (2023). Methods to prepare biosorbents and magnetic sorbents for water treatment: a review. Environmental Chemistry Letters, 21(4), 2337–2398.

Oumam, M., Abourriche, A., Mansouri, S., Mouiya, M., Benhammou, A., Abouliatim, Y., El Hafiane, Y., Hannache, H., Birot, M., Pailler, R., & Naslain, R. (2020). Comparison of chemical and physical activation processes at obtaining adsorbents from Moroccan oil shale. *Oil Shale*, 37(2), 139–157.

Partlan, E., Davis, K., Ren, Y., Apul, O. G., Mefford, O. T., Karanfil, T., & Ladner, D. A. (2016). Effect of bead milling on chemical and physical characteristics of activated carbons pulverized to superfine sizes. Water Research, 89, 161–170.

Patzak, M., Dostalek, P., Fogarty, R. V., Safarik, I., & Tobin, J. M. (1997). Development of magnetic biosorbents for metal uptake. Biotechnology Techniques, 11, 483–487.

Raji, Z., Karim, A., Karam, A., & Khalloufi, S. (2023). Adsorption of heavy metals: mechanisms, kinetics, and applications of various adsorbents in wastewater remediation—a review. Waste, 1(3), 775–805.

Rehman, A., Park, M., & Park, S. J. (2019). Current progress on the surface chemical modification of carbonaceous materials. Coatings, 9(2), 103.

Rinaudo, M. (2006). Chitin and chitosan: properties and applications. Progress in Polymer Science, 31(7), 603–632.

Satyam, S., & Patra, S. (2024). Innovations and challenges in adsorption-based wastewater remediation: a comprehensive review. Heliyon, 10(9), e29573.

Senturk, H. B., Ozdes, D., Gundogdu, A., Duran, C., & Soylak, M. (2009). Removal of phenol from aqueous solutions by adsorption onto organomodified Tirebolu bentonite: Equilibrium, kinetic and thermodynamic study. *Journal of hazardous materials*, 172(1), 353–362.

Sivaprasad, S. (2021). Hydrothermal carbonization of anaerobically digested effluent of sewage sludge to synthesize hydro char for remediation of methylene blue dye from water. Master's thesis. The Ohio State University.

Socías-Viciana, M. M., Ureña-Amate, M. D., González-Pradas, E., García-Cortés, M. J., & López-Teruel, C. (2008). Nitrate removal by calcined hydrotalcite-type compounds. Clays and Clay Minerals, 56(1), 2–9.

Sultana, M., Rownok, M. H., Sabrin, M., Rahaman, M. H., & Alam, S. N. (2022). A review on experimental chemically modified activated carbon to enhance dye and heavy metals adsorption. Cleaner Engineering and Technology, 6, 100382.

Tan, X., Liu, Y., Zeng, G., Wang, X., Hu, X., Gu, Y., & Yang, Z. (2015). Application of biochar for the removal of pollutants from aqueous solutions. Chemosphere, 125, 70–85.

Tee, G. T., Gok, X. Y., & Yong, W. F. (2022). Adsorption of pollutants in wastewater via biosorbents, nanoparticles and magnetic biosorbents: a review. Environmental Research, 212, 113248.

Tural, B., Ertaş, E., Enez, B., Fincan, S. A., & Tural, S. (2017). Preparation and characterization of a novel magnetic biosorbent functionalized with biomass of *Bacillus subtilis*: kinetic and isotherm studies of biosorption processes in the removal of Methylene Blue. Journal of Environmental Chemical Engineering, 5(5), 4795–4802.

Ugbebor, B. (2019). Feasibility of activated carbon via hydrothermally carbonized sewage sludge.

Usman, M., & Cheng, S. (2024). Recent trends and advancements in green synthesis of biomass-derived carbon dots. Eng, 5(3), 2223–2263.

Vajpayee, M., Singh, M., & Ledwani, L. (2021). Non-thermal plasma treatment of cellulosic biopolymer to enhance its surface property for various applications: a review. Materials Today: Proceedings, 43, 3250–3255.

Wang, J., & Chen, C. (2009). Biosorbents for heavy metals removal and their future. Biotechnology Advances, 27(2), 195–226.

Yang, Y., Li, X., Wan, C., Zhang, Z., Cao, W., Wang, G., & Wu, Y. (2024). A comprehensive review of cellulose nanomaterials for adsorption of wastewater pollutants: focus on dye and heavy metal Cr adsorption and oil/water separation. Collagen and Leather, 6(1), 35.

Yin, C. Y., Aroua, M. K., & Daud, W. M. A. W. (2007). Review of modifications of activated carbon for enhancing contaminant uptakes from aqueous solutions. Separation and Purification Technology, 52(3), 403–415.

Zhang, Q., Lu, T., Bai, D. M., Lin, D. Q., & Yao, S. J. (2016). Self-immobilization of a magnetic biosorbent and magnetic induction heated dye adsorption processes. Chemical Engineering Journal, 284, 972–978.

Zheng, C., Zheng, H., Hu, C., Wang, Y., Wang, Y., Zhao, C., Ding, W., & Sun, Q. (2020). Structural design of magnetic biosorbents for the removal of ciprofloxacin from water. Bioresource Technology, 296, 122288.

4

Nanotechnology-based Nano-biosorbents

4.1 Introduction

Global pollution levels have surged significantly due to rapid urbanization and the expansion of the industrial sector. Environmental assessments reveal that numerous pollutants are adversely affecting human health, wildlife, plant life, and the integrity of ecosystems. Addressing this challenge requires innovative, cost-effective, and environmentally sustainable methods for pollutant removal. Among emerging technologies, nano-biosorbents have received significant attention due to their remarkable efficiency and renewable origins. Derived from diverse natural sources, these nano-biosorbents show great promise for the removal of contaminants in the soil, water, and air, while aligning with environmental safety and economic feasibility (Khan et al., 2022a). Nanotechnology offers an effective strategy for heavy metal remediation owing to the exceptionally higher surface area and surface energy of nanoparticles. Among various treatment methods, adsorption remains the most widely utilized method for wastewater purification (Yadav et al., 2020).

The green synthesis (GS) of nanomaterials gives an environmentally friendly, reproducible, and scalable replacement to common chemical and physical methods. This approach encompasses the production of various nanomaterials, including metal nanoparticles, carbon nanoparticles, carbon nanofibers, nanocomposites, carbon dots, and quantum dots (Gupta et al., 2023). These nanoparticles exhibit unique properties such as chemical stability, high surface area, small size, and superior adsorption capacity, making them both cheap and efficient for pollutant removal. Currently, nanoparticles are produced from a various kinds of natural and synthetic sources. In particular, plants and agricultural wastes have gained attention as sustainable and abundant precursors due to their availability as environmental residues (Sachan et al., 2021). Plants act as ideal candidates for the GS of nanoparticles, offering an environmentally friendly and sustainable approach. Phytochemicals present in various plant parts—such as roots, stems, leaves, seeds, flowers, and latex—serve as natural capping and reducing agents in the biological synthesis of titanium dioxide (TiO_2) nanoparticles (Shahzadi et al., 2025). TiO_2 nanoparticles are among the most widely used engineered nanomaterials, finding applications across a broad range of nanotechnology-based consumer products. This review highlights the plant-based synthesis of TiO_2 nanoparticles across different taxonomic groups and explores their multifaceted applications (Sunny et al., 2022).

Biomass sources such as leaves, wood, and plant byproducts like husks act as renewable precursors in the GS of carbon nanoparticles (Debroy et al., 2023). This environmentally friendly approach can replace traditional metal catalysts in carbon nanoparticle production. Depending on the precursor type and processing conditions, carbon nanoparticles can form various morphologies, including nanofibers, nanotubes, or nanoporous structures. For example, cellulose subjected to hydrothermal treatment at 900 °C produces graphitic carbon coils, while coconut coir processed via hydrothermal carbonization and pyrolysis yields carbon nanosheets. Additionally, studies have explored the effects of clay minerals and the fabrication of hollow carbon nanostructures from cellulose husks through high-temperature pyrolysis and acid digestion. Food and agricultural wastes rich in proteins and lignin are also commonly employed for the GS of carbon nanoparticles (Qasim et al., 2023).

Magnetic nanoparticles have high surface area-to-volume ratio and the ease of separation via external magnetic fields. Among these, iron oxide nanoparticles are widely studied because of their cost-effectiveness and environmental compatibility. Functionalizing magnetic nanoparticles with biopolymers enhances their adsorption capacity and pollutant selectivity by introducing diverse functional groups. Biopolymers offer advantages such as sustainability, low toxicity, and biodegradability. The ideal biosorbent combines specificity, high efficiency, rapid adsorption, affordability, and easy recovery—qualities that the integration of magnetic iron oxides with biopolymers can fulfill. Advances in nanotechnology facilitate precise surface modifications, optimizing the efficiency and reusability of these composites for targeted water treatment applications (Soares et al., 2020).

This chapter highlights the innovative use of nano-biosorbents synthesized via green routes from renewable biomass, offering a sustainable alternative for pollutant removal. It emphasizes the unique properties arising from the nano–bio interface and discusses magnetic nano-biosorbents enabling easy recovery and reuse. By exploring diverse types, synthesis methods, and adsorption mechanisms, this chapter bridges nanotechnology and environmental science to advance water remediation strategies. Importantly, it also addresses environmental safety and toxicological aspects to promote responsible application. Overall, this chapter provides a comprehensive and novel perspective on nano-biosorbents as promising, eco-friendly materials for efficient and sustainable pollution control.

4.2 Nano-biosorbents

Nanotechnology focuses on materials ranging in size from 1 to 100 nm that have distinct chemical, biological, and physical characteristics from bulk materials. These nanoscale materials find applications across biology, electronics, environmental remediation, and catalysis. A major global challenge is water contamination caused by industrial effluents laden with heavy metals, dyes, and pharmaceutical residues. Owing to their chemical stability, resistance to biodegradation, and tendency to bioaccumulate, these pollutants are difficult to remove. Adsorption using bioadsorbents has emerged as an effective and promising method for water purification (Dutta and Charingia, 2020). Nano-biosorbents are advanced materials designed for the efficient elimination of harmful pollutants, due to their high surface area, small size, and enhanced stability. Frequently engineered at the nanoscale with zero-valent metals, they possess unique optical, electrical, and magnetic properties. These features enable greater reactivity with contaminants, faster chemical reactions, and reduced waste generation during treatment processes. Recognized as next-generation adsorbents, nano-biosorbents demonstrate exceptional performance in environmental purification, particularly in water and wastewater treatment. Their nontoxic nature further makes them an attractive and sustainable option for green environmental applications (Khan et al., 2022b). Table 4.1 illustrates the comparative evaluation of conventional adsorbents and nano-biosorbents for environmental pollutant removal. Figure 4.1 depicts the schematic representation of Nano-biosorbent-based water treatment process.

Magnetic composites derived from banana peel (BP) and pine needles (PNs) were synthesized and effectively utilized as biosorbents for the elimination of methylene blue (MB) dye from textile and aqueous wastewaters. Scanning electron microscopy (SEM) and X-ray diffraction (XRD) analyses revealed rough, heterogeneous surfaces and successful magnetite loading. Fourier-transform infrared spectroscopy (FTIR) confirmed the presence of functional groups responsible for dye adsorption. Thermogravimetric analysis indicated characteristic weight loss patterns associated with thermal degradation. Key parameters influencing the adsorption process included adsorbent dosage, pH, temperature, contact time, and initial pollutant concentration. Adsorption equilibrium data were best fitted by the Freundlich isotherm for PN composites and the Temkin isotherm for BP composites. Thermodynamic studies indicated endothermic adsorption behavior, while kinetic analysis followed a pseudo-second-order (PSO) model. Notably, PN

Table 4.1 Comparative evaluation of conventional adsorbents and nano-biosorbents for environmental pollutant removal.

Parameter	Conventional adsorbents	Nano-biosorbents	Reference
Adsorption efficiency	Moderate; limited surface area	High; enhanced surface area and reactivity	Zhang et al. (2016); Sajid et al. (2024)
Reusability	Limited cycles due to fouling	High; stable over multiple adsorption–desorption cycles	Ngoc et al. (2022); Sharma et al. (2009)
Cost	Generally low	Moderate; reduced with green synthesis	Batool and Valiyaveettil (2021); Das et al. (2022)
Eco-friendliness	May generate secondary pollutants	Biodegradable, low toxicity if synthesized via green routes	Sajid et al. (2024); Das et al. (2022)
Selectivity	Nonspecific to pollutants	High selectivity for dyes, metals, microplastics	Zhang et al. (2016); Batool and Valiyaveettil (2021)
Surface functionalization	Difficult to modify	Easily functionalized (e.g., with β-cyclodextrin, PEI)	Sajid et al. (2024); Batool and Valiyaveettil (2021)
Kinetic behavior	Often fits first-order kinetics	Often fits pseudo-second-order and Langmuir isotherms	Sharma et al. (2009); Ngoc et al. (2022)
Particle aggregation	Less prone	Prone; requires stabilizers (e.g., polymers, carbon nanotubes)	Sajid et al. (2024); Das et al. (2022)
Mechanism of adsorption	Physical adsorption dominant	Electrostatic interaction, hydrogen bonding, chemical coordination	Zhang et al. (2016); Batool and Valiyaveettil (2021)
Thermal stability	Moderate	High; especially for metal oxide nanomaterials	Ngoc et al. (2022)
Toxicological concerns	Low	Possible risks to ecosystems and health if not well controlled	Das et al. (2022); Bakand and Hayes (2016)
Application versatility	Limited to specific pollutants	Versatile; applicable to dyes, metals, microplastics, Cr(VI), arsenic	Deedar and Aslam (2009); Sharma et al. (2009); Zhang et al. (2016)

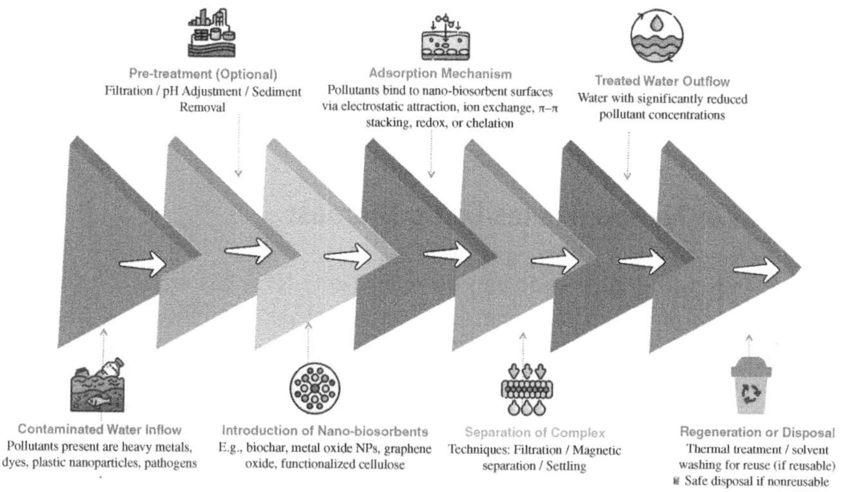

Figure 4.1 Schematic representation of Nano-biosorbent-based water treatment process.

composites exhibited the highest adsorption capacity, highlighting their potential as low-cost, eco-friendly biosorbents for sustainable wastewater treatment (Yaqub et al., 2022).

Lemon peel nanoparticles (LPNPs), sized between 40 and 60 nm, have demonstrated outstanding biosorption efficiency for Fe^{2+} and Ni^{2+} ions, achieving removal rates exceeding 93% under optimal pH and contact time conditions. Compared to conventional lemon peel adsorbents, LPNPs offer superior performance due to their higher surface area, enhanced reactivity, and cost-effective, eco-friendly synthesis. The nanoscale production of LPNPs enables uniform particle stabilization, significantly improving adsorption characteristics. Key synthesis parameters are temperature, pH, and reaction time—strongly influence particle morphology and performance. Additionally, LPNPs contribute to a better mechanistic understanding of molecular electrostatic interactions during adsorption. Further research focusing on the reusability, metal ion selectivity, and scale-up potential of LPNPs could facilitate their development as sustainable solutions for water purification (Karami et al., 2017).

Hazelnut skin-derived nano-adsorbents have shown highest removal of heavy metal ions, particularly Pb^{2+}, from contaminated water. Key operational parameters such as pH, adsorbent dosage, contact time, temperature, and sample volume significantly influence the adsorption process. The adsorption of Pb^{2+} ions was best described by the Langmuir isotherm model and followed PSO kinetics, indicating monolayer adsorption and chemisorption as the rate-limiting step. Under optimal conditions, the maximum adsorption capacity was 91 mg/g. Notably, the nano-biosorbent demonstrated exceptional recycling throughout many adsorption–desorption cycles, highlighting its promise as an environmentally renewable and cost-effective material for water purification applications (Karami and Kazemi, 2023).

4.3 GS of Nanoparticles Using Biomass

The biogenic GS of zinc oxide nanoparticles (ZnO NPs) has attained considerable attention owing to its cheap, environmental compatibility, and minimal utilization of hazardous chemicals. Unlike conventional chemical and physical methods, GS employs biological entities such as fungi, plant extracts, algae, and bacteria, which act as natural reducing and stabilizing agents. These biological substrates offer a renewable and nontoxic alternative, aligning with green chemistry principles. While large-scale production remains a technical challenge, recent advancements in the understanding of the chemical composition and reactivity of biological extracts offer promise for scalable and solvent-free synthesis routes. Such eco-friendly approaches significantly reduce environmental hazards and are paving the way for the broader application of ZnO NPs across diverse sectors, including environmental remediation, biomedicine, agriculture, and catalysis (Zeghoud et al., 2022). Various metal NPs have been successfully produced using plant extracts as efficient reducing agents, including silver nanoparticles (AgNPs) sized 6–15 nm produced with sesame oil cake, zero-valent iron nanoparticles (40–70 nm) prepared via *Nephrolepis auriculata*, and ZnO NPs measuring 70–75 nm using *Tecoma castanifolia* extracts. This biosynthesis generally involves a bio-reduction process wherein metal salt solutions are combined with plant extracts, facilitating reduction and nanoparticle formation. To enhance nanoparticle stability and broaden their functional applications, an additional step employing exogenous stabilizers is often incorporated into the synthesis protocol (Choudhary et al., 2020). Table 4.2 summarizes the different types of nano-biosorbents from biomass and their applications in environmental remediation.

In a recent study, lipid-extracted biomass from the thermotolerant microalga *Acutodesmus dimorphus*, cultivated in dairy effluent, was employed to biosynthesize spherical AgNPs with diameters ranging between 2 and 20 nm. These biosynthesized AgNPs exhibited notable free radical scavenging activity in ABTS (2,2′-azino-bis(3-ethylbenzothiazoline-6-sulfonic acid)) and DPPH (2,2-diphenyl-1-picrylhydrazyl) assays, indicating their potential as effective antiradical agents. However, additional study is necessary to build controlled synthesis methods capable of consistently producing nanoparticles with defined sizes and morphologies using specific microalgal strains (Chokshi et al., 2016).

Table 4.2 Summary of different types of nano-biosorbents and their applications in environmental remediation.

Biomass source	Type	Nanoparticles produced	Synthesis conditions	Nanoparticle size (nm)	Applications	Reference
Sisymbrium Irio	Plant	ZnO nanoparticle	Leaf extract, 60–70 °C, pH ~12	100	Anticancer activity	Maher et al. (2023)
Xylose	Natural monosaccharide	Au nanoparticle	Leaf extract, 37 °C, pH ~7	15 ± 5	Antibacterial agents	Badwaik et al. (2013)
Aspergillus fumigatus	Fungus	Au nanoparticle	Fungus extract, 37 °C, pH ~2.5	85.1–210	Antibacterial agents	Vasanthi Bathrinarayanan et al. (2013)
Wheat	Plant	Au nanoparticle	Aqueous solution, 95 °C, pH ~2–6	10–30	Antibacterial agents	Armendariz et al. (2004)
Oil palm empty fruit bunch	Lignin	Lignin nanoparticle	Aqueous solution, 50–80 °C, pH ~7	220	Food packing	Perera et al. (2023)
Botryosphaeria rhodina	Fungus	Ag nanoparticle	Sucrose extract, 27 °C, pH ~7	19.29	Biosorption of Ag(I)	Muñoz et al. (2021)
Botryosphaeria rhodina	Fungus	Ag nanoparticle	Yeast extract, 27 °C, pH ~7	24.9	Biosorption of Ag(I)	Muñoz et al. (2021)
Aloe vera	Plant	Ag nanoparticle	Plant extract, 27 °C, pH ~7	33.47±0.04	Antimicrobial activity	Muraro et al. (2022)
Phoenix dactylifera L.	Plant	ZnO nanoparticle	Plant extract, 60 °C, pH ~9.87	12.8	Biomedical and environmental applications	Doudi et al. (2024)

(Continued)

Table 4.2 *(Continued)*

Biomass source	Type	Nanoparticles produced	Synthesis conditions	Nanoparticle size (nm)	Applications	Reference
Phoenix dactylifera L.	Plant	Mg-Ag ZnO nanoparticle	Plant extract, 60 °C, pH ~9.87	22.0	Biomedical and environmental applications	Doudi et al. (2024)
Saccharum ravannae	Plant	Silica nanoparticle	Plant extract, 60–70 °C, pH ~2.5	3.56	Adsorption of Pb^{2+} and Cu^{2+}	Sachan et al. (2021)
Saccharum officinarum	Plant	Silica nanoparticle	Plant extract, 60–70 °C, pH ~2.5	39.47	Adsorption of Pb^{2+} and Cu^{2+}	Sachan et al. (2021)
Oryza sativa	Plant	Silica nanoparticle	Plant extract, 60–70 °C, pH ~2.5	29.13	Adsorption of Pb^{2+} and Cu^{2+}	Sachan et al. (2021)
Acutodesmus dimorphus	Microalgae	Ag nanoparticle	Algae extract, 4–60 °C, pH ~7	2–20	Antioxidant agents	Chokshi et al. (2016)
Aspergillus terreus	Fungus	Au nanoparticle	Fungus extract, 4–32 °C, pH ~2.5	186	Chemical, and biological sensors and as catalysts	Gurunathan et al. (2014)
Moringa oleifera	Plant	Fe_3O_4 nanoparticle	Plant extract, 25 °C, pH ~2.5–9.0	3.73	Adsorption of Pb(II) ions	Ngulube et al. (2024)
Moringa oleifera	Plant	Co-doped Fe_3O_4 nanoparticle	Plant extract, 160 °C, pH ~2.5–9.0	3.90	Adsorption of Pb(II) ions	Ngulube et al. (2024)
Moringa oleifera	Plant	Mn-doped Fe_3O_4 nanoparticle	Plant extract, 120 °C, pH ~2.5–9.0	3.90	Adsorption of Pb(II) ions	Ngulube et al. (2024)
Moringa oleifera	Plant	Co- and Mn-doped Fe_3O_4 nanoparticle	Plant extract, 120 °C, pH ~2.5–9.0	2.72	Adsorption of Pb(II) ions	Ngulube et al. (2024)

Al-Tamimi (2021), presents an eco-friendly approach to synthesizing manganese dioxide (MnO_2) and copper oxide nanoparticles (CuO NPs) using oat biomass for the fabrication of polymeric electrochemical sensors targeting cephalexin (CPX) and cefazolin (CFZ). The drugs were complexed with phosphomolybdic acid (PM) and phosphotungstic acid (PT) to form electroactive species—CPX-PM and CFZ-PT. These species were incorporated into nanoparticle-modified sensors, CPX-PM-MnO_2NPs and CFZ-PT-CuONPs, respectively. The modified sensors exhibited significantly enhanced conductivity and sensitivity compared to conventional sensors. Remarkably low detection limits were achieved—5.6×10^{-9} mol/L for CPX and 5.7×10^{-11} mol/L for CFZ—in both commercial and biological sample analyses. This GS approach underscores the promising role of biogenic nanoparticles in the advancement of high-performance electrochemical sensors.

In Massironi et al. (2019), ulvan—a sulfated polysaccharide derived from *Ulva* species of green algae—was employed as both a reducing and stabilizing agent in the GS of AgNPs. The ulvan-mediated synthesis produced AgNPs with significantly enhanced colloidal stability under physiological conditions, outperforming conventional stabilizers like citric acid owing to the presence of a robust polysaccharide shell. These ulvan-stabilized AgNPs demonstrated potent and rapid antibacterial activity against both Gram-positive and Gram-negative bacterial strains. This environmentally sustainable approach, leveraging renewable algal biomass, offers a promising pathway to the development of biocompatible antimicrobial nanoparticles for applications in cosmetics and biomedical fields. Figure 4.2 shows the GS pathways of Nanoparticles.

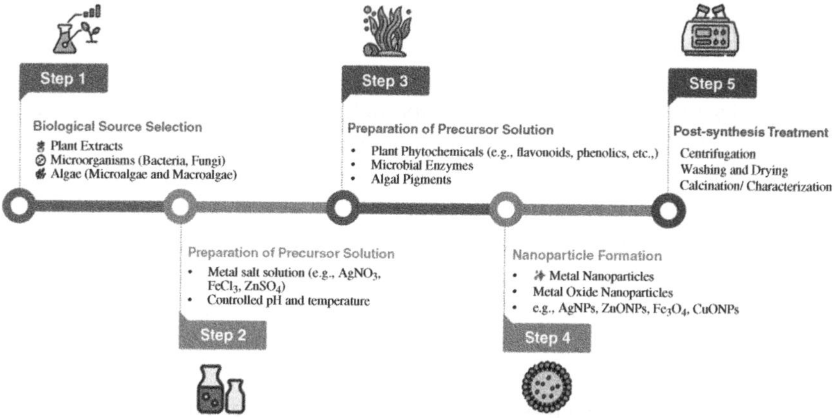

Figure 4.2 Green synthesis pathways of nanoparticles.

4.3.1 Plant Extract-mediated Synthesis

Bioactive compounds present in plant extracts—such as flavonoids, phenolics, and alkaloids—act as natural reducing and stabilizing agents during nanoparticle formation. These green-synthesized nanoparticles exhibit potent biocidal properties and have demonstrated effectiveness against multidrug-resistant bacterial strains. Unlike traditional antibiotics, their antibacterial mechanisms commonly involve the production of reactive oxygen species and rupture of microbial membranes, thus providing a novel approach to pathogen control (Küünal et al., 2018).

GS of nanoparticles, which employs fewer harmful chemicals, offers promising applications across medicinal and environmental fields (Jadoun et al., 2021). Nanoparticles derived from plant extracts like silver, gold, palladium, TiO_2, copper, and zinc oxide (ZnO) are extensively studied for their potential uses in imaging, catalysis, water treatment, medicine, sensors, textile engineering, and drug delivery. The unique phytochemicals in plants accelerate reactions and aid in nanoparticle synthesis. Despite the substantial potential of this green technology, concerns remain regarding the long-term ecological and health impacts on ecosystems, wildlife, and humans. To fully harness the benefits of biogenic nanoparticles, future research should prioritize assessing their environmental accumulation and safety policies.

Nanoparticles can be efficiently synthesized by simply mixing plant extracts with metal salt solutions at ambient temperature, with the reaction typically completing within few minutes. Several factors influence the synthesis and characteristics of the nanoparticles, including the metal salt concentration, pH, type and concentration of the plant extract, temperature, and reaction duration. Several research showed the production of Ag and AuNPs sized between 16 and 58 nm using leaf extracts from plants such as *Polyalthia longifolia, Cinnamomum camphora, and Ficus benghalensis.* Phytochemicals like phenolics, terpenoids, flavones, and polysaccharides play a crucial role in reducing metal ions and stabilizing the nanoparticles. The remarkable antibacterial and anticancer activities exhibited by these biosynthesized nanoparticles underscore the importance of plant-mediated GS in advancing nanotechnology (Mittal et al., 2013).

Magnolia leaf extract was employed as a natural reducing agent for the biosynthesis of copper nanoparticles. Upon addition of the extract to an aqueous solution of $CuSO_4{\cdot}5H_2O$, stable copper nanoparticles were formed, as confirmed by UV-visible spectroscopy. Characterization techniques including ICP, EDS, XPS, SEM, and HR-TEM revealed nanoparticle sizes ranging from 40 to 100 nm. The antibacterial efficacy of these nanoparticles was tested using shaking flasks containing latex foams coated with the copper NPs, by measuring viable *Escherichia coli* counts after 24 hours. The copper nanoparticle-coated foams demonstrated significantly enhanced antibacterial activity compared to uncoated foams. This biogenic synthesis method presents a sustainable and environmentally friendly alternative to conventional chemical and physical nanoparticle production techniques (Lee et al., 2011).

Plant extracts offer several advantages over chemical, physical, and microbial methods for AgNP synthesis. Unlike chemical approaches, plant-based synthesis does not require radiation, high temperatures, or hazardous reducing and capping agents. Compared to microbial synthesis, it avoids slow production rates, extensive culture maintenance, and contamination risks. Plant-mediated synthesis produces nanoparticles that are faster to generate, more stable, and more uniformly distributed, making them well-suited for biomedical applications. Additionally, reaction parameters such as temperature, pH, time, and reactant ratios can be easily adjusted to control nanoparticle size and shape—something often challenging with microbial techniques. The widespread availability, affordability, and straightforward preparation of plant extracts further enable environmentally friendly, scalable production of AgNPs. Nonetheless, advancements in genetic engineering and metabolite isolation may further enhance nanoparticle quality and production efficiency (Borase et al., 2014).

4.3.2 Microbial and Algal Mediated Synthesis

Due to their rapid growth, ease of cultivation, and rich content of secondary metabolites, algae have emerged as promising candidates for the GS of nanoparticles (NPs). In this environmentally friendly and sustainable approach, algae act as both reducing and capping agents, facilitating the formation of metal, metal oxide, or bimetallic nanoparticles through extracellular or intracellular mechanisms. Extensive research has been conducted using various algal types, including red, brown, micro and macro green, and blue-green, algae. This GS method is not only safe and cost-effective but also scalable, making it a viable alternative to conventional chemical and physical synthesis techniques. Algal-derived nanoparticles exhibit unique physicochemical properties and excellent biocompatibility, showing great potential in biomedical and environmental applications such as antibacterial, antifungal, anticancer, antifouling, bioremediation, and biosensing (Chaudhary et al., 2020). Algal extracts, rich in secondary metabolites, act as natural reducing and stabilizing agents in the formation of metal, metal oxide, and bimetallic NPs. Depending on the interaction site between algal biomolecules and metal ions, nanoparticle synthesis can occur either extracellularly or intracellularly. Commonly synthesized nanoparticles through this method include silver (Ag), gold (Au), copper oxide (CuO), cuprous oxide (Cu_2O), and ZnO. Due to their multifunctionality and excellent biocompatibility, these algal-derived nanoparticles hold significant promise for biomedical and environmental applications such as bioremediation, biosensing, anticancer therapies, and antibacterial, antifungal, and antifouling activities. Consequently, this GS route offers a sustainable alternative to conventional nanoparticle production techniques (Jain, 2022).

Microbial preparation of NPs using fungi, bacteria, and microalgae offers a low-energy, cost-effective, and environmentally friendly alternative to conventional methods. This green approach eliminates the use of hazardous chemicals and enables the formation of unique nanostructures such as nanowires, nanocellulose, and biomineralized materials (Qureshi et al., 2021). Members of the Chlorophyceae class, such as *Chlorella* and *Ulva*, have been reported to produce gold (AuNPs) and AgNPs with potent antibacterial, antifungal, antiprotozoal, and anticancer properties. Similarly, cyanobacteria such as *Spirulina* and *Microcoleus* are known for synthesizing AgNPs with broad-spectrum antimicrobial activity. Brown algae like *Sargassum* spp. have been utilized for the synthesis of a wide range of nanoparticles—including AuNPs, AgNPs, ZnO NPs, and TiO_2 NPs—suitable for industrial, biomedical, and agricultural applications. Furthermore, green algae such as *Scenedesmus* have been used in the GS of cadmium (Cd) NPs, particularly for bioremediation purposes. Despite these promising applications, large-scale production remains challenged by biological variability and limited mechanistic understanding. Thus, standardized protocols and advanced characterization techniques are essential to fully harness the potential of algal-mediated nanoparticle synthesis (Khanna et al., 2019).

Microbial cells offer a scalable, eco-friendly, and rapid platform for the biosynthesis of NPs, capable of producing a diverse array of nanostructures due to their ease of cultivation and adaptability. However, achieving consistency in nanoparticle size, shape, and composition remains a challenge, as these characteristics are highly sensitive to environmental factors. Despite significant progress, the commercial-scale production of microbial-derived nanoparticles is limited by concerns over scalability, toxicity, and biocompatibility. Genetically engineered microorganisms present a promising alternative, enabling more precise and efficient nanoparticle synthesis. Nevertheless, critical issues such as biosafety, public acceptance, and reproducibility must be addressed. Current research is increasingly focused on optimizing both microbial and recombinant techniques to enable sustainable nanoparticle production with broad applications in industry, agriculture, and medicine (Koul et al., 2021).

4.4 Types of Nano-biosorbents

A novel class of materials known as nano-biosorbents has been created to overcome the inherent limitations of conventional biosorbents, such as low mechanical strength, limited selectivity, and poor regeneration potential. These materials integrate biological components with nanomaterials and are available in various forms, including metal oxide nanoparticles, carbon-based nano-biosorbents, polymer-functionalized nanocomposites, and magnetic nano-biosorbents. Owing to their nanoscale dimensions, they exhibit enhanced surface area, superior adsorption capacity, and reduced diffusion resistance. Functionalization further improves their selectivity and recyclability. These advancements represent a significant step forward in creating scalable, efficient, and economically viable biosorption technologies (Mahamadi, 2019).

4.4.1 Carbon Nanomaterials (Biochar, Graphene Oxide)

Nano zero-valent iron (nZVI) and graphene oxide-enhanced biochar have emerged as promising nano-bioadsorbents for immobilizing copper (Cu) in contaminated soils. In this study, graphene oxide (GO) was utilized to modify corn stalk-derived biochar pyrolyzed at 450 °C, followed by impregnation with nZVI to produce corn stack biochar (CTBC) supported GO. Characterization confirmed enhanced surface properties, functional group diversity, and thermal stability. Application of the composite resulted in a significant reduction (>65%) of bioavailable Cu in the soil within 14 days. Sequential extraction analysis indicated the transformation of mobile Cu into more stable, less bioavailable fractions, thereby reducing its environmental toxicity. These findings highlight CTBC-GO/nZVI as an effective soil amendment for heavy metal immobilization and ecological restoration (Mandal et al., 2020).

Carbon-based nanocomposites such as SG–PySA–CNT and SG–PySA–GO, derived from sweetgum (SG) biomass through slow pyrolysis at 600 °C and integrated with carbon nanotubes (CNTs) or GO, exhibit superior sorption capacities for Pb(II) and Cd(II) ions. Compared to pristine biochar, these nanocomposites demonstrate significantly enhanced metal adsorption owing to their increased surface area and the uniform dispersion of CNTs or GO within the biochar matrix. Notably, SG–PySA–GO achieved maximum adsorption capacities exceeding 40 mg/g for Pb(II) and 10 mg/g for Cd(II). The study by Liu et al. (2016) highlights the simplicity of the synthesis process, high sorptive performance, and promising potential of these nanocomposites in wastewater treatment and environmental remediation applications.

Graphene nanoparticles demonstrated superior or comparable adsorption capacities for 17β-estradiol and 17α-ethynyl estradiol when evaluated against multi-walled carbon nanotubes (MWCNTs), single-walled CNTs, biochars, and both powdered and granular activated carbons (PAC and GAC). Although all adsorbents experienced a reduction in estrogen adsorption along with the natural organic matter (NOM), graphene exhibited the smallest decrease. This resilience is attributed to its large surface area and strong $\pi - \pi$ interactions with the estrogen molecules. The degree of adsorption was also influenced by the hydrophobicity of the estrogens. While graphene presents a best option for the efficient elimination of estrogenic compounds from solution, concerns remain regarding its environmental release due to its strong contaminant-binding affinity (Jiang et al., 2017).

GO-coated biochar was synthesized by thermally treating carbonaceous industrial waste at 800 °C, offering a value-added reuse pathway for waste materials. Comprehensive characterization using FTIR, FE-SEM, XRD, TEM, Raman spectroscopy, and AFM confirmed the successful fabrication and morphological features of the composite. The GO-coated biochar exhibited superior phenol adsorption capacity from wastewater compared to pristine GO. Kinetic modeling indicated that the adsorption followed the PSO model, while isotherm analysis conformed to the Langmuir model, suggesting monolayer adsorption. Intraparticle diffusion played a significant role in the adsorption process. Thermodynamic analyses revealed that the phenol adsorption was spontaneous and endothermic. Notably, the adsorbent maintained effective removal performance even under saline conditions, highlighting its potential for wastewater treatment applications in diverse environmental contexts (Manna et al., 2019).

Compared to pristine biochar, ball-milled biochar exhibits altered physicochemical properties that significantly influence microbial responses. Specifically, it demonstrates increased toxicity toward *Streptomyces coelicolor*, reducing cell viability to 68.2%, a level notably lower than that observed with GO (90%) and MWCNTs. In contrast, pristine biochar at 10 mg/L displayed minimal cytotoxicity. Interestingly, ball-milled biochar also enhanced the biosynthesis of antibiotics by significantly upregulating the expression of key regulatory genes (*redD*, *redZ*, and *actII-ORF4*), thereby increasing the production of undecylprodigiosin (RED) and actinorhodin (ACT). Despite its potential as a biochar-based additive to stimulate antibiotic production, caution is warranted due to the unknown environmental and health implications associated with its use (Liu et al., 2019).

4.4.2 Metal/Metal Oxide-loaded Biosorbents

Although unmodified biochar often exhibits limited sorption capacity, it remains a sustainable and promising adsorbent for water purification applications. Arar (2025), study highlights the significant improvements in performance achieved through the incorporation of nanoscale metal oxides into biochar. Such modifications substantially enhance adsorption kinetics, contaminant selectivity, and overall removal efficiency—particularly for inorganic anions and other challenging pollutants. These engineered nanocomposites act as high-capacity, selective sorbents, enabling rapid and effective contaminant removal. The findings suggest that nano-metal oxide-modified biochar represents a viable, eco-friendly strategy for the remediation of contaminated water systems.

A novel cellulose–nanoscale manganese oxide composite (C–NMOC) was synthesized via an in situ soft chemical method for the efficient removal of Pb(II) from aqueous solutions. Unlike the control nanoscale manganese oxide (NMO), which exhibited Mn(IV), the composite predominantly featured Mn(III). The C–NMOC achieved over 90% Pb(II) removal within just 10 minutes. Notably, C–NMOC outperformed both commercial β-MnO_2 and NMO in terms of Pb(II) adsorption efficiency per unit of manganese content. Kinetic and isotherm studies confirmed that physisorption was the dominant adsorption mechanism. According to Maliyekkal et al. (2010), this hybrid material presents a promising, cost-effective sorbent for water purification applications.

To effectively remove MB from wastewater, a $Cu(OH)_2$ nanowires–polyvinyl alcohol–activated carbon ($Cu(OH)_2$-NWs-PVA-AC) nanocomposite was synthesized via a simple room-temperature precipitation method. The nanocomposite demonstrated a rapid adsorption performance, achieving 96–99% MB removal within just 10 minutes. Adsorption kinetics followed the PSO model, indicating that chemisorption governed the process through active site availability. Equilibrium data were best described by the Langmuir and Langmuir–Freundlich isotherms, suggesting monolayer adsorption behavior. Thermodynamic parameters confirmed the adsorption process was spontaneous and exothermic. This environmentally benign nanocomposite offers superior speed and efficiency compared to many conventional adsorbents, positioning it as a promising material for wastewater treatment (Lakkaboyana et al., 2021).

A ceramsite-supported nZVI composite was produced and applied in a fixed-bed reactor system for the removal of malachite green (MG) dye from aqueous solutions. The ceramsite, produced by sintering industrial solid waste, provided a stable and magnetic support matrix for the immobilization of nZVI, enhancing its applicability in continuous flow systems. Breakthrough curve analysis indicated that increased temperature or reduced influent dye concentration and flow rate effectively delayed the breakthrough and saturation points. Under optimal conditions (10 mg/L MG, 25 °C, 5 mL/min), breakthrough and saturation occurred at 230 and 515 hours, respectively. Among the kinetic models evaluated, the Yoon–Nelson model best described the system's performance, with R^2 values exceeding 0.96 (Li et al., 2022). Amorphous transition metal oxide nanoparticles—specifically those of Fe, Co, and Ni—have been successfully synthesized via a novel laser irradiation in liquid technique. These amorphous nanostructures exhibit exceptional performance in the adsorption of dye pollutants such as MB. Notably, amorphous NiO demonstrated an extraordinarily high adsorption capacity of 10,584.6 mg/g for MB, representing one of the highest values reported to date. The superior adsorption efficiency is attributed to the high surface reactivity and structural disorder inherent in the amorphous form, which enhances interaction with dye molecules. These findings highlight the promising potential of amorphous metal oxides as highly efficient adsorbents for environmental remediation applications (Li et al., 2015).

CuO NPs were synthesized through a green method utilizing orange peels and mint leaves as natural reducing agents. These eco-friendly nanoparticles (designated as CuO NPs-1 and CuO NPs-2) were extensively characterized and evaluated for their adsorption capacity toward heavy metals—Pb(II), Ni(II), and Cd(II)—from aqueous solutions. Under optimized conditions (pH 6, adsorbent dose 0.33 g/L), the maximum adsorption capacities were 88.80 mg/g for Pb(II), 54.90 mg/g for Ni(II), and 15.60 mg/g for Cd(II), following the order: Pb(II) > Ni(II) > Cd(II). The adsorption kinetics conformed well to the PSO model ($R^2 > 0.96$), indicating chemisorption as the rate-limiting step, while the Freundlich isotherm model ($R^2 > 0.97$) described the equilibrium data, suggesting multilayer adsorption on a heterogeneous surface. Equilibrium was achieved within 60 minutes, demonstrating the potential of these green-synthesized CuO NPs as efficient adsorbents for heavy metal remediation (Mahmoud et al., 2021).

CuO NPs were biosynthesized for the first time using a green chemistry approach employing *Psidium guajava* (guava) leaf extract. The phytochemicals present in the extract served as natural reducing, stabilizing, and capping agents, facilitating a low-toxicity synthesis route. FTIR and XRD analyses confirmed the biomolecule-mediated synthesis and crystalline structure of the nanoparticles. Field emission scanning electron microscopy (FE-SEM) and high-resolution transmission electron microscopy (HR-TEM) revealed spherical particles ranging in size from 40 to 150 nm, while energy-dispersive X-ray spectroscopy (EDS) and dynamic light scattering verified the elemental composition and particle size distribution. The CuO NPs exhibited notable antibacterial activity, particularly against Gram-negative bacteria such as *Pseudomonas aeruginosa* and *E. coli*, which was attributed to interactions with their thinner peptidoglycan layers. Furthermore, the biosynthesized CuO nanoparticles demonstrated significant dye degradation capabilities, achieving 89% removal of MB and 81% of Congo red (CR), thereby indicating their strong potential for applications in both biomedical fields and wastewater treatment (Sathiyavimal et al., 2021).

4.5 Role of Nanomaterials in Environmental Remediation

Due to its simplicity, cost-effectiveness, and environmental compatibility, adsorption is widely preferred for dye removal from aqueous solutions. However, conventional adsorbents often suffer from drawbacks such as particle aggregation and limited adsorption capacity. To address these limitations, nanomaterials—particularly metal sulfides like MoS_2, ZnS, and NiS—have gained attention owing to their high surface area and surface reactivity. Bimetallic sulfides, such as Ni–Zn–S, supported on functional materials like β-cyclodextrin, exhibit enhanced adsorption performance through mechanisms such as hydrogen bonding and electrostatic interactions. These surface modifications not only prevent nanoparticle agglomeration but also improve adsorption efficiency. Consequently, nano-enhanced adsorbents present an economical, sustainable, and highly effective strategy for water purification (Sajid et al., 2024).

Cobalt chromite ($CoCrO_4$) nanoparticles synthesized via the solvothermal method have demonstrated excellent adsorption capabilities for organic dyes, particularly MB and CR. Adsorption behavior conformed to the Langmuir isotherm and PSO kinetic model, with maximum adsorption capacities of 43.8 mg/g for CCO@450 and 59.4 mg/g for CCO@550. Notably, the MB removal efficiency increased to 81% when present in a binary dye mixture, suggesting potential synergistic adsorption interactions. Enhanced dye removal was attributed to smaller particle sizes and improved dispersibility. These properties make $CoCrO_4$ nanoparticles promising candidates for effective and robust wastewater treatment applications involving organic dye contaminants (Ngoc et al., 2022).

The adsorption efficiency of zinc oxide (ZnO) nanoparticles in removing both cationic (malachite green, MG) and anionic dyes (acid fuchsin, AF; Congo red, CR) from aqueous solutions was systematically investigated. Thermodynamic analysis indicated that the adsorption process was exothermic, spontaneous, and influenced by temperature and pH. Kinetic studies revealed that the adsorption followed a PSO model, while isotherm analysis was best described by the Langmuir model, suggesting monolayer adsorption. The maximum adsorption capacities achieved were 2963 mg/g for MG, 3307 mg/g for AF, and 1554 mg/g for CR. The adsorption mechanisms involved chemical precipitation, hydrogen bonding, and electrostatic interactions. Notably, ZnO nanoparticles exhibited selective dye adsorption and retained high efficiency over five reuse cycles, demonstrating their potential as robust and recyclable adsorbents for dye removal in wastewater treatment (Zhang et al., 2016).

The effectiveness of sol–gel-synthesized iron NPs for the elimination of hexavalent chromium [Cr(VI)] from aqueous solutions was systematically evaluated. XRD and TEM analyses confirmed the successful synthesis and nanoscale structure of the particles. Batch adsorption experiments revealed that at an acidic pH of 2.0, 100% Cr(VI) removal was achieved. Furthermore, adsorption efficiency increased with decreasing initial Cr(VI) concentrations. Kinetic studies indicated that the adsorption process followed intraparticle diffusion, first-order, and PSO models, suggesting multiple rate-controlling mechanisms. Thermodynamic analysis, based on parameters such as ΔG°, ΔH°, and ΔS°, confirmed that the process was feasible and spontaneous. These findings underscore the potential of nanoiron as an efficient adsorbent for the detoxification of chromium-contaminated industrial effluents (Sharma et al., 2009).

The potential of sol–gel-synthesized TiO_2 nanoparticles for the removal of arsenic from drinking water was systematically investigated. Both pure and iron-doped TiO_2 nanoparticles immobilized on sand were evaluated under various light and atmospheric conditions. XRD confirmed the anatase crystalline phase, while iron doping was found to reduce particle size and enhance photocatalytic activity under visible light. Adsorption behavior conformed to both the Langmuir and Freundlich isotherm models, with iron-doped TiO_2 demonstrating superior arsenic removal efficiency due to increased surface reactivity. Among the tested configurations, TiO_2-coated sand exhibited the highest removal efficiency, particularly for arsenate [As(V)]. Additionally, solution pH significantly influenced the adsorption performance. These results affirm the efficacy of TiO_2-based nanomaterials, especially iron-doped variants, for arsenic remediation in contaminated water systems (Deedar and Aslam, 2009).

Surface-functionalized cellulose fibers (PEI@CE) were developed from plantain pith and modified by grafting polyethyleneimine (PEI) to enhance their adsorption capacity for polymer nanoparticles (PNPs). The incorporation of amino groups via PEI significantly increased electrostatic interactions with negatively charged PNPs, resulting in high adsorption efficiencies: 97% for poly(methyl methacrylate) (PMMA), 98% for poly(vinyl chloride) (PVC), and 99% for poly(vinyl acetate) (PVAc). FTIR, SEM, and zeta potential analyses confirmed the successful modification and adsorption mechanism. The PEI@CE fibers also effectively removed microplastics from a diluted commercial body scrub, demonstrating their real-world applicability. These stable, renewable, and efficient adsorbents offer a sustainable solution for the removal of micro- and nanoplastics from aquatic environments (Batool and Valiyaveettil, 2021). Table 4.3 gives the types of nano-biosorbents and their pollutant removal capabilities.

4.6 Environmental and Toxicological Considerations

Although nanoparticles have demonstrated high efficiency in pollutant removal, their large-scale environmental application raises critical toxicological concerns. The accumulation of nanoparticles and associated waste from industrial production and usage poses risks of environmental contamination. These particles can disrupt biological functions, exhibiting toxicity not only to humans but also to microorganisms, flora, and fauna. Such ecological hazards hinder the direct translation of laboratory successes to real-world applications. Achieving a sustainable balance necessitates a deeper understanding of nanoparticle interactions with

Table 4.3 Types of nano-biosorbents and their pollutant removal capabilities.

Type of nano-biosorbent	Source/modifier	Pollutants removed	Adsorption capacity/ efficiency (mg/g)	Adsorbent dose (g/L)	pH range of application	Initial concentration of pollutant	Reference
Superparamagnetic nanoparticles	Sludge biomass	Cr (VI)	80.21	5	2	100 mg/L	Ramya et al. (2022)
Nanoparticles	*Rhodotorula mucilaginosa*-Live	Cu (II)	12.7	16.67	5	50 mg/L	Salvadori et al. (2014)
Nanoparticles	*Rhodotorula mucilaginosa*-Dead	Cu (II)	26.2	16.67	5	50 mg/L	Salvadori et al. (2014)
Iron oxide nanoparticle	*Agrobacterium fabrum*	Methylene blue	91	1	11	200 mg/L	Sharma et al. (2018)
Zerovalent iron nanoparticles	*Macrocystis pyrifera*	Reactive Black 5	39.9	6	3	100 mg/L	García et al. (2018)
Magnetic nanoparticles	Aminopropyl silane	Hg (II)	42650	3	7	100 mg/L	Marimón-Bolívar et al. (2018)
Magnetic nanoparticles	Peptone	Hg (II)	278.63	3	7	100 mg/L	Marimón-Bolívar et al. (2018)
Magnetic nanoparticles	Yam peel biomass	Hg (II)	215.6	3	8	100 mg/L	Marimón-Bolívar et al. (2018)
Copper oxide nanoparticles	Sal seed deoiled cake	Erichrome Black	76.92	3	10	100 mg/L	Varshney and Gupta (2024)
Copper oxide nanoparticles	Sal seed deoiled cake	Congo Red	111.11	3	10	100 mg/L	Varshney and Gupta (2024)

(Continued)

Table 4.3 *(Continued)*

Type of nano-biosorbent	Source/modifier	Pollutants removed	Adsorption capacity/ efficiency (mg/g)	Adsorbent dose (g/L)	pH range of application	Initial concentration of pollutant	Reference
Copper oxide nanoparticles	Sal seed deoiled cake	Reactive Violet 1	87.72	3	10	100 mg/L	Varshney and Gupta (2024)
Copper oxide nanoparticles	*Solanum tuberosum*	Azo direct red 80	13.09	1.25	11	10 mg/L	Hamami and Javanbakht (2021)
Ultrasonic-assisted Copper oxide nanoparticles	*Solanum tuberosum*	Azo direct red 80	14.07	1.25	11	10 mg/L	Hamami and Javanbakht (2021)
Titanium dioxide nanoparticles	Orange peel	Nickel	79.36%	5	6	100 mg/L	Herrera-Barros et al. (2019)
Alumina nanoparticles	Orange peel	Nickel	87.85%	5	6	100 mg/L	Herrera-Barros et al. (2019)
Alumina nanoparticles	Orange peel	Cadmium	95.6%	5	6	100 mg/L	Herrera-Barros et al. (2019)
Titanium dioxide nanoparticles	Orange peel	Cadmium	85%	5	6	100 mg/L	Herrera-Barros et al. (2019)
Iron nanoparticles	Coffee husk	Losartan	98.26 %	1.5	-	0.1–0.4 mmol/L	Costa et al. (2024)
Iron nanoparticles	Coffee husk	Diclofenac	98.42 %	1.5	-	0.1–0.4 mmol/L	Costa et al. (2024)
Iron nanoparticles	*Leptolyngbya valderiana*	Cr (VI)	99%	10	3	50 mg/L	Banerjee et al. (2021)
Magnetic nanocomposite	*Palm fiber powder*	As (III)	16.23 mg/g	2	4.0–9.0	10 mg/L	Cui et al. (2019)

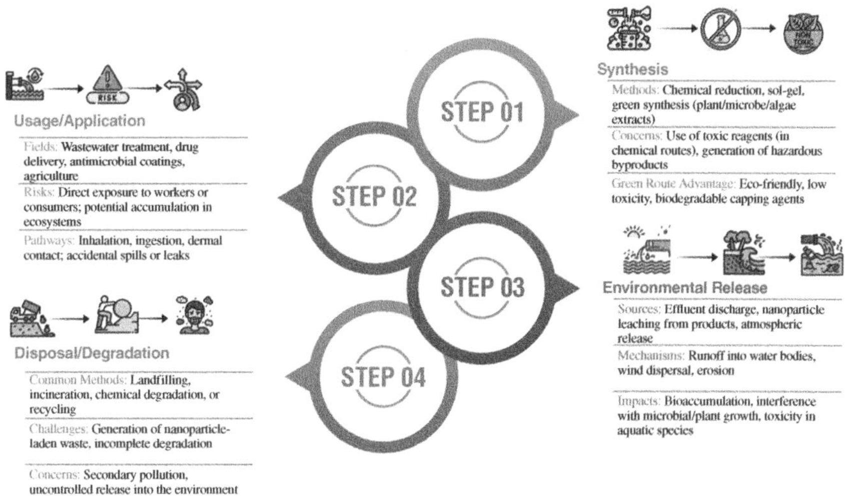

Figure 4.3 Environmental risk assessment framework for nanoparticles.

living systems and the implementation of effective waste management strategies during their synthesis and deployment (Das et al., 2022). Figure 4.3 depicts the environmental risk assessment framework for nanoparticles.

To mitigate the health risks associated with the uncertain nature of nanoparticle (NP) exposure—especially in occupational settings—several risk management strategies have been proposed. Given the lack of comprehensive exposure data and long-term toxicological profiles, a precautionary, conservative approach is essential. Effective reduction of unintentional human exposure to NPs requires applying the hierarchy of controls, which encompasses elimination, substitution, engineering controls, administrative controls, and personal protective equipment. While in vitro toxicological testing provides an initial basis for assessing NP safety, in vivo studies remain essential for accurate risk evaluation. Regulatory frameworks must evolve in parallel with the rapid advancement of nanotechnology to embed safety-by-design principles into the innovation pipeline. Furthermore, interdisciplinary research is imperative to bridge the gap between health risk mitigation and continued NP development (Bakand and Hayes, 2016).

4.7 Future Prospects

Despite their proven effectiveness, the application of nano-biosorbents in water treatment faces significant challenges related to cost and safety. The high expense associated with nanoparticle synthesis and implementation has limited large-scale adoption. Strategies such as the reuse and regeneration of nano-biosorbents, as well as the utilization of low-cost, low-purity precursors, have been proposed to reduce costs and improve economic feasibility. The global environmental nanotechnology market was projected to reach €1.6 trillion by 2020, highlighting its rapid growth and potential for substantial economic and employment benefits. Nevertheless, concerns surrounding nanotoxicity persist due to the nanoscale particles' high surface area and reactivity, which may induce oxidative stress, DNA damage, and cytotoxic effects. At present, regulatory oversight for the safe production, handling, and disposal of these materials remains inadequate. Consequently, the advancement of environmentally friendly synthesis methods, effective end-of-life disposal practices, and robust regulatory frameworks is essential to ensure the safe and sustainable deployment of nano-biosorbents in environmental remediation (Ali et al., 2020).

4.8 Conclusion

The use of nano-biosorbents represents a promising and sustainable strategy for addressing water pollution, particularly through the effective removal of heavy metals and dyes. This chapter has highlighted the advantages of GS methods—leveraging plant extracts, microorganisms, and algae—which enable the environmentally friendly production of nanoparticles with high surface area, tunable properties, and enhanced adsorption performance. Various nano-biosorbents, such as carbon-based materials and metal/metal oxide-loaded composites, have demonstrated superior adsorption capacities. However, the potential environmental and toxicological risks associated with their use—such as toxicity, persistence, and bioaccumulation—necessitate a balanced and cautious approach. The future of nano-biosorbents in environmental remediation hinges on integrating safety-by-design practices, thorough eco-toxicological assessment, and robust regulatory oversight. With responsible development, these greenly synthesized nanomaterials offer a viable, cost-effective, and scalable solution for sustainable pollution control. Their success depends on aligning innovation with environmental and public health safeguards.

References

Ali, M. E., Hoque, M. E., Safdar Hossain, S. K., & Biswas, M. C. (2020). Nanoadsorbents for wastewater treatment: next generation biotechnological solution. International Journal of Environmental Science and Technology, 17(9), 4095–4132.

Al-Tamimi, S. A. (2021). Biogenic green synthesis of metal oxide nanoparticles using oat biomass for ultrasensitive modified polymeric sensors. Green Chemistry Letters and Reviews, 14(2), 166–179.

Arar, Ö. (2025). Preparation of nano-metal-oxide incorporated biochar for remediation of water pollution. Current Nanomaterials, 10(2), 156–166.

Armendariz, V., José Yacamán, M., Duarte Moller, A., Peralta Videa, J. R., Troiani, H. E., Herrera, I., & Gardea Torres, J. L. (2004). HRTEM characterization of gold nanoparticles produced by wheat biomass. Revista Mexicana de Fisica, 50, 7–11.

Badwaik, V. D., Willis, C. B., Pender, D. S., Paripelly, R., Shah, M., Kherde, Y. A., Vangala, L. M., Gonzalez, M. S., & Dakshinamurthy, R. (2013). Antibacterial gold nanoparticles-biomass assisted synthesis and characterization. Journal of Biomedical Nanotechnology, 9(10), 1716–1723.

Bakand, S., & Hayes, A. (2016). Toxicological considerations, toxicity assessment, and risk management of inhaled nanoparticles. International Journal of Molecular Sciences, 17(6), 929.

Banerjee, S., Banerjee, I., Dutta, M., & Pal, R. (2021). Fabrication of iron nanoparticles using *Leptolyngbya valderiana* and investigation of its Cr (VI) removal potential in the free and biomass associated forms. Algal Research, 58, 102373.

Batool, A., & Valiyaveettil, S. (2021). Surface functionalized cellulose fibers–a renewable adsorbent for removal of plastic nanoparticles from water. Journal of Hazardous Materials, 413, 125301.

Borase, H. P., Salunke, B. K., Salunkhe, R. B., Patil, C. D., Hallsworth, J. E., Kim, B. S., & Patil, S. V. (2014). Plant extract: a promising biomatrix for ecofriendly, controlled synthesis of silver nanoparticles. Applied Biochemistry and Biotechnology, 173, 1–29.

Chaudhary, R., Nawaz, K., Khan, A. K., Hano, C., Abbasi, B. H., & Anjum, S. (2020). An overview of the algae-mediated biosynthesis of nanoparticles and their biomedical applications. Biomolecules, 10(11), 1498.

Chokshi, K., Pancha, I., Ghosh, T., Paliwal, C., Maurya, R., Ghosh, A., & Mishra, S. (2016). Green synthesis, characterization and antioxidant potential of silver nanoparticles biosynthesized from de-oiled biomass of thermotolerant oleaginous microalgae *Acutodesmus dimorphus*. RSC Advances, 6(76), 72269–72274.

Choudhary, S., Kumar, R., Dalal, U., Tomar, S., & Reddy, S. N. (2020). Green synthesis of nanometal impregnated biomass–antiviral potential. Materials Science and Engineering: C, 112, 110934.

Costa, H. P., Duarte, E. D., da Silva, F. V., da Silva, M. G., & Vieira, M. G. (2024). Green synthesis of carbon nanotubes functionalized with iron nanoparticles and coffee husk biomass for efficient removal of losartan and diclofenac: adsorption kinetics and ANN modeling studies. Environmental Research, 251, 118733.

Cui, J., Jin, Q., Li, Y., & Li, F. (2019). Oxidation and removal of As (III) from soil using novel magnetic nanocomposite derived from biomass waste. Environmental Science: Nano, 6(2), 478–488.

Das, P. K., Mohanty, C., Purohit, G. K., Mishra, S., & Palo, S. (2022). Nanoparticle assisted environmental remediation: applications, toxicological implications and recommendations for a sustainable environment. Environmental Nanotechnology, Monitoring & Management, 18, 100679.

Debroy, A., Joshi, S., Yadav, M., & George, N. (2023). Green synthesis of nanoparticles from bio-waste for potential applications: current trends, challenges, and prospects. In: Bio-Based Materials and Waste for Energy Generation and Resource Management (pp. 431–466). Elsevier.

Deedar, N. A. B. I., & Aslam, I. (2009). Evaluation of the adsorption potential of titanium dioxide nanoparticles for arsenic removal. *Journal of Environmental Sciences*, 21(3), 402–408.

Doudi, D., Mahboub, N., Gheraissa, N., Laib, I., Cherrada, N., Messai, R., & Slimani, N. (2024). Eco-friendly synthesis and characterization of ZnO and Mg-Ag-doped ZnO nanoparticles using *Phoenix dactylifera* L. seeds: exploring biological activity and structural properties. Biomass Conversion and Biorefinery, 15, 1–21.

Dutta, M. M., & Charingia, A. (2020). Nanotechnology-based nano-biosorbents. In: Handbook of Research on Emerging Developments and Environmental Impacts of Ecological Chemistry (pp. 386–408). IGI Global.

García, F. E., Plaza-Cazón, J., Montesinos, V. N., Donati, E. R., & Litter, M. I. (2018). Combined strategy for removal of Reactive Black 5 by biomass sorption on *Macrocystis pyrifera* and zerovalent iron nanoparticles. Journal of Environmental Management, 207, 70–79.

Gupta, D., Boora, A., Thakur, A., & Gupta, T. K. (2023). Green and sustainable synthesis of nanomaterials: recent advancements and limitations. Environmental Research, 231, 116316.

Gurunathan, B., Bathrinarayanan, P. V., Muthukumarasamy, V. K., & Thangavelu, D. (2014). Characterization of intracellular gold nanoparticles synthesized by biomass of *Aspergillus terreus*. Acta Metallurgica Sinica (English Letters), 27, 569–572.

Hamami, Z., & Javanbakht, V. (2021). Biosynthesis of copper oxide nanoparticles using biomass, peel, and extract polysaccharides of *Solanum tuberosum* for ultrasound-assisted adsorption of azo direct red 80 contaminants. Ceramics International, 47(17), 24170–24181.

Herrera-Barros, A., Tejada-Tovar, C., Villabona-Ortíz, A., González-Delgado, Á., & Mejía-Meza, R. J. (2019). Assessment of the effect of Al2O3 and TiO2 nanoparticles on orange peel biomass and its application for Cd (II) and Ni (II) uptake. Transactions of the ASABE, 62(1), 139–147.

Jadoun, S., Arif, R., Jangid, N. K., & Meena, R. K. (2021). Green synthesis of nanoparticles using plant extracts: a review. Environmental Chemistry Letters, 19(1), 355–374.

Jain, A. (2022). Algae-mediated synthesis of biogenic nanoparticles. Advances in Natural Sciences: Nanoscience and Nanotechnology, 13(4), 043001.

Jiang, L., Liu, Y., Liu, S., Zeng, G., Hu, X., Hu, X., Guo, Z., Tan, X., Wang, L., & Wu, Z. (2017). Adsorption of estrogen contaminants by graphene nanomaterials under natural organic matter preloading: comparison to carbon nanotube, biochar, and activated carbon. Environmental Science & Technology, 51(11), 6352–6359.

Karami, H., & Kazemi, S. K. (2023). Mechanically produced hazelnut skin micro and nano biosorbents for the fast removal of lead ions from polluted waters. Iranian Journal of Analytical Chemistry, 10(2), 1–15.

Karami, H., Moosakhani, Z., & Goli, F. (2017). Walnut skin nanoparticles as a novel and fast nano-biosorbent for the fast removal of lead ions from polluted water. Desalination and Water Treatment, 60, 291–300.

Khan, A., Malik, S., Ali, N., Yang, Y., Akhter, M. S., & Bilal, M. (2022a). Introduction to nano-biosorbents. In: Nano-Biosorbents for Decontamination of Water, Air, and Soil Pollution (pp. 29–43). Elsevier.

Khan, S., Zeb, S., Vega-Chacón, J., Torres, S. E., Martínez, S. Q., López, R., Mejía, I. M. M., Hernández, C. R. J., Vila, J. L., Cornejo, E. J., Aquino, C. P., Mortari, B., Borges, L. F. T., Ruiz-Córdova, G. A., Carhuas, F. L. R., Sotomayor, M. D. P. T., & Picasso, G. (2022b). Non-toxic nature of nano-biosorbents as a positive approach toward green environment. In: Nano-Biosorbents for Decontamination of Water, Air, and Soil Pollution (pp. 187–226). Elsevier.

Khanna, P., Kaur, A., & Goyal, D. (2019). Algae-based metallic nanoparticles: synthesis, characterization and applications. Journal of Microbiological Methods, 163, 105656.

Koul, B., Poonia, A. K., Yadav, D., & Jin, J. O. (2021). Microbe-mediated biosynthesis of nanoparticles: applications and future prospects. Biomolecules, 11(6), 886.

Küünal, S., Rauwel, P., & Rauwel, E. (2018). Plant extract mediated synthesis of nanoparticles. In: Emerging Applications of Nanoparticles and Architecture Nanostructures (pp. 411–446). Elsevier.

Lakkaboyana, S. K., Soontarapa, K., Asmel, N. K., Kumar, V., Marella, R. K., Yuzir, A., & Wan Yaacob, W. Z. (2021). Synthesis and characterization of Cu (OH) 2-NWs-PVA-AC nano-composite and its use as an efficient adsorbent for removal of methylene blue. Scientific Reports, 11(1), 5686.

Lee, H. J., Lee, G., Jang, N. R., Yun, J. H., Song, J. Y., & Kim, B. S. (2011). Biological synthesis of copper nanoparticles using plant extract. Nanotechnology, 1(1), 371–374.

Li, L. H., Xiao, J., Liu, P., & Yang, G. W. (2015). Super adsorption capability from amorphousization of metal oxide nanoparticles for dye removal. Scientific Reports, 5(1), 9028.

Li, T., Chen, H., Wang, Z., Ding, H., Xiao, C., Li, C., & Sun, T. (2022). Malachite green dye removal using ceramsite-supported nanoscale zero-valent iron in a fixed-bed reactor. Green Processing and Synthesis, 11(1), 275–286.

Liu, T., Gao, B., Fang, J., Wang, B., & Cao, X. (2016). Biochar-supported carbon nanotube and graphene oxide nanocomposites for Pb (II) and Cd (II) removal. RSC Advances, 6(29), 24314–24319.

Liu, X., Tang, J., Wang, L., Liu, Q., & Liu, R. (2019). A comparative analysis of ball-milled biochar, graphene oxide, and multi-walled carbon nanotubes with respect to toxicity induction in *Streptomyces*. Journal of Environmental Management, 243, 308–317.

Mahamadi, C. (2019). Will nano-biosorbents break the Achilles' heel of biosorption technology?. Environmental Chemistry Letters, 17(4), 1753–1768.

Maher, S., Nisar, S., Aslam, S. M., Saleem, F., Behlil, F., Imran, M., Assiri, M. A., Nouroz, A., Naheed, N., Khan, Z. A., & Aslam, P. (2023). Synthesis and characterization of ZnO nanoparticles derived from biomass (*Sisymbrium irio*) and assessment of potential anticancer activity. ACS Omega, 8(18), 15920–15931.

Mahmoud, A. E. D., Al-Qahtani, K. M., Alflaij, S. O., Al-Qahtani, S. F., & Alsamhan, F. A. (2021). Green copper oxide nanoparticles for lead, nickel, and cadmium removal from contaminated water. Scientific Reports, 11(1), 12547.

Maliyekkal, S. M., Lisha, K. P., & Pradeep, T. (2010). A novel cellulose–manganese oxide hybrid material by in situ soft chemical synthesis and its application for the removal of Pb (II) from water. Journal of Hazardous Materials, 181(1–3), 986–995.

Mandal, S., Pu, S., He, L., Ma, H., & Hou, D. (2020). Biochar induced modification of graphene oxide & nZVI and its impact on immobilization of toxic copper in soil. Environmental Pollution, 259, 113851.

Manna, S., Prakash, S., & Das, P. (2019). Synthesis of graphene oxide nano-materials coated bio-char using carbonaceous industrial waste for phenol separation from water. Colloids and Surfaces A: Physicochemical and Engineering Aspects, 581, 123818.

Marimón-Bolívar, W., Tejeda-Benítez, L., & Herrera, A. P. (2018). Removal of mercury (II) from water using magnetic nanoparticles coated with amino organic ligands and yam peel biomass. Environmental Nanotechnology, Monitoring & Management, 10, 486–493.

Massironi, A., Morelli, A., Grassi, L., Puppi, D., Braccini, S., Maisetta, G., Esin, S., Batoni, G., Della Pina, C., & Chiellini, F. (2019). Ulvan as novel reducing and stabilizing agent from renewable algal biomass: application to green synthesis of silver nanoparticles. Carbohydrate Polymers, 203, 310–321.

Mittal, A. K., Chisti, Y., & Banerjee, U. C. (2013). Synthesis of metallic nanoparticles using plant extracts. Biotechnology Advances, 31(2), 346–356.

Muñoz, A. J., Espínola, F., Ruiz, E., Barbosa-Dekker, A. M., Dekker, R. F., & Castro, E. (2021). Biosorption mechanisms of Ag (I) and the synthesis of nanoparticles by the biomass from *Botryosphaeria rhodina* MAMB-05. Journal of Hazardous Materials, 420, 126598.

Muraro, P. C. L., Pinheiro, L. D. S. M., Chuy, G., Vizzotto, B. S., Pavoski, G., Espinosa, D. C. R., Rech, V. C., & da Silva, W. L. (2022). Silver nanoparticles from residual

biomass: biosynthesis, characterization and antimicrobial activity. Journal of Biotechnology, 343, 47–51.
Ngoc, P. K., Mac, T. K., Nguyen, H. T., Viet, D. T., Thanh, T. D., Van Vinh, P., Phan, B. T., Duong, A. T., & Das, R. (2022). Superior organic dye removal by $CoCr_2O_4$ nanoparticles: adsorption kinetics and isotherm. Journal of Science: Advanced Materials and Devices, 7(2), 100438.
Ngulube, R., Pillay, L., & Nombona, N. (2024). Synthesis and characterization of electrospun composite nanofibers from *Moringa oleifera* biomass and metal oxide nanoparticles as potential adsorbents for the removal of lead ions. Chemical Papers, 78(1), 599–611.
Perera, U. P., Foo, M. L., & Chew, I. M. L. (2023). Synthesis and characterization of lignin nanoparticles isolated from oil palm empty fruit bunch and application in biocomposites. Sustainable Chemistry for Climate Action, 2, 100011.
Qasim, M., Clarkson, A. N., & Hinkley, S. F. (2023). Green synthesis of carbon nanoparticles (CNPs) from biomass for biomedical applications. International Journal of Molecular Sciences, 24(2), 1023.
Qureshi, A., Blaisi, N. I., Abbas, A. A., Khan, N. A., & Rehman, S. (2021). Prospectus and development of microbes mediated synthesis of nanoparticles. In: Microbial Nanotechnology: Green Synthesis and Applications (pp. 1–15). Springer.
Ramya, V., Murugan, D., Lajapathirai, C., Meenatchisundaram, S., & Arumugam, S. (2022). A composite adsorbent of superparamagnetic nanoparticles with sludge biomass derived activated carbon for the removal of chromium (VI). Journal of Cleaner Production, 366, 132853.
Sachan, D., Ramesh, A., & Das, G. (2021). Green synthesis of silica nanoparticles from leaf biomass and its application to remove heavy metals from synthetic wastewater: a comparative analysis. Environmental Nanotechnology, Monitoring & Management, 16, 100467.
Sajid, M., Sharma, A., Shukla, S., Khan, A., & Ali Chaudhry, S. (2024). Nano-enhanced water purification: uncovering the dye adsorption efficiency of Ni-Zn-S@Cyclodextrine nanomaterial. ChemistrySelect, 9(33), e202400093.
Salvadori, M. R., Ando, R. A., Oller do Nascimento, C. A., & Corrêa, B. (2014). Intracellular biosynthesis and removal of copper nanoparticles by dead biomass of yeast isolated from the wastewater of a mine in the Brazilian Amazonia. PLoS One, 9(1), e87968.
Sathiyavimal, S., Vasantharaj, S., Veeramani, V., Saravanan, M., Rajalakshmi, G., Kaliannan, T., Al-Misned, F. A., & Pugazhendhi, A. (2021). Green chemistry route of biosynthesized copper oxide nanoparticles using *Psidium guajava* leaf extract and their antibacterial activity and effective removal of industrial dyes. Journal of Environmental Chemical Engineering, 9(2), 105033.
Shahzadi, S., Fatima, S., Shafiq, Z., & Janjua, M. R. S. A. (2025). A review on green synthesis of silver nanoparticles (SNPs) using plant extracts: a multifaceted

approach in photocatalysis, environmental remediation, and biomedicine. RSC Advances, 15(5), 3858–3903.

Sharma, S., Hasan, A., Kumar, N., & Pandey, L. M. (2018). Removal of methylene blue dye from aqueous solution using immobilized *Agrobacterium fabrum* biomass along with iron oxide nanoparticles as biosorbent. Environmental Science and Pollution Research, 25, 21605–21615.

Sharma, Y. C., Srivastava, V., Weng, C. H., & Upadhyay, S. N. (2009). Removal of Cr (VI) from wastewater by adsorption on iron nanoparticles. *The Canadian Journal of Chemical Engineering*, 87(6), 921–929.

Soares, S. F., Fernandes, T., Trindade, T., & Daniel-da-Silva, A. L. (2020). Recent advances on magnetic biosorbents and their applications for water treatment. Environmental Chemistry Letters, 18(1), 151–164.

Sunny, N. E., Mathew, S. S., Chandel, N., Saravanan, P., Rajeshkannan, R., Rajasimman, M., Vasseghian, Y., Rajamohan, N., & Kumar, S. V. (2022). Green synthesis of titanium dioxide nanoparticles using plant biomass and their applications-a review. Chemosphere, 300, 134612.

Varshney, S., & Gupta, A. (2024). Forest industrial biomass residue-mediated green synthesized multifunctional copper oxide nanoparticles for efficient wastewater treatment and biomedical applications. Journal of Cleaner Production, 434, 140109.

Vasanthi Bathrinarayanan, P. R. A. N. A. V., Thangavelu, D., Muthukumarasamy, V. K., Munusamy, C., & Gurunathan, B. (2013). Biological synthesis and characterization of intracellular gold nanoparticles using biomass of *Aspergillus fumigatus*. Bulletin of Materials Science, 36, 1201–1205.

Yadav, V. K., Choudhary, N., Khan, S. H., Malik, P., Inwati, G. K., Suriyaprabha, R., & Ravi, R. K. (2020). Synthesis and characterisation of nano-biosorbents and their applications for waste water treatment. In: Handbook of Research on Emerging Developments and Environmental Impacts of Ecological Chemistry (pp. 252–290). IGI Global.

Yaqub, A., Ajab, H., Almas, A., Syed, S. M., Azam, A., Khan, M. I., Awais, M., Muhammad, I., Galal, A. M., & Alshahrani, M. Y. (2022). Utilization of nano-biosorbents based on pine needles and banana peel for methylene blue removal: equilibrium, kinetics, thermodynamic study, and application. Biomass Conversion and Biorefinery, 12(5), 1787–1802.

Zeghoud, S., Hemmami, H., Seghir, B. B., Amor, I. B., Kouadri, I., Rebiai, A., Messaoudi, M., Ahmed, S., Pohl, P., & Simal-Gandara, J. (2022). A review on biogenic green synthesis of ZnO nanoparticles by plant biomass and their applications. Materials Today Communications, 33, 104747.

Zhang, F., Chen, X., Wu, F., & Ji, Y. (2016). High adsorption capability and selectivity of ZnO nanoparticles for dye removal. Colloids and Surfaces A: Physicochemical and Engineering Aspects, 509, 474–483.

5

Characterization Techniques of Bioadsorbents

5.1 Introduction

The effective removal of pollutants from aqueous matrices is critical for environmental protection, technological advancement, and public health. Adsorption has evolved as a highly effective and frequently used treatment method. The efficacy of this approach is heavily reliant on the qualities of the adsorbent, which should have a wide surface area, stability, reusability, and a high adsorption capacity. To improve these features, researchers created composite adsorbents by incorporating functional elements into basic matrices, which improve selectivity and cost-effectiveness. Characterizing these composites at various phases is critical for determining their structural integrity and adsorption efficiency (Nazari et al., 2022). Depending on the substance and the instruments that are available, different characterization procedures are used to collect complete physicochemical data on adsorbent materials.

Fourier Transform Infrared Spectroscopy (FTIR), Raman Spectroscopy, X-ray Photoelectron Spectroscopy (XPS), Energy Dispersive X-ray Spectroscopy (EDX), X-ray Diffraction (XRD), Scanning Electron Microscopy (SEM) and Field Emission Scanning Electron Microscopy (FESEM), Transmission Electron Microscopy (TEM), Atomic Force Microscopy (AFM), Vibrating Sample Magnetometry (VSM), Dynamic Light Scattering (DLS), Thermogravimetric Analysis (TGA), and Differential Scanning Calorimetry (DSC) are typical methods for identifying chemical components and analyzing structural, morphological, magnetic, and thermal properties. These techniques are critical for assessing the synthesis, surface changes, and performance of adsorbents. They also aid in understanding adsorption mechanisms by showing the interactions between adsorbent and adsorbate, allowing for the development of effective materials for environmental applications (Danya et al., 2024).

SEM and TEM are effective techniques for thorough characterization of two-dimensional (2D) and three-dimensional (3D) materials. With excellent spatial resolution ranging from nanometers to micrometers, they provide crucial information about surface topography, grain size, and local chemical composition. These technologies supplement other nondestructive characterization tools by exploiting electron—matter interactions to produce a variety of relevant signals for imaging and analysis. SEM and TEM are critical tools for analyzing microstructural aspects, allowing researchers to properly analyze material quality, performance, and structural modifications (Inkson, 2016). Their broad applications make them indispensable in academic and industrial materials research.

The BET method is commonly used to calculate the specific surface area of solid powders using gas adsorption isotherms. This approach provides useful information about surface properties by estimating two critical parameters: the BET C constant and the monolayer adsorption capacity. The theoretical foundation of the approach highlights the necessity of correctly identifying the monolayer point, as it effects surface area calculations and isotherm comparisons. Experimental approaches, such as single-point methods, are also considered for practical applications. The BET equation remains an important tool in surface analysis, particularly for understanding adsorption behavior in porous materials (Dollimore et al., 1976).

XRD has been a standard technique in material characterization since the discovery of X-rays and their diffraction by crystals. XRD, which is widely used in materials science and engineering, can identify phases, measure residual stress, and analyze the texture of polycrystalline materials. This technique uses the interaction of X-rays with a crystal's atomic planes to provide comprehensive structural information. Epp (2016) discusses the principles of X-ray production, diffraction theory, instrumentation, and practical applications. Advances in XRD continue to improve its precision and scope, making it essential for research and industry quality control.

FTIR is a strong analytical tool for studying the chemical composition and surface chemistry of membranes. It is critical for determining which functional groups are responsible for changes in membrane behavior and performance. FTIR analysis allows researchers to track changes in membrane surface properties, determine chemical stability, and assess durability over time. These findings are critical for understanding the changes in physicochemical properties that influence membrane efficiency in a variety of applications (Mohamed et al., 2017). XPS, also known as Electron Spectroscopy for Chemical Analysis (ESCA), is an important technique for surface characterization, especially for biomedical polymers. It offers quantitative elemental analysis (excluding hydrogen and helium) and chemical bonding data from the top 10–200 Å of a solid surface. XPS is highly regarded for its accuracy, interpretability, and comprehensive chemical insights, making it one of the most informative surface analysis techniques available. Despite the need for expensive gear, its broad availability in academic, industrial, and commercial labs has made it a popular technique for studying polymer surfaces in biomedical research (Andrade, 1985).

TGA is a popular analytical technique that measures the mass change of a sample when it is heated to 1600 °C with regulated gas flow. TGA gives significant data on decomposition, oxidation, volatilization, and other mass-changing processes, making it useful for understanding kinetics, mechanisms, and conversions in chemical engineering. However, TGA cannot identify non-mass changes like as phase transitions or polymorphic transformations. Coupling with DSC or differential thermal analysis (DTA) aids in the differentiation between chemical and physical changes. TGA is also combined with evolved gas analysis to enable real-time monitoring of emitted gases (Saadatkhah et al., 2020). EDX chemical mapping is a strong, incoherent imaging technology that provides direct chemical information about a variety of elements using their unique X-ray peaks. One significant advantage of EDX mapping is its capacity to detect larger energy-loss peaks with better spatial localization. When paired with Z-contrast imaging or electron energy-loss spectroscopy (EELS), EDX improves the analytical capabilities of scanning transmission electron microscopy (STEM). This integration enables extensive structural, chemical, and functional analysis at atomic resolution, making it extremely useful for researching a wide range of innovative materials (Allen et al., 2012).

The pH-dependent surface charging of solid particles in aqueous solutions, especially in the presence of 1–1 electrolytes, is an important element of colloid and interface research. The point of zero charge (PZC) and isoelectric points (IEPs) are important characteristics for describing this behavior. In pristine conditions—typically in dilute solutions of alkali halides, nitrates, or chlorates—these values are stable and irrespective of salt type or concentration, allowing for reliable classification. However, the presence of certain adsorbing ions, such as sulfates, can alter PZC and IEP values, making them less universally relevant (Kosmulski, 2014). Figure 5.1 illustrates the classification of characterization techniques.

This chapter uniquely consolidates a comprehensive suite of modern characterization techniques tailored specifically for bioadsorbents, highlighting their relevance in understanding structure–function relationships crucial for adsorption applications. By systematically integrating morphological, structural, surface chemistry, thermal, and surface charge analyses, this chapter provides a multidisciplinary perspective that is rarely addressed in a single resource.

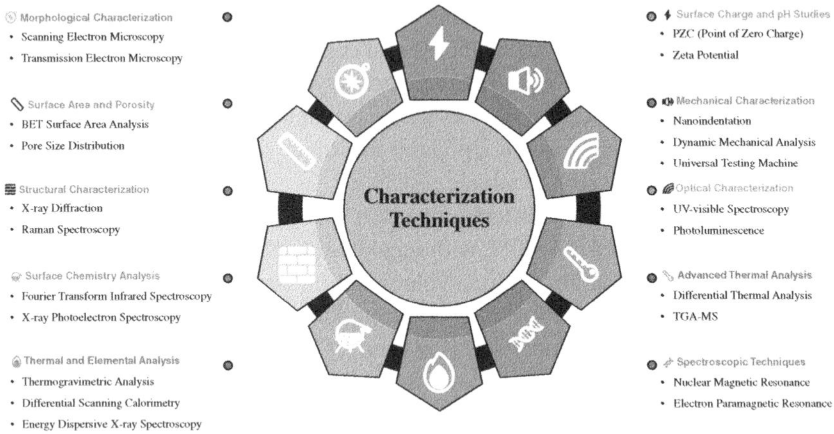

Figure 5.1 Classification of characterization techniques.

5.2 Morphological Characterization

Morphological characterization is critical for understanding the structure–property relationships between nanomaterials and bioadsorbents. Nanomaterials can take on a variety of morphologies, including nanospheres, nanowires, nanocubes, and nanotubes, depending on the synthesis method. These various morphologies have a substantial impact on the optical, electrical, and catalytic properties of the materials. SEM, TEM, and AFM offer extensive information about surface topography, particle size, and microstructural characteristics. Such analysis is critical for improving material performance in applications such as electronics, photonics, and environmental remediation (Mayeen et al., 2018). Understanding morphology allows us to design nanomaterials to specific functional and technological requirements.

5.2.1 Scanning Electron Microscopy

SEM is a strong analytical technique used to investigate the shape and surface properties of organic and inorganic materials at the micro- to nanoscale levels. SEM offers high-resolution imaging at magnifications of 5× to 300,000×, and even up to 1,000,000× in advanced versions. When combined with EDX, SEM provides elemental composition analysis. This combination allows for qualitative and semi-quantitative examination of sample microstructure and composition beyond what is feasible with traditional lab procedures. SEM is very useful in material research since it can handle a wide range of samples, including metals, polymers, and

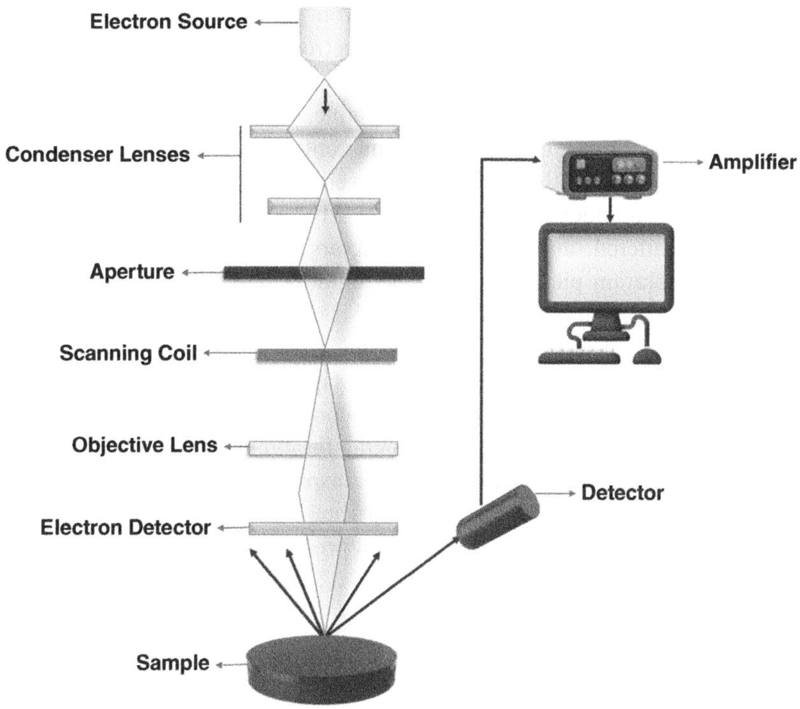

Figure 5.2 Schematic representation of SEM instrumentation.

biological specimens, with minimal pretreatment (Mohammed and Abdullah, 2018). Figure 5.2 depicts the schematic representation of SEM instrumentation.

The SEM instrument comprises of an electrical console and an electron column. It produces high-resolution photographs by scanning a specimen with a concentrated electron beam under vacuum. Primary signals comprise secondary and backscattered electrons observed within the specimen chamber. SEM is frequently combined with EDX to do elemental analysis using point, line, or area scans. This integration enables the imaging and compositional mapping of materials like metals, ceramics, and polymers. SEM provides information about shape, topography, composition, and microstructure (Ural, 2021). SEM is particularly good in visualizing exposed surface facets in nanoparticles and requires less sample preparation than TEM. Samples can usually be dispersed directly onto a SEM stub. While SEM only shows surface morphology, cross-sectional imaging can provide insight into interior structures. This is also beneficial for finding spatial distributions of components in structured materials, such as locating active phases in catalyst extrudates or eggshell morphologies with functional layers at the surface (Datye and DeLaRiva, 2023).

Key components of a SEM include an electron gun, condenser lenses, scan coils, specimen chamber, and detectors. The electron gun, which can be thermionic (tungsten or LaB_6) or field emission (cold or Schottky), generates a focused beam of electrons accelerated toward the sample. Condenser lenses are comprised of electromagnetic coils that shape and control the beam's diameter and intensity. Scan coils deflect the beam in x–y directions across the sample surface. The specimen is held under a high vacuum inside the sample chamber. Everhart–Thornley, backscattered, and solid-state detectors gather emitted signals and produce high-resolution pictures and elemental data. The SEM runs in a high vacuum to allow electrons to move freely, and it uses electromagnetic lenses to focus and control the beam. Scan coils move the beam in a raster pattern, while detectors capture emitted electrons to produce images. Brightness in SEM images is determined by the amount and energy of secondary electrons that reach the detector, with slanted or edged surfaces looking brighter due to higher electron escape. SEM is useful for evaluating big, nontransparent, bulk specimens and is frequently utilized in materials science, biology, and nanotechnology. While SEM does not achieve atomic resolution like TEM, it provides better sample flexibility, faster imaging, and easier interpretation (Ma et al., 2006).

5.2.2 Transmission Electron Microscopy

TEM is widely utilized for detailed examination of membrane materials, revealing shape, crystalline structure, and elemental composition. The approach uses high-energy electron beams that are passed through ultra-thin specimens to allow for atomic-level imaging. TEM primarily functions in two modes: bright-field and dark-field, which enable contrast formation for structural visualization. These modes aid in detecting flaws, grain boundaries, and crystalline orientations. Furthermore, elemental analysis can be performed by combining TEM with techniques such as EDX. TEM is particularly useful in membrane research for examining nanostructures and validating nanoscale synthesis or functionalization techniques (Tang and Yang, 2017). Figure 5.3 describes the schematic representation of TEM instrumentation.

TEM is made up of an electron gun, magnetic lenses, apertures, detectors, and cameras. Field emission guns (FEGs), particularly Schottky and cold-FEGs, are preferred for their brightness and excellent energy resolution, making them ideal for EELS investigation. TEM modes include diffraction, scanning TEM, and high-resolution TEM (HRTEM). Electrons scattered or transmitted through substances produce bright-field pictures or diffraction patterns. Annular detectors, EDX, and EELS allow for elemental and structural analysis. Various sample holders (e.g., tilt, cryo, tomography, and environmental)

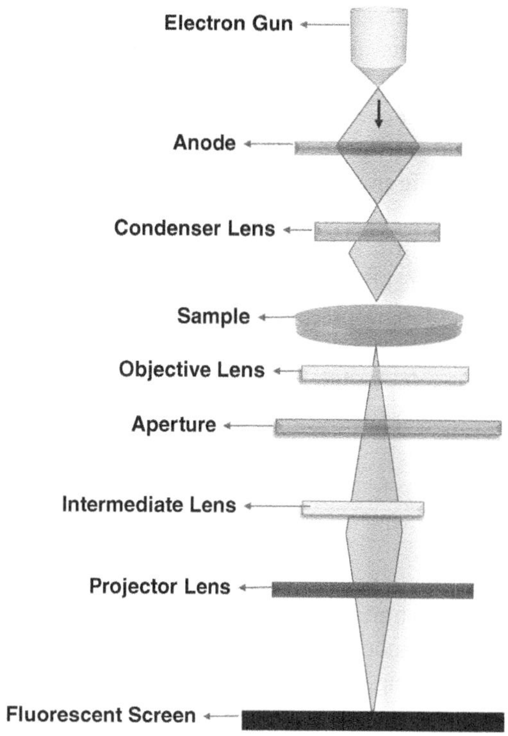

Figure 5.3 Schematic representation of TEM instrumentation.

broaden application possibilities, including operando catalysis experiments. TEM offers remarkable resolution and compositional insights into nanoscale materials (Braidy et al., 2020).

5.3 Surface Area and Porosity Analysis

Surface area and porosity are important factors that affect biochar effectiveness in applications like as wastewater treatment and soil remediation. The biomass feedstock type, particularly lignocellulosic materials like wood, and the pyrolysis temperature (400–700 °C) have a significant impact on pore development. Engineered biochar with increased porosity and surface area is increasingly highly desired. Among treatment approaches, chemical activation is the most effective at improving these properties. Additional processes like as ball milling, templating, and carbonaceous material coating improve porosity. Future research should focus on creating integrated treatment methods that improve the structure and functioning of biochar, allowing for broader and more efficient environmental uses (Leng et al., 2021).

Surface area is an important feature of metal-organic frameworks (MOFs), since it influences their performance in gas storage and separation applications. Accurate surface area measurements aid in the optimization of MOF design and application. Two widely used methods for determining surface area are: (1) the isotherm-based method, which uses the Brunauer–Emmett–Teller (BET) theory to estimate monolayer gas coverage, and (2) the geometry-based method, which calculates accessible surface area by rolling a probe molecule over atomic surfaces. While BET is commonly used, it has limits when working with microporous materials such as MOFs. As a result, modifications and careful analytical method selection are required for accurate surface area characterization (Sinha et al., 2019).

5.3.1 BET Surface Area Analysis

Surface area is an important parameter for nanomaterials, since it influences their reactivity, adsorption capacity, and environmental behavior. The BET method is the most often used surface area measurement methodology, which calculates a material's specific surface area by evaluating the quantity of gas adsorbed on its surface. This approach is highly applicable to nanomaterials with high surface-to-volume ratios, such as nano-aluminum oxide and nano-graphene. Accurate BET analysis needs precise sample preparation and a thorough understanding of instrument-specific needs. Brame and Griggs (2016) created a standard operating procedure that explains the theory, constraints, and step-by-step operation of the BET method, allowing for reproducible and useful surface area data. Table 5.1 gives the BET surface area and pore size data of various nanomaterials.

The BET approach, which employs low-temperature adsorption of nitrogen or krypton gas, is the most often used method for experimental SSA measurement. The total accessible internal and external surface area is calculated using the adsorption isotherm and the adsorptive gas's cross-sectional area. ISO 9277:2010, for example, provides recommendations for using the BET technique on both mesoporous and nonporous materials. Certified reference materials and specialist devices improve its accuracy and dependability (Gibson et al., 2020). Despite developments in density functional theory, the BET method remains a cornerstone of surface area analysis for current porous materials, having initially been devised for non-microporous surfaces in the 1930s. BET surface areas are frequently employed in the energy, medicinal, and environmental domains. However, estimating BET areas from adsorption isotherms might be difficult due to manual

Table 5.1 BET surface area and pore size data of various nanomaterials.

Nanomaterial	BET surface area (m^2/g)	Pore volume (cm^3/g)	Average pore diameter (nm)	Reference
Activated carbon	7.9168	0.027785	14.039	Alkhawaja and Abdul-Hameed (2025)
Orange peel	3.879	0.01789	18.448	Alkhawaja and Abdul-Hameed (2025)
Molecular sieves	100	0.22	1.33	Ilavarasi Jeyamalar and Kannan (2025)
Quercetin	27.83	0.157	22.51	Febriantini et al. (2025)
Silica gel powders	200	0.5	10	Danchova et al. (2025)
Silicon carbide (SiC) nanoparticles	37.6	0.124975	126.1853	Al Banna et al. (2018)
Capecitabine	419.36 ± 6.98	0.73 ± 0.21	8.12 ± 0.43	Narayan et al. (2021)
Eggshells nanocatalyst	5.54 ± 0.48	0.016	18.57 ± 2.16	Masime et al. (2022)
SBA-15	705.42	0.114	5.14	Ghosh et al. (2015)
2.5 Cu @SBA-15	338.9	0.028	6.45	Ghosh et al. (2015)
5 Cu @SBA-15	291.83	0.025	6.82	Ghosh et al. (2015)
7.5 Cu @SBA-15	252.89	0.021	8.21	Ghosh et al. (2015)
10 Cu @SBA-15	235.3	0.019	8.22	Ghosh et al. (2015)
12.5 Cu @SBA-15	195.54	0.015	8.38	Ghosh et al. (2015)
0 MPa—Chitin	4.98	0.00949	7.62	Tian et al. (2018)
1.6 MPa—Chitin	12.69	0.0328	10.34	Tian et al. (2018)

(Continued)

Table 5.1 *(Continued)*

Nanomaterial	BET surface area (m^2/g)	Pore volume (cm^3/g)	Average pore diameter (nm)	Reference
1.8 MPa—Chitin	11.66	0.0284	9.73	Tian et al. (2018)
2.0 MPa—Chitin	12.18	0.0333	10.92	Tian et al. (2018)
2.2 MPa—Chitin	12.58	0.0319	10.14	Tian et al. (2018)
2.4 MPa—Chitin	12.39	0.0309	9.96	Tian et al. (2018)
Rice husk—CS	653	0.9	2.6	Rumiyanti et al. (2025)
Rice husk—400	182	0.2	2.3	Rumiyanti et al. (2025)
Rice husk—600	113	0.1	2.7	Rumiyanti et al. (2025)
Silica nanoparticles	694.29	0.278503	2.45	Shiekh et al. (2022)
Natural clay	74.78	0.0994	5.31595	Zarime et al. (2018)
Natural clay-nZVI	5.6418	0.0226	16.02891	Zarime et al. (2018)
Boron nitride (BN) nano/microrods	1109.11	0.454	2.60	Zhao et al. (2021)

calculation errors. Osterrieth et al. (2022) underlined that while new computational techniques such as BETSI can increase reproducibility and transparency, they cannot completely address theoretical limits. Researchers should always publish isotherms in detail, including pressure range and data points, because the isotherm represents the genuine experimental result, not the derived BET area.

The BET approach works best for mesoporous and macroporous materials because it assumes multilayer adsorption on a flat, homogeneous surface with no lateral interactions. However, for microporous materials (pore diameter < 2 nm), this assumption fails, resulting in merely the perceived surface area rather than the genuine geometrically accessible area. For MOFs and mixed zeolites with complicated or multimodal pore architectures, BET frequently fails to offer linear fitting throughout any pressure zone. Although the consistency criterion can be used to determine an acceptable pressure range, it does not ensure an accurate representation of the surface area. As a result, when interpreting BET results for microporous materials, extreme caution is advised (Tian and Wu, 2018).

Although the BET method is traditionally discouraged for materials with ultra-micropores (<7 Å), Bae et al. (2010) demonstrated that accurate BET surface areas can be obtained for MOFs and zeolites if carried out within a pressure range defined by published consistency criteria. Surprisingly, when these conditions are implemented, BET surface areas obtained from nitrogen or argon isotherms closely match accessible surface areas computed from crystal structures—despite fundamental methodological differences. The study also demonstrates that utilizing the usual BET pressure range often results in an overestimate of surface areas, as adsorbate filling in ultra-micropores occurs at substantially lower pressures. As a result, for such materials, argon adsorption at 87 K and precise pressure range selection provide a more reliable BET measurement.

5.3.2 Pore Size Distribution

The origin of pores in solid materials can be used to classify them into two general categories: interparticle and intraparticle. In order to customize material properties for uses like adsorption and catalysis, it is essential to comprehend the structural features of these pores. Numerous methods, including as mercury intrusion porosimetry, small-angle X-ray scattering, nuclear magnetic resonance, thermoporosimetry, and molecular adsorption methods (molecular resolution porosimetry), are used to assess the pore size distribution (Nimmo, 2004). Designing useful porous materials is greatly aided by these methods, which offer comprehensive insights into pore structures across a range of scales (Kaneko, 1994).

Changes in pore structures under different compaction levels were found in pore-size distribution experiments utilizing mercury porosimetry on statically compacted clay specimens, such as kaolinite, illite, and Boston Blue Clay. Larger pores in kaolinite were diminished by first compaction, whereas finer pores stayed unaltered. While finer pores remained intact, intermediate pores (>0.14 μm) were impacted by ongoing compaction. When kaolinite was made via sedimentation or slurry, the pore structures varied considerably. Larger pores were specifically eliminated by drying shrinkage. According to these results, pore-size distribution is a crucial indicator for assessing the engineering behavior of fine-grained soils and is extremely sensitive to microstructural changes (Sridharan et al., 1971).

5.4 Structural Characterization

A popular nondestructive method for characterizing structures, XRD provides crucial information about material phases, crystallinity, crystallite size, lattice strain, and crystal defects. With little sample preparation, it can be used with bulk materials, polycrystalline thin films, and multilayer structures. XRD is an essential technique in materials science and engineering since it aids in the determination of orientation parameters and texture coefficients. Pandey et al. (2021) assert that XRD offers basic structural information that is essential for developing and refining sophisticated functional materials, especially for thin-film, electronic, and nanotechnology applications.

5.4.1 X-ray Diffraction

A crucial method for analyzing the structural characteristics of both crystalline and noncrystalline materials is XRD. Lattice parameters, phases, crystallite size, crystallinity, and crystal defects are all determined by it. Additionally, strain, roughness, and electron density are revealed via sophisticated XRD techniques. Coherent scattering by periodically organized atoms produces diffraction patterns; polycrystals produce ring patterns, whereas single crystals make spot patterns. The crystal structure of the material is shown by the location and strength of the diffraction peaks. According to Khan et al. (2020), XRD provides comprehensive insight into both short- and long-range atomic order, making it an essential tool for materials characterization.

The distinct fingerprints of the atomic arrangement found in crystalline materials are XRDs patterns, which allow for accurate phase identification and structural study. According to Bunaciu et al. (2015), XRD is particularly useful for measuring strain and crystallinity as well as for identifying preferred orientations (texture). When X-rays interact with particular lattice planes, constructive interference produces the diffraction peaks, whose intensity correspond to the positions of the atoms. Thus, this method makes it easier to examine microstructural characteristics that are essential for comprehending material behavior, such as average grain size and lattice defects. Because of its depth of structural knowledge and dependability, XRD is still a fundamental technique in materials research.

XRD occurs when X-rays (usually ~1 Å) contact with periodic electron clouds in crystalline materials. Cassetta (2014) explains that when X-rays strike an ordered atomic array, secondary waves scatter in all directions. The superposition of these waves produces constructive or destructive interference, depending on the atomic configuration. Constructive interference creates different diffracted beams at certain angles, whose intensities are determined by atomic number, crystal symmetry, and spatial atomic distribution. This diffraction pattern provides critical information about the crystal's 3D structure, making XRD an essential tool in material characterization.

5.4.2 Raman Spectroscopy

Raman spectroscopy is a powerful vibrational technique based on inelastic light scattering that provides molecular-level information about material structure and composition. It detects shifts in photon frequency when they interact with chemical bonds, providing a unique vibrational fingerprint of materials. According to Rostron et al. (2016), its great sensitivity and low sample preparation requirements make it excellent for detecting trace chemicals and corrosion products. Raman's application in structural research has extended across numerous domains due to technological developments, allowing for the identification of phases and functional groups in solids, liquids, and thin films.

Historically, Raman spectroscopy was limited by poor scattering efficiency and high instrumentation costs. According to Kudelski (2008), the usual Raman cross-section (~10^{-29} cm^2/molecule) is substantially lower than that of UV or IR absorption, which requires large analyte concentrations and limits early application. Surface-enhanced Raman scattering enables detection of single molecules with cross-sections of ~10^{-14} cm^2. Furthermore, the development of tiny, portable, and low-cost Raman spectrometers has increased their relevance for on-site and distant sensing applications, including space exploration, transforming Raman spectroscopy into a highly versatile and powerful structural characterization tool. The development of small, cost-effective, and portable Raman equipment has extended its utility beyond laboratories to field and space missions. Furthermore, Raman spectroscopy contributes to corrosion investigations, environmental sensing, and material characterization by providing comprehensive vibrational information about molecular structures (Rostron et al., 2016; Moon et al., 2023), making it a valuable tool in modern materials and chemistry research.

5.5 Surface Chemistry Analysis

Surface chemistry analysis is critical for determining the outermost atomic layers of materials, which are typically within the first few nanometers. Techniques including XPS, Auger electron spectroscopy (AES), and time-of-flight secondary ion mass spectrometry (ToF-SIMS) can reveal important information on elemental composition, chemical states, and molecular interactions. These techniques are critical for evaluating surface contaminants, oxidation, corrosion, and thin film coatings. Advanced features like depth profiling and high-resolution imaging facilitate the examination of surface heterogeneity and layered structures, which is critical for understanding interfacial behavior and material performance (Jayawardena et al., 2021).

5.5.1 Fourier Transform Infrared Spectroscopy

FTIR spectroscopy is an effective instrument for studying the vibrational properties of amino acids, cofactors, and water molecules, which are extremely sensitive to small structural changes. Its broad applicability allows for the investigation of a wide range of functional groups, whereas reaction-induced FTIR difference spectroscopy permits the selective detection of vibrations from certain chemical groups. Isotope tagging, site-directed mutagenesis, and hydrogen/deuterium exchange improve spectrum interpretation. When combined with theoretical normal mode calculations, FTIR offers critical insights into structural and functional dynamics in biological systems (Berthomieu and Hienerwadel, 2009).

FTIR spectroscopy produces information in the form of "absorption versus wavenumber" spectra, which are commonly used to characterize the structural properties of organic materials. The mid-IR region (400–4000/cm) is widely examined. It is separated into four unique regions: single bond (2500–4000/cm), triple bond (2000–2500/cm), double bond (1500–2000/cm), and fingerprint area (600–1500/cm). Each of these locations has distinct absorption bands corresponding to particular functional groups, which aids in compound identification. The mid-IR spectrum interpretation aids in determining the molecular structure and functional groups present in samples (Nandiyanto et al., 2019). Table 5.2 depicts the functional group assignments in FTIR spectra.

A significant feature of FTIR spectroscopy is its universal applicability—nearly all chemical substances display characteristic absorption or emission in the infrared range, allowing for qualitative and quantitative study. This feature makes FTIR the favored approach in a variety of fields. One of its most significant

Table 5.2 Functional group assignments in FTIR spectra.

Adsorbent source	Type of particle	Wavenumber (cm^{-1})	Type of vibration	Functional group	Reference
Peels and pulp of *Solanum lycopersicum*	Phytochemical	3630.4, 99.694	O—H stretching	Alcohol	Kainat et al. (2022)
Peels and pulp of *Solanum lycopersicum*	Phytochemical	3503.7, 99.836	O—H stretching	Alcohol	Kainat et al. (2022)
Peels and pulp of *Solanum lycopersicum*	Phytochemical	3451.5, 99.702	O—H stretch, H-bonded	Alcohols, phenols	Kainat et al. (2022)
Peels and pulp of *Solanum lycopersicum*	Phytochemical	3339.7, 64.344	O—H stretch, H-bonded	Alcohols, phenols	Kainat et al. (2022)
Peels and pulp of *Solanum lycopersicum*	Phytochemical	3332.2, 69.03	O—H stretch, H-bonded	Alcohols, phenols	Kainat et al. (2022)
Peels and pulp of *Solanum lycopersicum*	Phytochemical	3013.23	C—H stretch	Aromatics	Kainat et al. (2022)
Peels and pulp of *Solanum lycopersicum*	Phytochemical	2974.4, 75.264	C—H stretch	Alkanes	Kainat et al. (2022)
Peels and pulp of *Solanum lycopersicum*	Phytochemical	2817.9, 99.861	C—H stretching	Aldehyde	Kainat et al. (2022)
Peels and pulp of *Solanum lycopersicum*	Phytochemical	2571.9, 99.825	S—H stretching	Thiol	Kainat et al. (2022)
Peels and pulp of *Solanum lycopersicum*	Phytochemical	2288.6, 99.784	N=C=O stretching	Isocyanate	Kainat et al. (2022)

(Continued)

Table 5.2 *(Continued)*

Adsorbent source	Type of particle	Wavenumber (cm^{-1})	Type of vibration	Functional group	Reference
Peels and pulp of *Solanum lycopersicum*	Phytochemical	879.9, 63.683	C=C bending	Alkene	Kainat et al. (2022)
TiO_2	Nanoparticle	3400	O—H stretching (broad)	Hydroxyl group	Hussain et al. (2010)
TiO_2	Nanoparticle	1650	C=O stretching	Carbonyl group	Hussain et al. (2010)
Elephantopus scaber L.	Nanoparticle	3415	O—H stretching (broad)	Hydroxyl group	Kharat and Mendhulkar (2016)
Elephantopus scaber L.	Nanoparticle	2865	C—H stretch	Alkanes groups	Kharat and Mendhulkar (2016)
Elephantopus scaber L.	Nanoparticle	2933	C—H stretch	Alkanes groups	Kharat and Mendhulkar (2016)
Elephantopus scaber L.	Nanoparticle	1611–1400	C—H stretch	Aromatic rings	Kharat and Mendhulkar (2016)
Elephantopus scaber L.	Nanoparticle	1109	O—H stretching (broad)	Hydroxyl group	Kharat and Mendhulkar (2016)
Elephantopus scaber L.	Nanoparticle	1109	N—H group	Primary and secondary amides	Kharat and Mendhulkar (2016)
Elephantopus scaber L.	Nanoparticle	827	C—H stretch	Alkyl halides	Kharat and Mendhulkar (2016)
Iresine herbstii leaf	Nanoparticle	~3271	N—H or O—H stretching	Hydrogen-bonded —OH (alcohols, phenols) or N—H (amines, amides)	Dipankar and Murugan (2012)

Iresine herbstii leaf	Nanoparticle	~2916	C—H asymmetric stretching	Aliphatic C—H (CH_2, CH_3)	Dipankar and Murugan (2012)
Iresine herbstii leaf	Nanoparticle	~2847	C—H symmetric stretching	Aliphatic C—H (CH_2, CH_3)	Dipankar and Murugan (2012)
Iresine herbstii leaf	Nanoparticle	~2360	O=C=O asymmetric stretching	CO_2 (often from atmospheric contamination)	Dipankar and Murugan (2012)
Iresine herbstii leaf	Nanoparticle	~1643	C=O stretching/ N—H bending	Amide I (proteins), conjugated C=O, or adsorbed water	Dipankar and Murugan (2012)
Iresine herbstii leaf	Nanoparticle	~1527	N—H bending and C—N stretching	Amide II (proteins), aromatic nitro compounds	Dipankar and Murugan (2012)
Iresine herbstii leaf	Nanoparticle	~1033	C—O stretching	Alcohols, ethers, carboxylic acids, esters	Dipankar and Murugan (2012)
Iresine herbstii leaf	Nanoparticle	~827	C—H out-of-plane bending	Aromatic (para-substituted benzene rings)	Dipankar and Murugan (2012)
FePt	Nanoparticle	2854	C—H asymmetric stretching	Aliphatic C—H (CH_2, CH_3)	Shukla et al. (2003)
FePt	Nanoparticle	2922	C—H asymmetric stretching	Aliphatic C—H (CH_2, CH_3)	Shukla et al. (2003)
FePt	Nanoparticle	3006	=C—H stretching	Aromatic or alkene C—H stretch	Shukla et al. (2003)
FePt	Nanoparticle	1648	C=C or C=O stretching	Conjugated C=C (alkenes or aromatics), or amide I band	Shukla et al. (2003)

applications has been in the detection of atmospheric trace gases. The fast progress of FTIR instrumentation during the 1970s has significantly improved remote and laboratory-based atmospheric monitoring. As a result, FTIR has emerged as a trustworthy technique for environmental research, notably for determining air composition and trace gases (Bacsik et al., 2004).

FTIR spectroscopy is a sensitive vibrational technique commonly used to study molecular changes in sick tissues. It identifies functional groups, bonding types, and molecular conformations using narrow, molecule-specific absorption peaks. While FTIR is best suited for nonaqueous materials due to water's significant IR absorption, it does provide direct biochemical composition information. Compared to Raman spectroscopy, FTIR lacks confocal imaging and has lesser spatial resolution (10–20 μm vs. 1–2 μm). FTIR signals result from changes in dipole moment, whereas Raman spectra emerge from changes in polarizability, making the techniques complimentary for structural and biological tissue research (Movasaghi et al., 2008).

5.5.2 X-ray Photoelectron Spectroscopy

X-ray Photoelectron Spectroscopy (XPS), also called Electron Spectroscopy for Chemical Analysis (ESCA), is a surface-sensitive technique based on Einstein's 1905 photoelectric effect theory. It has advanced greatly since the 1950s, becoming an important analytical technique for investigating the elemental composition and electrical structure of materials. Kai Siegbahn's creation of high-resolution XPS, which earned him the Nobel Prize in 1981, represented a significant milestone. Continuous advances in instrumentation and theoretical techniques have expanded XPS capabilities, allowing for more precise surface characterization and deeper insights into chemical states and bonding conditions (Fadley, 2010).

XPS is frequently utilized for surface characterization due to its high sensitivity (~10 nm) and ability to discriminate chemical states. It detects all elements except hydrogen and helium, providing quantitative analysis without standards and detection limits of 0.1–1%. Although XPS has lower lateral resolution than techniques like as AES or EDX, it is comparatively nondestructive. Surface contamination and sample charging can be managed using correct methods. Careful interpretation and detailed reporting of instrument and analysis conditions are required for accurate and reproducible results (Stevie and Donley, 2020). Modern XPS equipment require ultra-high vacuum (UHV) settings ($\sim 10^{-9}$ mbar) for accurate surface investigation, particularly of reactive materials. The X-ray system consists of an Al Kα anode with a monochromator, a motorized sample stage for accurate placement and tilting, and electron optics with extraction lenses, an energy analyzer (often hemispherical), and a detector. These components allow for high-resolution measurements by examining photoelectron kinetic energies.

Additional techniques, such as ion guns for sputtering and flood guns for charge compensation, facilitate depth profiling and investigation of insulating materials (Greczynski et al., 2023).

XPS is an important method for elemental detection and chemical state research because it is sensitive to electron binding energy alterations. Each element has distinct peaks; differences in peak position, known as chemical shifts, indicate interactions with electronegative or electropositive atoms. Narrow scans allow for comprehensive investigation of peak forms and splitting patterns, such as shake-up satellites or multiplet effects. These spectrum patterns represent the complicated electron interactions that occur during photoionization. Quantitative analysis involves integrating peak regions while accounting for inelastic scattering tails and background signals (Shard, 2020).

5.6 Thermal and Elemental Characterization

Thermal and elemental characterization procedures are critical for determining a material's structural integrity, stability, and composition. TGA measures weight changes with temperature, exposing thermal transitions via slope variations in the thermogram. DSC enhances TGA by discriminating between endothermic and exothermic processes. When combined with elemental analysis, which quantifies elemental composition, these techniques provide a complete picture of a sample's physical and chemical properties. Zhang et al. (2015) proved the value of integrating these methods for assessing combustion chamber deposits, revealing compositional variations and thermal properties.

5.6.1 Thermogravimetric Analysis

TGA is a quantitative thermal technique that analyzes mass changes in materials at temperatures up to 1600 °C under regulated gas atmospheres. It is used to study decomposition, evaporation, oxidation, and thermal stability, providing information on reaction kinetics and processes. While TGA is good in identifying processes involving mass change, it cannot detect phase transitions or polymorphic shifts where mass remains constant. Therefore, connecting TGA with DSC or DTA helps separate physical from chemical phenomena. Evolved gas analysis improves TGA by identifying gaseous products in real time (Saadatkhah et al., 2020). TGA is useful in evaluating biomass and other materials for thermochemical conversion because it analyzes mass changes during thermal processing. Key TGA setups allow for the assessment of thermal degradation behavior based on positive or negative mass variations. The derivative thermogravimetry (DTG) curve, which is derived from TGA data, improves resolution by indicating the rate of mass loss. TGA combined with DSC or DTA allows for the simultaneous calculation of weight change and heat flow, resulting in a more complete thermal profile. This integrated method aids in distinguishing between overlapping thermal processes, such as decomposition and phase transitions (De Blasio, 2019).

The thermobalance is the heart of a TGA system, with critical subunits including an electronic microbalance, sample holder, furnace, temperature programmer, and recorder. Accurate measurements are dependent on precise balance arm handling, particularly the return to zero after sample placement. Stability is critical, so the system is usually allowed to equilibrate for 15–20 minutes before beginning analysis. Mishandling, such as transporting the instrument with the clamp engaged, might result in inaccurate mass readings. Proper setup and calibration are thus required for accurate thermogravimetric data (Loganathan et al., 2017).

TGA is widely used to study the thermal behavior of biomass materials, providing exact measures of mass loss at various heating rates and atmospheres. Its simplicity, consistency, and small sample size (<20 mg) make it excellent for biomass characterization. TGA helps to estimate kinetic parameters like as activation energy and pre-exponential factors, which are critical for optimizing pyrolysis operations. Furthermore, combining TGA with techniques like FTIR, mass spectrometry (MS), gas chromatography–mass spectrometry (GC/MS), or TGA allows for a more comprehensive investigation of volatile compounds produced during decomposition. This makes TGA an important technique for identifying acceptable biomass for industrial-scale bioenergy applications (Escalante et al., 2022).

5.6.2 Differential Scanning Calorimetry

DSC is a popular thermal analytic technique for determining the thermodynamic properties of biomacromolecules including proteins, nucleic acids, and lipid assemblies. It permits accurate monitoring of unfolding and phase transitions, revealing information on the thermodynamic stability and structural transitions of these molecules. DSC also enables the characterization of intermediate states in melting processes. Advances in instrumentation and software have substantially increased its sensitivity and resolution, allowing for extensive research of structural and environmental influences on biomolecular behavior. Thus, DSC is critical in molecular biology for determining stability and conformational dynamics (Spink, 2008).

DSC and modulated DSC (MDSC) are commonly used in industry to measure key thermal transitions such as melting, crystallization, and glass transition temperatures. These thermodynamic characteristics are essential for product development, stability testing, and process optimization. MDSC provides improved resolution and sensitivity, allowing for distinct discrimination between reversible and irreversible transitions. In the food sector, these procedures are used to evaluate protein denaturation and lipid behavior, whereas in medicines, they help to analyze medication component stability and compatibility. Their dependability makes them essential tools for thermal characterization in industrial settings (Leyva-Porras et al., 2019).

DSC is a versatile technique for determining the physical properties of polymers, particularly semicrystalline kinds. It allows for exact measurements of melting, crystallization, mesomorphic transitions, and glass transitions, as well as the accompanying enthalpy and entropy changes. DSC provides a wide range of heating and cooling speeds, from 1 $\mu K\ s^{-1}$ to more than 1 MK/s, making it ideal for researching rate-dependent phase transitions in nonequilibrium systems. DSC was introduced in the 1960s and has since proven indispensable in polymer science due to its simplicity, universality, and ability to deliver energetically significant data (Schick, 2009).

5.6.3 Energy Dispersive X-ray Spectroscopy

EDX has become an important method for determining the elemental composition of solid materials, particularly nanoparticles. When combined with SEM or STEM-in-SEM, EDX allows qualitative elemental analysis at spatial resolutions of around 10 nm. Technological advances in detector sensitivity have permitted quick analysis of individual nanoparticles, supplementing surface morphological data collected by electron microscopy. This makes EDX extremely useful in nanoparticle studies. While EDX provides meaningful data, it has limits in terms of sensitivity, quantification accuracy, and interference from overlapping spectral lines (Hodoroaba, 2020). SEM/EDX is widely used in elemental microanalysis and can quantify almost all elements except hydrogen, helium, and lithium. Even with peak overlaps, advanced silicon drift detectors (SDD-EDX) allow SEM/EDX to attain precision comparable to Wavelength Dispersive Spectrometry (WDS). However, limitations remain, such as misidentification of elemental peaks (particularly for trace elements), mistakes in standardless quantification, and geometric distortions caused by complicated sample topography. These issues can result in severe systematic mistakes unless suitable methods, such as k-ratio referencing and careful specimen preparation, are followed (Newbury and Ritchie, 2013).

EDX is a common analytical technique for determining and measuring elemental compositions in nanomaterials. It produces a spectrum with specific peaks for the elements present. This chapter discusses the fundamentals, benefits, limitations, and most recent advances in EDX, including its combination with electron microscopy for nanoscale investigation. EDX provides quick, localized compositional mapping, making it especially useful in nanotechnology research. Recent advancements have enhanced spatial resolution and detection sensitivity, enabling more accurate characterization of complex nanostructures. These results highlight the importance of EDX in the advancement of nanomaterial science (Mishra et al., 2017).

5.7 Surface Charge and pH Studies

Surface charge behavior cannot be separated to pH, as proven by considerable research on colloids, oxides, and soils. Surface charge in oxides results from the protonation/deprotonation of surface hydroxyl groups, and potentiometric acid–base titration is the basic method for estimating pH-dependent charge density and the PZC. Recent study suggests that even "permanently charged" soils and clay minerals have pH-dependent surface charge numbers (SCNs), which decrease with lower pH due to H^+ adsorption polarization and organic matter interaction. For variable-charge minerals like kaolinite and goethite, experimental variables such as pH, ion concentration ratios, and electrolyte addition procedures have a significant impact on measurement accuracy (Szekeres and Tombácz, 2012).

5.7.1 PZC Determination

The PZC is the pH at which a sorbent's net surface charge equals zero, with equal concentrations of positive and negative surface sites. This value is critical for estimating the surface charge behavior and interaction potential of sorbents with ions. At pH > PZC, surfaces are negatively charged, encouraging cation adsorption, while at pH < PZC, anions interact more. Bourikas et al. proposed the potentiometric mass titration (PMT), which is commonly used to estimate PZC in oxides and hydroxides (Fiol and Villaescusa, 2009).

The PZC is an important characteristic that affects adsorption effectiveness and varies between materials. Untreated algal biomass with a PZC of 5 demonstrated excellent fluoride adsorption at pH 5, while biochar from Colocasia esculenta stem had PZC values of 6.6 (raw) and 8.3 (activated), reaching maximum fluoride removal at pH 4. Al-Maliky et al. (2021) used potentiometric titration to measure PZC in concrete particles with various ionic strengths. The ΔpH approach, which involves KCl additions and dual pH measurements, allows graphing of initial pH versus ΔpH to find PZC using curve projection.

5.7.2 Zeta Potential

The zeta potential (ZP) measures the electrostatic potential at the electrical double layer of nanoparticles in solution, and it is an important indicator of colloidal stability and surface charge behavior. Nanoparticles having ZPs between −10 and +10 mV are neutral, but values greater than −30 mV suggest strong anionic or cationic properties. This charge affects interactions with biological membranes, with cationic particles causing more membrane rupture and toxicity. Clogston and Patri (2010) showed this technique by contrasting colloidal gold (anionic) and amine-terminated PAMAM dendrimers (cationic) in terms of biological interaction and stability.

Accurate nanoparticle characterization, particularly particle size and surface charge, is crucial for developing effective nanoformulations in drug delivery. DLS and ZP are widely adopted techniques due to their simplicity and reproducibility. DLS determines hydrodynamic diameter, while ZP evaluates colloidal stability and surface charge interactions. However, challenges in sample preparation, principle comprehension, and data interpretation often hinder accurate application. As Bhattacharjee (2016) notes, these techniques originate from colloid chemistry and require interdisciplinary understanding, which is often lacking in nanomedicine research, emphasizing the need for clear methodological guidance.

The ZP is an important indication of colloidal stability because it reflects electrochemical interactions between nanoparticles and their surrounding medium. It is especially important in assessing dispersion stability of nanomaterials utilized in medicinal, environmental, and industrial applications. Although theoretical models like Smoluchowski's and Debye-Hückel's provide a core understanding of electrokinetic events, they have shortcomings, such as failing to account for surface conductivity or counterion concentration effects. According to Lunardi et al. (2021), accurate ZP measurements necessitate precise control over a variety of factors such as voltage, electrophoretic mobility, temperature, viscosity, particle shape, and concentration, and pH. Table 5.3 gives the overview of characterization techniques—advantages, disadvantages, and applications.

5.8 Conclusion

This chapter provides a complete overview of physicochemical characterization techniques required to understand the morphology, surface, structure, thermal behavior, and surface charge properties of nanomaterials and adsorbents. SEM and TEM are useful tools for understanding surface texture and particle size at the

Table 5.3 Overview of characterization techniques – advantages, disadvantages, and applications.

Technique	Advantages	Disadvantages	Applications	Reference
SEM	High-resolution imaging of surface morphology	Requires conductive coating; vacuum needed	Morphology, particle shape, size	Afanasyev et al. (2019)
TEM	Atomic-scale resolution, internal structure imaging	Expensive; complex sample prep; thin samples required	Crystallography, lattice defects, core-shell structures	Wang (2003)
BET surface area analysis	Quantifies surface area accurately	Requires degassing; assumes model-dependent interpretation	Surface area of adsorbents, catalysts	Sinha et al. (2019)
Pore Size Distribution (BJH)	Determines mesoporous structure	Less accurate for micropores; limited to certain size ranges	Porous material evaluation (e.g., activated carbon, MOFs)	Shi et al. (2021)
XRD	Identifies crystalline phases, lattice parameters	Poor for amorphous materials; overlapping peaks can confuse analysis	Phase identification, crystallinity, particle size (Debye–Scherrer)	Bunaciu et al. (2015)
Raman spectroscopy	Nondestructive, provides molecular fingerprint	Fluorescence interference; weak signal for some materials	Carbon materials, biomolecules, polymer crystallinity	Eberhardt et al. (2015)
FTIR	Identifies functional groups, chemical bonding	Overlapping bands; water interference	Functional group analysis, polymer degradation, sorption studies	Simonescu (2012)

XPS	Surface-sensitive (10 nm), chemical state identification	Cannot detect H or He; expensive; vacuum required	Surface chemistry, oxidation states, elemental quantification	Ketenoglu (2022)
TGA	Measures weight loss, decomposition temperatures	No info on structure or phase transition alone	Thermal stability, composition, reaction kinetics	Nickle et al. (1987)
DSC	Determines Tg, Tm, crystallization temps, enthalpy	Cannot detect mass changes; baseline drift possible	Polymers, biomolecules, food/pharma thermal behavior	Ahmadi Khoshooei et al. (2019)
EDX (with SEM or TEM)	Elemental analysis, quick and nondestructive	Cannot detect light elements (H, He, Li); qualitative if standardless	Composition analysis, elemental mapping	Boyes (2003)
PZC determination	Simple; helps predict surface charge behavior	Requires careful pH control; limited to aqueous systems	Adsorption, catalyst surface interactions	Bakatula et al. (2018)
Zeta potential	Assesses colloidal stability, surface charge	Affected by ionic strength, concentration, pH	Nanoparticle dispersion, drug delivery, wastewater treatment	Lunardi et al. (2021)

nanoscale. BET and pore size distribution investigations provide for exact surface area and porosity determinations, which are crucial in adsorption and catalytic applications. Structural investigation using XRD and Raman spectroscopy reveals crystallinity and molecular ordering, whereas FTIR and XPS disclose chemical functionalities and elemental states on the surface. TGA, DSC, and EDX are effective tools for assessing thermal stability and compositional information, offering a comprehensive picture of material behavior under thermal stress. Furthermore, PZC and ZP investigations provide critical information about surface charge properties, which influence dispersion stability and adsorption efficacy. Together, these strategies provide an extensive toolkit for modifying material characteristics to specific environmental, biological, and industrial needs.

References

Afanasyev, S. S., Kychkina, T. V., & Savvinova, L. N. (2019). Scanning electron microscope (advantages and disadvantages). Colloquium-Journal, 2-2(26), 25–27.

Ahmadi Khoshooei, M., Fazlollahi, F., & Maham, Y. (2019). A review on the application of differential scanning calorimetry (DSC) to petroleum products: characterization and kinetic study. Journal of Thermal Analysis and Calorimetry, 138, 3455–3484.

Al Banna, L., Salem, N., Ghrair, A. M., & Habash, S. S. (2018). Impact of silicon carbide nanoparticles on hatching and survival of soil nematodes *Caenorhabditis elegans* and *Meloidogyne incognita*. Applied Ecology and Environmental Research, 16(3), 2651–2662.

Alkhawaja, E. S., & Abdul-Hameed, H. M. (2025). Novel synthesis and characterization of nano-activated carbon derived from agricultural orange peel waste. International Journal of Design & Nature and Ecodynamics, 20(1), 31–41.

Allen, L. J., D'Alfonso, A. J., Freitag, B., & Klenov, D. O. (2012). Chemical mapping at atomic resolution using energy-dispersive x-ray spectroscopy. MRS Bulletin, 37(1), 47–52.

Al-Maliky, E. A., Gzar, H. A., & Al-Azawy, M. G. (2021, September). Determination of point of zero charge (PZC) of concrete particles adsorbents. In: IOP Conference Series: Materials Science and Engineering (Vol. 1184, No. 1, p. 012004). IOP Publishing.

Andrade, J. D. (1985). X-ray photoelectron spectroscopy (XPS). In: Surface and Interfacial Aspects of Biomedical Polymers: Volume 1 Surface Chemistry and Physics (pp. 105–195). Springer.

Bacsik, Z., Mink, J., & Keresztury, G. (2004). FTIR spectroscopy of the atmosphere. I. Principles and methods. Applied Spectroscopy Reviews, 39(3), 295–363.

Bae, Y. S., Yazaydın, A. O., & Snurr, R. Q. (2010). Evaluation of the BET method for determining surface areas of MOFs and zeolites that contain ultra-micropores. Langmuir, 26(8), 5475–5483.

Bakatula, E. N., Richard, D., Neculita, C. M., & Zagury, G. J. (2018). Determination of point of zero charge of natural organic materials. Environmental Science and Pollution Research, 25, 7823–7833.

Berthomieu, C., & Hienerwadel, R. (2009). Fourier transform infrared (FTIR) spectroscopy. Photosynthesis Research, 101, 157–170.

Bhattacharjee, S. (2016). DLS and zeta potential–what they are and what they are not?. Journal of Controlled Release, 235, 337–351.

Boyes, E. (2003). Pros and cons of low voltage SEM EDX elemental analysis. Microscopy Today, 11(1), 5–11.

Braidy, N., Béchu, A., de Souza Terra, J. C., & Patience, G. S. (2020). Experimental methods in chemical engineering: transmission electron microscopy—TEM. The Canadian Journal of Chemical Engineering, 98(3), 628–641.

Brame, J. A., & Griggs, C. S. (2016). Surface area analysis using the Brunauer-Emmett-Teller (BET) method: scientific operation procedure series: SOP-C. https://erdc-library.erdc.dren.mil/server/api/core/bitstreams/81b728f8-a50f-4ef8-e053-411ac80adeb3/content

Bunaciu, A. A., UdriŞTioiu, E. G., & Aboul-Enein, H. Y. (2015). X-ray diffraction: instrumentation and applications. Critical Reviews in Analytical Chemistry, 45(4), 289–299.

Cassetta, A. (2014). X-ray diffraction (XRD). In: Encyclopedia of Membranes (pp. 1–3). Berlin, Heidelberg: Springer.

Clogston, J. D., & Patri, A. K. (2010). Zeta potential measurement. In: Characterization of Nanoparticles Intended for Drug Delivery (pp. 63–70). Totowa, NJ: Humana press.

Danchova, N., Shandurkov, D., Tsekov, R., Mihaylov, L., Spassov, T., & Gutzov, S. (2025). Porous silica gels doped with gold nanoparticles–preparation, microstructure, optical and textural properties. Gels, 11(6), 454.

Danya, U., Sagayaraj, C., Yaazhini, J., Ramachandran, B., Aravind, J., & Muthukumaran, P. (2024). Characterization techniques for carbon-based adsorbents and carbon composites. Physical Sciences Reviews, 9(11), 3387–3407.

Datye, A., & DeLaRiva, A. (2023). Scanning electron microscopy (SEM). In: Springer Handbook of Advanced Catalyst Characterization (pp. 359–380). Cham: Springer International Publishing.

De Blasio, C. (2019). Thermogravimetric analysis (TGA). In: Fundamentals of Biofuels Engineering and Technology (pp. 91–102). Springer.

Dipankar, C., & Murugan, S. (2012). The green synthesis, characterization and evaluation of the biological activities of silver nanoparticles synthesized from *Iresine herbstii* leaf aqueous extracts. Colloids and Surfaces B: Biointerfaces, 98, 112–119.

Dollimore, D., Spooner, P., & Turner, A. J. S. T. (1976). The bet method of analysis of gas adsorption data and its relevance to the calculation of surface areas. Surface Technology, 4(2), 121–160.

Eberhardt, K., Stiebing, C., Matthäus, C., Schmitt, M., & Popp, J. (2015). Advantages and limitations of Raman spectroscopy for molecular diagnostics: an update. Expert Review of Molecular Diagnostics, 15(6), 773–787.

Epp, J. (2016). X-ray diffraction (XRD) techniques for materials characterization. In: Materials Characterization Using Nondestructive Evaluation (NDE) Methods (pp. 81–124). Woodhead Publishing.

Escalante, J., Chen, W. H., Tabatabaei, M., Hoang, A. T., Kwon, E. E., Lin, K. Y. A., & Saravanakumar, A. (2022). Pyrolysis of lignocellulosic, algal, plastic, and other biomass wastes for biofuel production and circular bioeconomy: a review of thermogravimetric analysis (TGA) approach. Renewable and Sustainable Energy Reviews, 169, 112914.

Fadley, C. S. (2010). X-ray photoelectron spectroscopy: progress and perspectives. Journal of Electron Spectroscopy and Related Phenomena, 178, 2–32.

Febriantini, D., Rianto, A. K., Pebriani, S., Apriandanu, D. O. B., & Yulizar, Y. (2025). Transformation of novel quercetin-facilitated ZrO_2 into ZrO_2@SiO_2–BF_3 nanoparticles: synthesis, characterization, and catalytic behavior. Surfaces and Interfaces, 67, 106589.

Fiol, N., & Villaescusa, I. (2009). Determination of sorbent point zero charge: usefulness in sorption studies. Environmental Chemistry Letters, 7, 79–84.

Ghosh, B. K., Hazra, S., Naik, B., & Ghosh, N. N. (2015). Preparation of Cu nanoparticle loaded SBA-15 and their excellent catalytic activity in reduction of variety of dyes. Powder Technology, 269, 371–378.

Gibson, N., Kuchenbecker, P., Rasmussen, K., Hodoroaba, V. D., & Rauscher, H. (2020). Volume-specific surface area by gas adsorption analysis with the BET method. In: Characterization of Nanoparticles (pp. 265–294). Elsevier.

Greczynski, G., Haasch, R. T., Hellgren, N., Lewin, E., & Hultman, L. (2023). X-ray photoelectron spectroscopy of thin films. Nature Reviews Methods Primers, 3(1), 40.

Hodoroaba, V. D. (2020). Energy-dispersive X-ray spectroscopy (EDS). In: Characterization of Nanoparticles (pp. 397–417). Elsevier.

Hussain, M., Ceccarelli, R., Marchisio, D. L., Fino, D., Russo, N., & Geobaldo, F. (2010). Synthesis, characterization, and photocatalytic application of novel TiO_2 nanoparticles. Chemical Engineering Journal, 157(1), 45–51.

Ilavarasi Jeyamalar, J., & Kannan, C. (2025). novel nanoporous material. International Journal of Current Rese Key words.

Inkson, B. J. (2016). Scanning electron microscopy (SEM) and transmission electron microscopy (TEM) for materials characterization. In: Materials Characterization Using Nondestructive Evaluation (NDE) Methods (pp. 17–43). Woodhead Publishing.

Jayawardena, H. S. N., Liyanage, S. H., Rathnayake, K., Patel, U., & Yan, M. (2021). Analytical methods for characterization of nanomaterial surfaces. Analytical Chemistry, 93(4), 1889–1911.

Kainat, S., Gilani, S. R., Asad, F., Khalid, M. Z., Khalid, W., Ranjha, M. M. A. N., Bangar, S. P., & Lorenzo, J. M. (2022). Determination and comparison of phytochemicals, phenolics, and flavonoids in *Solanum lycopersicum* using FTIR spectroscopy. Food Analytical Methods, 15(11), 2931–2939.

Kaneko, K. (1994). Determination of pore size and pore size distribution: 1. Adsorbents and catalysts. Journal of Membrane Science, 96(1–2), 59–89.

Ketenoglu, D. (2022). A general overview and comparative interpretation on element-specific X-ray spectroscopy techniques: XPS, XAS, and XRS. X-Ray Spectrometry, 51(5–6), 422–443.

Khan, H., Yerramilli, A. S., D'Oliveira, A., Alford, T. L., Boffito, D. C., & Patience, G. S. (2020). Experimental methods in chemical engineering: X-ray diffraction spectroscopy—XRD. The Canadian Journal of Chemical Engineering, 98(6), 1255–1266.

Kharat, S. N., & Mendhulkar, V. D. (2016). Synthesis, characterization and studies on antioxidant activity of silver nanoparticles using *Elephantopus scaber* leaf extract. Materials Science and Engineering: C, 62, 719–724.

Kosmulski, M. (2014). The pH dependent surface charging and points of zero charge. VI. Update. Journal of Colloid and Interface Science, 426, 209–212.

Kudelski, A. (2008). Analytical applications of Raman spectroscopy. Talanta, 76(1), 1–8.

Leng, L., Xiong, Q., Yang, L., Li, H., Zhou, Y., Zhang, W., Jiang, S., Li, H., & Huang, H. (2021). An overview on engineering the surface area and porosity of biochar. Science of the Total Environment, 763, 144204.

Leyva-Porras, C., Cruz-Alcantar, P., Espinosa-Solís, V., Martínez-Guerra, E., Piñón-Balderrama, C. I., Compean Martínez, I., & Saavedra-Leos, M. Z. (2019). Application of differential scanning calorimetry (DSC) and modulated differential scanning calorimetry (MDSC) in food and drug industries. *Polymers*, 12(1), 5.

Loganathan, S., Valapa, R. B., Mishra, R. K., Pugazhenthi, G., & Thomas, S. (2017). Thermogravimetric analysis for characterization of nanomaterials. In: Thermal and Rheological Measurement Techniques for Nanomaterials Characterization (pp. 67–108). Elsevier.

Lunardi, C. N., Gomes, A. J., Rocha, F. S., De Tommaso, J., & Patience, G. S. (2021). Experimental methods in chemical engineering: zeta potential. The Canadian Journal of Chemical Engineering, 99(3), 627–639.

Ma, H., Shieh, K. J., & Qiao, T. X. (2006). Study of transmission electron microscopy (TEM) and scanning electron microscopy (SEM). Nature and Science, 4(3), 14–22.

Masime, J. O., Ogur, E. O., Mbatia, B. N., Ochieng'Aluoch, A., & Otieno, G. (2022). Characterization of eggshells nanocatalyst: synthesized by bottom-up technology. Walisongo Journal of Chemistry, 5(2), 202–211.

Mayeen, A., Shaji, L. K., Nair, A. K., & Kalarikkal, N. (2018). Morphological characterization of nanomaterials. In: Characterization of Nanomaterials (pp. 335–364). Woodhead Publishing.

Mishra, R. K., Zachariah, A. K., & Thomas, S. (2017). Energy-dispersive X-ray spectroscopy techniques for nanomaterial. In: Microscopy Methods in Nanomaterials Characterization (pp. 383–405). Elsevier.

Mohamed, M. A., Jaafar, J., Ismail, A. F., Othman, M. H. D., & Rahman, M. A. (2017). Fourier transform infrared (FTIR) spectroscopy. In: Membrane Characterization (pp. 3–29). Elsevier.

Mohammed, A., & Abdullah, A. (2018, November). Scanning electron microscopy (SEM): a review. In: Proceedings of the 2018 International Conference on Hydraulics and Pneumatics—HERVEX, Băile Govora, Romania (Vol. 2018, pp. 7–9).

Moon, J., Li, M., Ramirez-Cuesta, A. J., & Wu, Z. (2023). Raman spectroscopy. In: Springer Handbook of Advanced Catalyst Characterization (pp. 75–110). Cham: Springer International Publishing.

Movasaghi, Z., Rehman, S., & ur Rehman, D. I. (2008). Fourier transform infrared (FTIR) spectroscopy of biological tissues. Applied Spectroscopy Reviews, 43(2), 134–179.

Nandiyanto, A. B. D., Oktiani, R., & Ragadhita, R. (2019). How to read and interpret FTIR spectroscope of organic material. Indonesian Journal of Science and Technology, 4(1), 97–118.

Narayan, R., Gadag, S., Cheruku, S. P., Raichur, A. M., Day, C. M., Garg, S., Manandhar, S., Pai, K. S. R., Suresh, A., Mehta, C. H., Nayak, Y., Kumar, N., & Nayak, U. Y. (2021). Chitosan-glucuronic acid conjugate coated mesoporous silica nanoparticles: a smart pH-responsive and receptor-targeted system for colorectal cancer therapy. Carbohydrate Polymers, 261, 117893.

Nazari, M. T., Schnorr, C., Rigueto, C. V., Alessandretti, I., Melara, F., da Silva, N. F., Crestani, L., Ferrari, V., Vieillard, J., Dotto, G. L., Silva, L. F. O., & Piccin, J. S. (2022). A review of the main methods for composite adsorbents characterization. Environmental Science and Pollution Research, 29(59), 88488–88506.

Newbury, D. E., & Ritchie, N. W. (2013). Is scanning electron microscopy/energy dispersive X-ray spectrometry (SEM/EDS) quantitative?. Scanning, 35(3), 141–168.

Nickle, S. K., Meyers, K. O., & Nash, L. J. (1987, September). Shortcomings in the use of TGA/DSC techniques to evaluate in-situ combustion. In: SPE Annual Technical Conference and Exhibition (pp. SPE-16867). SPE.

Nimmo, J. R. (2004). Porosity and pore size distribution. Encyclopedia of Soils in the Environment, 3(1), 295–303.

Osterrieth, J. W., Rampersad, J., Madden, D., Rampal, N., Skoric, L., Connolly, B., Allendorf, M. D., Stavila, V., Snider, J. L., Ameloot, R., Marreiros, J., Ania, C., Azevedo, D., Vilarrasa-Garcia, E., Santos, B. F., Bu, X. H., Chang, Z., Bunzen, H.,

Champness, N. R., ... & Fairen-Jimenez, D. (2022). How reproducible are surface areas calculated from the BET equation?. Advanced Materials, 34(27), 2201502.
Pandey, A., Dalal, S., Dutta, S., & Dixit, A. (2021). Structural characterization of polycrystalline thin films by X-ray diffraction techniques. Journal of Materials Science: Materials in Electronics, 32, 1341–1368.
Rostron, P., Gaber, S., & Gaber, D. (2016). Raman spectroscopy, review. Laser, 21, 24.
Rumiyanti, L., Garcia, S., Manurung, P., Firdaus, I., Chotimah, C., & Suharyadi, E. (2025). Synthesis of rice husk–based–mesoporous silica nanoparticles by facile calcination. Functional Composites and Structures, 7, 015004.
Saadatkhah, N., Carillo Garcia, A., Ackermann, S., Leclerc, P., Latifi, M., Samih, S., Patience, G. S., & Chaouki, J. (2020). Experimental methods in chemical engineering: thermogravimetric analysis—TGA. The Canadian Journal of Chemical Engineering, 98(1), 34–43.
Schick, C. (2009). Differential scanning calorimetry (DSC) of semicrystalline polymers. Analytical and bioanalytical chemistry, 395(6), 1589–1611.
Shard, A. G. (2020). X-ray photoelectron spectroscopy. In: Characterization of Nanoparticles (pp. 349–371). Elsevier.
Shi, K., Santiso, E. E., & Gubbins, K. E. (2021). Current advances in characterization of nano-porous materials: pore size distribution and surface area. In: Porous Materials: Theory and Its Application for Environmental Remediation (pp. 315–340). Springer.
Shiekh, A., Mushtaq, A., Jabeen, U., Bashir, F., Zahra, M., Behlil, F., Hina, N., & Hafeez, I. (2022). Surface modification of mesoporous silica nanoparticles with hexamethyl disilazane as smart carriers for tocopherol acetate. Nano Biomedicine & Engineering, 14(3), 216–224.
Shukla, N., Liu, C., Jones, P. M., & Weller, D. (2003). FTIR study of surfactant bonding to FePt nanoparticles. Journal of Magnetism and Magnetic materials, 266(1–2), 178–184.
Simonescu, C. M. (2012). Application of FTIR spectroscopy in environmental studies. In: Advanced Aspects of Spectroscopy (pp. 49–84). IntechOpen.
Sinha, P., Datar, A., Jeong, C., Deng, X., Chung, Y. G., & Lin, L. C. (2019). Surface area determination of porous materials using the Brunauer–Emmett–Teller (BET) method: limitations and improvements. The Journal of Physical Chemistry C, 123(33), 20195–20209.
Spink, C. H. (2008). Differential scanning calorimetry. Methods in Cell Biology, 84, 115–141.
Sridharan, A., Altschaeffl, A. G., & Diamond, S. (1971). Pore size distribution studies. Journal of the Soil Mechanics and Foundations Division, 97(5), 771–787.
Stevie, F. A., & Donley, C. L. (2020). Introduction to x-ray photoelectron spectroscopy. Journal of Vacuum Science & Technology A, 38(6), 063204.

Szekeres, M., & Tombácz, E. (2012). Surface charge characterization of metal oxides by potentiometric acid–base titration, revisited theory and experiment. Colloids and Surfaces A: Physicochemical and Engineering Aspects, 414, 302–313.

Tang, C. Y., & Yang, Z. (2017). Transmission electron microscopy (TEM). In: Membrane Characterization (pp. 145–159). Elsevier.

Tian, Y., & Wu, J. (2018). A comprehensive analysis of the BET area for nanoporous materials. AIChE Journal, 64(1), 286–293.

Tian, Z., Wang, S., Hu, X., Zhang, Z., & Liang, L. (2018). Crystalline reduction, surface area enlargement and pore generation of chitin by instant catapult steam explosion. Carbohydrate Polymers, 200, 255–261.

Ural, N. (2021). The significance of scanning electron microscopy (SEM) analysis on the microstructure of improved clay: an overview. *Open Geosciences*, 13(1), 197–218.

Wang, Z. L. (2003). New developments in transmission electron microscopy for nanotechnology. Advanced Materials, 15(18), 1497–1514.

Zarime, N. A., Yaacob, W. Z. W., & Jamil, H. (2018, April). Removal of heavy metals using bentonite supported nano-zero valent iron particles. In: AIP Conference Proceedings (Vol. 1940, No. 1). AIP Publishing.

Zhang, X., Zeng, X., Sun, X., Deng, B., Peng, G., Xie, H., Wu, Z., & Xiao, T. (2015). Thermal and elemental analysis of the combustion chamber deposits in a large-scale two-stroke marine diesel engine. Science China Technological Sciences, 58, 1717–1725.

Zhao, Z., Bai, C., An, L., Zhang, X., Wang, F., Huang, Y., Qu, M., & Yu, Y. (2021). Biocompatible porous boron nitride nano/microrods with ultrafast selective adsorption for dyes. Journal of Environmental Chemical Engineering, 9(1), 104797.

6

Application of Bioadsorbents

6.1 Introduction

Hazardous pollutants are being released more frequently as a result of the chemical industry' explosive growth, endangering both human health and ecosystems. Adsorption has grown in popularity as a remediation technology due to its low cost and versatility in removing a wide range of contaminants. Super adsorbent polymers (SAPs), which have an extremely high adsorption capacity, are critical in removing harmful chemicals such as heavy metals, dyes, and organic compounds from wastewater. These materials are important in environmental cleaning efforts because of their efficiency and versatility (Xiang et al., 2022). Environmentally friendly biomaterials have arisen as long-term options for water filtration due to their abundance, low toxicity, biodegradability, and efficient adsorption. Recent research has highlighted the use of natural materials such as chitosan, luffa, sisal, and hydroxyapatite to cleanse water contaminated with dyes, heavy metals, and pharmaceutical residues (Khadir et al., 2021). These pollutants frequently coexist in real-world water matrices, necessitating multifunctional adsorbents capable of tackling many contaminants simultaneously. The incorporation of such biomaterials into water treatment systems not only improves pollutant removal effectiveness but also adheres to green chemistry principles, encouraging environmentally friendly and economically feasible wastewater management solutions.

Dye pollutants are among the most common contaminants in wastewater, coming mostly from the textile, leather, paper, and plastic sectors. With over 100,000 commercially accessible dyes generated annually at an estimated 700,000 tons, around 2% are discharged into aquatic systems, creating major ecological problems due to their great visibility, toxicity, and harmful impacts on aquatic creatures. Similarly, pharmaceuticals and personal care products (PPCPs) remain in water bodies even after traditional use and cleanup, causing endocrine disruption in living creatures. Heavy metals like Pb, As, Cd, and Hg also offer long-term environmental and health risks, necessitating effective remediation solutions (Khan et al., 2013).

Volatile organic compounds (VOCs) are harmful to both human health and the environment, demanding effective removal procedures. Adsorption has been widely used for over three decades due to its effectiveness and versatility (Li et al., 2020). Adsorption capabilities vary with adsorbents, with metal-organic frameworks (MOFs) outperforming activated carbon, hypercrosslinked polymeric resins (HPR), and zeolite. Adsorption mechanisms include electrostatic interactions, polarity-based affinities, and partitioning into porous structures. Increased surface area, pore volume, and functional groups all improve adsorption efficiency. Future research should concentrate on creating targeted, cost-effective adsorbents with tailored surface properties for better VOC abatement (Li et al., 2020). Activated carbon is commonly used for CO_2 adsorption due to its high surface area; however, its weak interaction with CO_2 molecules leads to low selectivity and poor performance at higher temperatures (Zulys, et al., 2020). To address these constraints, metal-organic frameworks (MOFs) have developed as improved adsorbents with superior structural features such as large surface area, configurable pore diameters, and customizable surface functions. Coordination polymers, made up of metal ions and organic ligands, have potential for CO_2 storage, catalysis, drug delivery, and sensing (Ghanbari et al., 2020).

The conversion of agricultural waste biomass into biochar is an environmentally friendly and cost-effective technique for pollution prevention and long-term soil restoration. Biochar made from agricultural waste is known for its high adsorption efficiency, environmental friendliness, and potential to improve soil health by lowering greenhouse gas emissions, nutrient leaching, and pollutant bioavailability (Van Nguyen et al., 2022). It also promotes bioremediation and phytoremediation by immobilizing cationic metal(loid)s in acidic soils, hence increasing soil characteristics and microbial diversity. However, there are limitations in stabilizing highly toxic or anionic metal(loid)s, highlighting the importance of customized biochar selection and additional study on biochar–soil–microbe interactions and field-scale performance (Van Nguyen et al., 2022).

However, despite the proven potential of adsorption technologies, significant research gaps remain. These include limited knowledge of adsorbent–pollutant interaction mechanisms under real environmental conditions, lack of scalability of novel materials, and reduced performance in multi-contaminant systems. This chapter aims to bridge these gaps by exploring the multifunctional applications of advanced adsorbents in water, air, and soil remediation. It also highlights recent innovations, identifies challenges, and proposes future directions to enhance environmental sustainability through adsorption-based approaches.

6.2 Water and Wastewater Treatment

The utilization of waste-derived bioadsorbents is bringing about a practical change in wastewater treatment, meeting the need for economical treatment techniques and the increase in waste generation globally. This approach encourages sustainability and resource efficiency by supporting the circular economy and energy recovery. A variety of waste materials are used either directly or following physical, chemical, thermal, or hydrothermal changes. These include biomass, municipal solid trash, industrial byproducts, and agricultural residues. Rice husk, orange peel, sawdust, eggshells, black gram husk, straw, sewage sludge, and microalgae are all effective adsorbents. These materials have significant promise for eliminating pollutants and turning waste into useful resources, hence lowering environmental burden and treatment costs (Hossain et al., 2020).

In recent years, algae have received a lot of attention for their effectiveness as bioadsorbents. They are eco-friendly, inexpensive, and readily available. Algae also have a high adsorption capability. Phytoremediation uses microalgae and macroalgae to remove heavy metals from wastewater. Compared to other adsorbents, algae provide a low-cost raw material, are easy to cultivate, have good metal selectivity, and cause no secondary contamination. Their unique construction allows for large-scale application. Algae often outperform traditional adsorbents. More study is required to optimize this procedure. pH, temperature, contact time, and ion concentration all have a substantial influence on performance. Understanding this allows us to scale up the biosorption process. Because industrial applications are limited, algae-based approaches should be integrated with other technologies such as membrane filtration, electrochemical treatment, and bioprecipitation (Ameri et al., 2020).

Algae also serve as natural ion exchangers. They have a huge surface area and a distinct cell wall composition. Brown algae include alginic acid and sulfated carbohydrates, which assist metals bind. Red algae include sulfated galactans, which aid in metal uptake. Green algae contain heteropolysaccharides containing carboxyl and sulfate groups. These groups increase the binding of hazardous substances. Chlorella, a green microalga, contains proteins, cellulose, and glycoproteins in its cell wall. It has several active sites, including —COOH, —OH, $—NH_2$, and $—SO_4^-$. These groups serve an important role in collecting metal ions. The diversity of functional groups among algal varieties renders them adaptable to various pollutants (Younas et al., 2021).

Sewage wastewater is a big issue. In a recent study, Carica papaya seed powder was utilized as a natural bioadsorbent. The material was examined using FTIR, SEM, DLS, XRD, XRF, BET, and zeta potential. Batch studies were conducted to assess various dosages and contact times. At 0.5 g/500 mL for 180 minutes, chemical oxygen demand (COD), biochemical oxygen demand (BOD), and total suspended solids (TSS) were reduced by 77, 79.2, and 95%, respectively. Total dissolved solids (TDS), total nitrogen (TN), and total phosphorus (TP) all decreased dramatically. The data agreed well with the Langmuir isotherm and pseudo-second-order kinetics. This demonstrates that papaya seed powder is effective and sustainable for treating sewage effluent (Nady et al., 2024).

Many studies have investigated seashells as low-cost adsorbents. The majority of studies focus on heavy metal removal from oyster, snail, and shrimp shells. These shells have unique functional groups such as $-NH_2$, $-CH$, $-OH$, and PO_4^{3-}, providing active binding sites. Oyster shells are high in calcium carbonate, which enhances adsorption. Shrimp shells are structurally similar to chitin and chitosan, which makes them effective biosorbents. Modifying these materials or modifying the pyrolysis conditions can increase their capacity even further. Performance is affected by parameters such as pH, temperature, metal ion concentration, and contact time. The adsorption was best described by using Langmuir or Freundlich isotherms and pseudo-second-order kinetics. Although promising, marine biosorbents are rarely exploited on an industrial basis. Hybrid systems that incorporate other technologies may enhance real-world applications (Tamjidi and Ameri, 2020).

Bio-based polymers provide a sustainable solution for wastewater treatment. Traditional synthetic polymers perform well but represent environmental problems due to their low biodegradability. Biopolymers like chitosan and other polysaccharide-based compounds are inexpensive and readily available. However, they require chemical modification to improve solubility and charge density. With the proper modifications, these materials might be effective, environmentally friendly alternatives. Future research should aim to make them multipurpose, simple to use, and reusable. Real-world testing, cost analysis, and regeneration studies are all vital. Collaboration between researchers and industry will accelerate the development of viable solutions. This transition could result in greener and more effective water treatment systems (Kolya and Kang, 2023).

EDTA-functionalized magnetic chitosan (EDTA-M-Cs) composite that removes Cu(II), Cd(II), and Zn(II) ions from water. EDTA modification considerably increased metal ion adsorption. Adsorption effectiveness decreased as the pH fell. The mechanism followed pseudo-second-order kinetics and was consistent with the Langmuir isotherm model. Thermodynamic studies indicated that the adsorption occurred spontaneously. Cu(II) demonstrated the best removal effectiveness of the studied metals, followed by Cd(II) and Zn(II). The high adsorption capacity and selectivity of EDTA-M-Cs indicate that it is a potential material for wastewater treatment applications (Karimi et al., 2022).

In industrial-scale dye wastewater treatment, it is critical to guarantee that bioadsorbents have adequate adsorption strength, reusability, site accessibility, and economic viability. Recent research emphasizes the need for bioadsorbents being both efficient and cost-effective throughout numerous cycles of use. A significant concern is the balance of performance and operational costs, and bio-adsorption has the potential to surpass traditional technologies such as distillation in terms of affordability and environmental effect. When adsorbents have high regeneration potential, bio-adsorption becomes a viable and long-term option for large-scale industrial applications (Li et al., 2019).

6.3 Adsorption of Inorganic and Organic Pollutants

Adsorption is a frequently used method for eliminating contaminants from wastewater. It is easy, inexpensive, and effective. Recent research has concentrated on biosorbents, nanoparticles, and magnetic biosorbents. Magnetic biosorbents are promising due to their high surface area, ease of recovery, and inexpensive cost. Chemical activation using acid, alkali, or salt increases porosity and performance. Inorganic contaminants are eliminated by ion exchange and electrostatic attraction. Organic pollutants are adsorbed by hydrogen bonding, π–π interaction, and hydrophobicity. The efficiency of removal is affected by factors such as pH, contact time, adsorbent dosage, and pollutant concentration. Too much adsorbent reduces capacity, whereas increased contact time or concentration enhances it. These materials provide a sustainable solution for large-scale wastewater treatment (Tee et al., 2022).

In recent years, biodegradable and environmentally friendly adsorbents have received increased interest for pollutant removal. Chitosan stands out for its biodegradability, functional groups, and availability. However, its original form frequently lacks mechanical strength and stability. To address these constraints, chemical and physical changes such as grafting and cross-linking have been investigated. These methods increase chitosan's adsorption capability for both organic (dyes and medicines) and inorganic (heavy metals) contaminants. Cross-linking improves structural stability but may limit adsorption sites. To address this, recent research has used clays, activated carbon, or grafting agents prior to cross-linking, resulting in enhanced performance. The overall adsorption efficiency is determined by modification methods, materials employed, and operational parameters (Saheed et al., 2021).

Biochar, a carbon-rich substance formed by the pyrolysis of organic waste, is increasingly being employed as a low-cost adsorbent for eliminating a variety of contaminants from water and soil. Its porous structure, large surface area, and surface functional groups increase its potential to adsorb both inorganic pollutants like heavy metals and organic contaminants like dyes and medicines. The efficiency of biochar is determined by the feedstock utilized, the pyrolysis temperature, and the technique of modification. In comparison to commercial adsorbents, biochar has demonstrated competitive performance. Despite its advantages, long-term field tests are required to demonstrate its stability and reusability under real-world situations. Overall, biochar is a potential and long-term alternative for environmental restoration (Jagadeesh and Sundaram, 2023).

Plant barks from Eucalyptus, Neem, Pine, Acacia, and Mango trees have been extensively researched for pollutant adsorption, notably in India and Nigeria, where water pollution is a serious issue. These biosorbents are categorized into five kinds based on their modification processes. They have high potential for eliminating heavy metals, dyes, pesticides, and other pollutants. Most adsorption studies use Langmuir or Freundlich models and pseudo-second-order kinetics. Adsorption is typically spontaneous and physical in nature, with a high reusability. However, there are still gaps in testing for pharmaceuticals and other new contaminants. Disposal procedures and cost assessments also require further study (Ighalo and Adeniyi, 2020).

6.3.1 Heavy Metal Removal

Heavy metal pollution poses a major hazard to both the environment and human health. There are various removal procedures, each with advantages and disadvantages. Biological adsorption is gaining popularity. It is inexpensive, efficient, and generates little sludge. It also supports metal recovery. Agricultural and food industry wastes show great promise as bioadsorbents. Materials such as potato peel, sawdust, citrus peels, mango peel, rice husk, wheat bran, and tea waste have demonstrated high potential. These adsorbents are inexpensive and readily available. The majority of investigations continue on a pilot stage. Industrial applications require upgrading. Using these wastes is both environmentally friendly and cost-effective. Research should focus on optimizing approaches for large-scale usage in water and food decontamination (Ziarati et al., 2018). Table 6.1 gives the efficiency of bioadsorbents in heavy metal removal.

Recent research investigated the use of ZnO-modified date pits (MDP) to remove Cu^{2+}, Ni^{2+}, and Zn^{2+} from water and industrial effluent. The adsorbent was evaluated by FT-IR, SEM, BET, and XRD. Key process variables such as pH, particle size, shaking speed, dose, and temperature were optimized. Adsorption followed the Langmuir and Freundlich models. Kinetics corresponded well to pseudo-second-order and intra-particle diffusion models. The maximum adsorption was 82.4, 71.9, and 66.3 mg/g for Cu^{2+}, Ni^{2+}, and Zn^{2+}, respectively. These results outperformed typical date pit adsorbents by more than four times. The process was spontaneous and exothermic, with adsorption occurring at both homogeneous and heterogeneous locations. The work focuses on an affordable, waste-derived adsorbent with improved performance (Hummadi et al., 2023).

Table 6.1 Efficiency of bioadsorbents in heavy metal removaL

Bioadsorbent	Heavy metal removed	Removal efficiency (%)	Optimum pH	Optimum dosage (g/L)	Optimum contact time	Adsorption capacity (mg/g)	References
Peanut husk	Cadmium	41.3	6	5	3 h	11.36	Abdelfattah et al. (2016)
Moringa aptera Gaertn	Chromium	91	Solution pH	10	40 min	5.497	Matouq et al. (2015)
Tea waste	Chromium	85	2	2	2 h	61.34	Babapoor et al. (2022)
Tangerine peel	Chromium	51	2	2	2 h	49.50	Babapoor et al. (2022)
Bovine gut	Chromium	46	2	2	2 h	30.86	Babapoor et al. (2022)
Sunflower seeds	Chromium	34	2	2	2 h	4.23	Babapoor et al. (2022)
Peanut husk	Cobalt	30	6	5	3 h	6.10	Abdelfattah et al. (2016)
Moringa aptera Gaertn	Copper	90	Solution pH	10	40 min	6.07	Matouq et al. (2015)
Fe_3O_4/montmorillonite	Copper	94.89	Solution pH	0.08/0.025	120 s	182.94	Kalantari et al. (2015)
Tea waste	Iron	96	4	2	2 h	20.83	Babapoor et al. (2022)
Tangerine peel	Iron	87	4	1	2 h	62.50	Babapoor et al. (2022)
Bovine gut	Iron	96	4	1	2 h	73.00	Babapoor et al. (2022)

(Continued)

Table 6.1 *(Continued)*

Bioadsorbent	Heavy metal removed	Removal efficiency (%)	Optimum pH	Optimum dosage (g/L)	Optimum contact time	Adsorption capacity (mg/g)	References
Sunflower seeds	Iron	56	4	2	2 h	13.00	Babapoor et al. (2022)
Fe_3O_4/montmorillonite	Lead	89.72	Solution pH	0.06 /0.025	120 s	510.16	Kalantari et al. (2015)
Peanut husk	Lead	100	6	5	3 h	27.03	Abdelfattah et al. (2016)
Peanut husk	Manganese	38.4	6	5	3 h	14.29	Abdelfattah et al. (2016)
Moringa aptera Gaertn	Nickel	68	Solution pH	10	30 min	5.53	Matouq et al. (2015)
Fe_3O_4/montmorillonite	Nickel	76.15	Solution pH	0.08/0.025	120 s	111.90	Kalantari et al. (2015)
Peanut husk	Nickel	23.6	6	5	3 h	56.82	Abdelfattah et al. (2016)
Coconut shell	Zinc	52	6	25	3 h	45.14	Amuda et al. (2007)
Coconut shell carbon with chitosan	Zinc	58	6	21	3 h	50.93	Amuda et al. (2007)
Acid-treated Coconut shell carbon with chitosan	Zinc	90	6	18	3 h	60.41	Amuda et al. (2007)

Nejadshafiee and Islami (2019) produced a magnetic bioadsorbent using Fe_3O_4 nanoparticles and 1,4-butane sultone functionalized activated carbon from pistachio shells. This sustainable and cost-effective material was studied utilizing TEM, VSM, FT-IR, and TGA methods. Its adsorption ability for Pb(II), As(III), and Cd(II) ions was assessed under ultrasound-assisted conditions using the Langmuir isotherm model. The maximal adsorption capacities were 147.05, 151.51, and 119.04 mg/g, respectively. The adsorbent had several advantages, including eco-friendly synthesis, recyclability, low production costs, and simple magnetic separation. The use of ultrasound dramatically increased removal effectiveness, indicating the magnetic nanocomposite's promise as a rapid, efficient, and long-term solution for heavy metal cleanup from contaminated water sources. Lee and Choi (2018) showed that persimmon leaf biowaste is an effective and environmentally acceptable method for removing heavy metals from water. Dried persimmon leaves (DPL) raw persimmon leaves (RPL) by 10–15% in terms of Pb, Cu, and Cd removal, with Pb having the highest affinity. Adsorption followed the Freundlich isotherm, indicating heterogeneous surface binding, and the kinetics were consistent with the pseudo-second-order model. The maximum adsorption capacities were 22.59, 19.42, and 18.26 mg/g for Pb, Cu, and Cd, respectively. Thermodynamic data confirm spontaneity (negative $\Delta G°$) and endothermic behavior (positive $\Delta H°$), with increased adsorption efficiency at higher temperatures. This waste-derived adsorbent combines sustainability, cost-effectiveness, and high performance, demonstrating its utility for heavy metal remediation in aqueous environments.

Baby et al. (2023) created a functionalized palm kernel shell (PKS-Sulfo) bioadsorbent to combat heavy metal contamination in water. Sulfuric acid oxidation introduced oxygenated and sulfonated groups into activated carbon produced from PKS. The adsorbent was tested for Cr^{6+}, Pb^{2+}, Cd^{2+}, and Zn^{2+} removal using optimum pH, metal ion concentration, contact time, and dosage. The results revealed 99% elimination of Cr^{6+} and Pb^{2+}, and 80 and 70% of Cd^{2+} and Zn^{2+}, respectively. Adsorption followed the pseudo-second-order model, with the Langmuir isotherm proving to be the most accurate fit. This low-cost, waste-derived adsorbent showed great promise for large-scale water treatment while also providing a sustainable solution to waste management and heavy metal contamination.

Ahmed et al. (2024) developed a biological treatment unit in El-Menoufia, Egypt, that uses agricultural residues—rice husk, plum leaves, banana peels, and green bean peels—as low-cost bioadsorbents for heavy metal removal from industrial wastewater. Batch studies revealed a chromium removal efficiency (RE) of 96–99.5%, followed by lead (86.79–95.90%), with a maximum RE at 45 minutes retention time. Continuous flow reduced lead levels to around 0.041 mg/L (97.89–97.94% RE), cadmium to 0.002 mg/L (99.42% RE), and chromium to

0.002–0.003 mg/L (98.99–99.33% RE). All post-treatment metal levels were within international guidelines. The system is economically viable, costing ~ 1226.86 USD/year for a 1 m^3/h discharge unit, making it ideal for large-scale industrial applications. Jiang et al. (2020) used free radical polymerization to create SA-PAM/GO hydrogel composites that were then evaluated by BET, SEM-EDS, FTIR, and TGA. The BET surface area more than tripled after alteration, consistent with SEM findings. The ideal conditions resulted in maximum adsorption capabilities of 68.76 mg/g for Cu^{2+} and 240.69 mg/g for Pb^{2+}. Kinetic and isotherm analysis revealed that the pseudo-second-order model and the Langmuir isotherm best fit the data. Adsorption occurred by physical adsorption, chemical adsorption, and electrostatic interactions, with ion exchange being the most common process. The composites maintained high efficiency after five adsorption-regeneration cycles, demonstrating their significant potential for effective heavy metal ion removal from water.

Chicken eggshell powder (CESP) was used as a low-cost bioadsorbent to remove lead, cadmium, and chromium from water. The BET, FE-SEM/EDX, XRD, and TG-DTA studies revealed good surface characteristics for adsorption. Batch studies showed that Pb^{2+} removal efficiency approached 99% at 2.5 g dosage and 4 h contact time, with adsorption capacity ranking $Pb^{2+} > Cr^{3+} > Cd^{2+}$. Pb^{2+} was quickly absorbed in mixed-metal systems. Kinetic data fit the pseudo-second-order model, and equilibrium matched the Langmuir isotherm, confirming monolayer adsorption. CESP is a viable and environmentally beneficial solution for heavy metal cleanup due to its large capacity, lack of pretreatment requirements, and ease of supply (Kim et al., 2019). Wang et al. (2017) developed a bioadsorbent made of sugarcane cellulose that effectively removes lead, copper, and zinc from aqueous solutions while also being environmentally friendly. Adsorption capacity was affected by initial ion concentration and temperature. Adsorption in single-component systems followed the Langmuir model, with maximal capacities of 558.9 mg/g for Pb^{2+}, 446.2 mg/g for Cu^{2+}, and 363.3 mg/g for Zn^{2+}. In binary systems, the competitive Langmuir model predicts that the addition of Pb^{2+} reduces Cu^{2+} uptake while leaving other ions unaltered. SEM-EDAX research revealed that adsorption occurs through coordination, ion exchange, and electrostatic interaction. These findings highlight the high potential of sugarcane cellulose-based adsorbents for efficient heavy metal removal in both single- and mixed-ion systems.

Ziarati et al. investigated the ability of *Rumex crispus* L. seeds to remove heavy metals from pharmaceutical effluent. To enhance bio-adsorption, dried seeds were evaluated at various pH levels, temperatures, dosages, contact times, and particle sizes. Lead, zinc, copper, cobalt, and nickel removal efficiency increased with higher adsorbent concentrations, peaking at 2% seed dosage. Adsorption varied from 83.5 to 91% after one week of agitation, with no notable change thereafter.

The study emphasizes the hitherto unexplored mechanisms of metal complexation at the bioadsorbent interface and proposes additional research on innovative herbal, agricultural, and food-industry waste-based materials for large-scale heavy metal remediation (Ziarati et al., 2017).

Kılıç et al. (2013) used almond shell biochar, a pyrolysis byproduct, as a low-cost adsorbent to remove Ni^{2+} and Co^{2+} from aqueous solutions. The optimal circumstances were determined by adjusting pH, dose, beginning metal concentration, contact time, and temperature. Adsorption followed the Langmuir isotherm, showing monolayer coverage, whereas kinetics matched the pseudo-second-order model, implying chemisorption. Thermodynamic analysis ($\Delta G°$, $\Delta H°$, $\Delta S°$) showed that the adsorption was spontaneous, endothermic, and chemically favorable. The findings suggest that almond shell biochar is an efficient and economical option for heavy metal remediation, offering possibilities for sustainable water treatment applications in place of more expensive conventional adsorbents.

6.3.2 Dye Removal

Dye wastewater is a major environmental contaminant, and bio-adsorption has emerged as a promising remedy due to its simplicity, adaptability, and effectiveness. Despite its benefits, large-scale applications are limited due to high running expenses. The chemical and physical structure of bioadsorbents is a key aspect influencing adsorption and desorption, since it determines performance efficiency. Li et al. (2019) underlined that no previous studies had systematically specified parameters for industrial-scale bio-adsorption use. They presented a cost-performance connection to determine feasibility and assist the selection of effective, low-cost bioadsorbents. Balancing adsorption efficiency and economic viability is critical for moving from laboratory research to real-world dye wastewater treatment, allowing for sustainable industrial-scale application while limiting environmental impact. Table 6.2 explores the adsorption performance of bioadsorbents for dye removal.

Sawdust, a low-cost bioadsorbent, was examined for its ability to remove methylene blue (MB) from water. Chemical activation was achieved through the use of sulfuric acid. In batch tests, contact time, dye concentration, and adsorbent dosage were investigated. At a neutral pH and 4 g/L dosage, treated sawdust had a removal efficiency of 98.12%, whereas raw sawdust had a removal efficiency of 97.18% at 100 mg/L dye concentration. Under varying settings, the difference in removal effectiveness between the two forms was less than 5%. Adsorption data fits well with the Freundlich isotherm model, demonstrating multilayer adsorption. The findings indicate that sawdust, even in its raw state, is an excellent, low-cost solution for dye removal from wastewater (Raghuvanshi et al., 2005). Khattri and Singh (2000) explored dye removal from water with neem sawdust,

Table 6.2 Adsorption performance of bioadsorbents for dye removal.

Bioadsorbent	Dye removed	Removal efficiency (%)	Optimum pH	Optimum dosage	Optimum contact time (min)	Adsorption capacity (mg/g)	References
Orange albedo	Methylene blue	99.15	Solution pH	1%	60	77.79	de Farias Silva et al. (2020)
Parthenium hysterophorus	Methylene blue	99.9	7	0.22 g	60	98.06	Chatterjee et al. (2012)
Leather shaving waste	Acid red 357	87.37	2.3	8.09 g/L	60	24.74	Gomes et al. (2016)
Cashew nutshell	Congo red	99.99	3.2	24.76 g/L	67	—	Ponnusamy and Subramaniam (2013)
Rice straw	Methylene blue	—	6	0.15 g/25 mL	120	32.6	Fathy et al. (2013)
Alkali-treated rice straw (ARS)	Methylene blue	97.1	6	0.15 g/25 mL	120	62.9	Fathy et al. (2013)
ARS-0.5C	Methylene blue	—	6	0.15 g/25 mL	120	95.2	Fathy et al. (2013)
ARS-1C	Methylene blue	99.9	6	0.15 g/25 mL	120	135.1	Fathy et al. (2013)
Arginine based composite	Methylene blue	—	8	20 mg/250 mL	120	219.9	Naushad et al. (2019)

Pistachio hull waste	Methylene blue	99.7	8	1.5 g/L	70	602	Moussavi and Khosravi (2011)
Acacia Nilotica sawdust	Indigo carmine	77.5	6.76	0.5 g/50 mL	120	31.01	Gupta and Lataye (2018)
Acacia Nilotica sawdust	Methylene blue	95.4	6.96	0.5 g/50 mL	90	57.36	Gupta and Lataye (2018)
Date seed	Malachite green	97.6	8	0.02 g/25 mL	60	91.22	Amode et al. (2016)
Metroxylon spp.	Methylene blue	—	Solution pH	5 g/L	60	83.5	Amode et al. (2016)
Potassium hydroxide-treated sago wastes	Methylene blue	—	Solution pH	5 g/L	60	212.8	Amode et al. (2016)
Phosphoric acid-treated sago wastes	Methylene blue	—	Solution pH	5 g/L	60	36.82	Amode et al. (2016)

a low-cost bioadsorbent. Tests on crystal violet, malachite green, methylene blue, and Rhodamine B at various concentrations, pH levels, and temperatures revealed high removal effectiveness, with crystal violet adsorption decreasing from 91.56 to 78.94% as concentration increased. Similar patterns were seen with various dyes. Langmuir isotherm modeling revealed monolayer adsorption, with adsorption capacity decreasing as temperature increased. Kinetic data fit the pseudo-first-order model, and thermodynamic characteristics validated the process' feasibility. Surface ionization and complexation were found to be responsible for pH effects. These findings show that neem sawdust is a viable and environmentally friendly solution for treating dye effluent.

Ali et al. (2022) investigated the use of raw Iraqi date seeds as a low-cost, natural bioadsorbent to remove methyl violet (MV) color from aqueous solutions. Before beginning adsorption investigations, the seeds were examined using XRD, SEM, FTIR, and BET analysis. The effects of contact time, initial dye concentration, and adsorbent dosage on dye removal efficiency were studied. The pseudo-second-order model $\left(R^2 = 0.9917\right)$ outperformed the pseudo-first-order model in accurately describing adsorption kinetics. Equilibrium data suit the Langmuir isotherm $\left(R^2 = 0.9837\right)$ better than the Freundlich model, resulting in a maximum adsorption capacity of 59.5 mg/g. Thermodynamic research demonstrated that the process was spontaneous and endothermic, demonstrating date seeds' efficacy and sustainability as a bioadsorbent for cationic dye removal. Grafting acrylic acid and acrylamide resulted in the creation of a porous cellulose-based bioadsorbent capable of removing anionic dye acid blue 93 (AB93) and cationic dye MB from single and binary systems. Adsorption efficacy was affected by initial dye concentration, dosage, contact time, pH, temperature, ionic strength, and surfactant content. Both dyes had a maximum adsorption capacity of 1372 mg/g at 2500 mg/L. Equilibrium data suited the Freundlich isotherm, kinetics followed the pseudo-second-order model, and electrostatic interactions were the primary mechanism. The adsorbent performed consistently across three reuse cycles, even in complex matrices, proving its efficacy and reusability for textile wastewater treatment (Liu et al., 2015).

Waste tea, a common agricultural byproduct, was treated with citric acid to increase its adsorption ability for MB dye. Batch tests revealed that citric acid-modified waste tea (CA-WT) had a maximum adsorption capacity of 212.77 mg/g at 298 K, which corresponded to the Langmuir isotherm model. Following dye removal, the spent adsorbent was regenerated with cold atmospheric plasma (CAP) at normal temperature and pressure. Even after six adsorption-regeneration cycles, the regeneration efficiency was still high, indicating stability and reusability. This technology provides a low-cost and environmentally friendly method for creating and replenishing bioadsorbents for dye removal in wastewater treatment (Zhou et al., 2016).

Corn cobs were activated with CO_2 and steam at 1173 K to remove MB. The effects of contact time, dosage, beginning concentration, and temperature were all studied in batches. Characterization methods included BET-N_2 adsorption, temperature-programmed breakdown, and thermogravimetric analysis. Adsorption followed the Langmuir isotherm, with maximal capacities of 100 mg/g (CCC) and 75.5 mg/g (SCC). The pseudo-second-order model accurately characterized kinetics. Thermodynamic characteristics revealed an endothermic and spontaneous adsorption, with negative ΔG° values indicating feasibility. The findings point to maize cob-derived activated carbon as an effective, low-cost adsorbent for dye removal from aqueous solutions (Reddy et al., 2016).

Activated carbon from *Catha edulis* stem was of good quality, with 83.65% fixed carbon. Malachite green removal was investigated using various adsorption conditions. The best results were obtained with a dye concentration of 10 mg/L, pH of 10, 0.5 g dosage, and a contact time of 60 minutes. Under these settings, removal efficiency was 98.8%. Data analysis revealed that the process matched the pseudo-second-order kinetic model. Equilibrium corresponded well to the Freundlich isotherm model. The adsorbent's high fixed carbon content indicates good stability and reusability. Using discarded stems makes the process less expensive. The technology is environmentally safe and appropriate for dye-contaminated water treatment. This approach demonstrates how agricultural waste may be a valuable instrument for wastewater treatment. It provides a simple, sustainable, and effective method for eliminating hazardous dyes (Abate et al., 2020).

CMC-T-CH is a biopolymer adsorbent made by crosslinking carboxymethyl cellulose and chitosan with triethylenetetramine. It was tested to remove Congo red and direct blue dyes from water. Characterization confirmed its advantageous structure. At 35 °C, Congo red was removed most effectively at pH 6 and direct blue at pH 5. Removal efficiencies reached 99 and 98.33%, respectively. The Langmuir isotherm described adsorption well. The pseudo-second-order kinetic model fit the process. Maximum adsorption capacities were 519.53 mg/g for Congo red and 534.25 mg/g for direct blue. The adsorbent might be reused for multiple cycles. A phytotoxicity test revealed improved seed germination following treatment, demonstrating its environmental safety and potential for wastewater treatment (Kumar et al., 2023). The textile dyeing industry's wastewater poses significant dangers to aquatic life and human health. To remove Congo red dye, a cross-linked chitosan composite was developed: chitosan/fly ash/polyvinyl alcohol (Ch/FA/PVA). To enhance performance, fly ash was added to chitosan at various ratios. Under ideal circumstances of 0.9 g/L dosage, 50 min contact time, and 40 mg/L dye concentration, the Ch/FA0.75/PVA composite removed 99.7% of the dye. SEM-EDX, BET, FTIR, and pHzpc analysis validated its adsorption capabilities. The Freundlich isotherm described the process and had a maximum capacity of 263.15 mg/g. Adsorption kinetics followed the pseudo-second-order model. Pilot-scale tests demonstrated significant removal of dye, COD, BOD_5, and TDS, indicating potential for industrial wastewater treatment (Elhady et al., 2024).

Hawthorn kernel (HK), an agricultural waste material, was employed as a low-cost bioadsorbent to remove MB. To increase its limited adsorption capability, a sulfonated hawthorn kernel (SHK) was synthesized using sulfuric acid activation under ideal conditions: 6 hours at 85 °C in 18 M acid with a 1 : 5 solid-to-liquid ratio. Characterization demonstrated that SHK had higher acidity, porosity, and thermal stability than HK. Adsorption occurred using the pseudo-second-order kinetic model and the Langmuir isotherm, with maximum capacities of 49.5 mg/g for HK and 151.5 mg/g for SHK. Thermodynamic analysis revealed spontaneous adsorption under various circumstances. SHK maintained 99% efficiency over five reuse cycles, indicating a high potential for wastewater treatment (Akköz et al., 2019).

Raw olive stone (OS) was studied as a low-cost, environmentally acceptable adsorbent for MB removal. SEM revealed well-developed pores, while FTIR validated functional groups that aid in adsorption. The BET surface area was 0.9 m^2/g. Optimal elimination (93.65%) occurred at 0.2 g OS for 90 minutes with 20 ppm MB. The Langmuir and Temkin isotherms best described equilibrium, with a maximum capacity of 44.5 mg/g each. The kinetic data fit the pseudo-second-order model $\left(R^2 = 1.0\right)$ and Elovich model $\left(R^2 = 0.935\right)$. Thermodynamic analysis revealed that the adsorption was endothermic and spontaneous. OS remained 78% efficient after five reuse cycles, demonstrating its potential as a low-cost, long-term wastewater treatment option (Alardhi et al., 2023).

A chitosan-pectin composite (Cs-Pc) was evaluated to remove MB from synthetic effluent. SEM, EDX, FTIR, and pHpzc revealed a porous, rough structure with active adsorption sites. Optimal conditions (50 mg/L MB, pH ~ 11, 1.5 g/L Cs-Pc, 180 min) resulted in 97.77% elimination. At 500 mg/L MB, removal rate was 98.67%, with a capacity of 328.02 mg/g. Kinetics followed the pseudo-second-order model, while equilibrium matched the Freundlich isotherm. pH had a greater influence than adsorbent dosage. Cs-Pc outperformed chitosan and pectin separately. After four reuse cycles, elimination remained over 96%. According to Mohrazi and Ghasemi-Fasaei (2023), Cs-Pc is a reliable and reusable choice for dye removal due to its strong electrostatic and π–π interactions.

6.3.3 Emerging Contaminants Removal

Emerging contaminants, such as pesticides, pharmaceutical and personal care products (PPCPs), and endocrine-disrupting chemicals (EDCs), have been found in water sources, raising significant environmental and health issues. Activated carbon adsorption is a potential approach for eliminating these contaminants because of its high efficiency and low cost. Adsorption performance is determined by the features of activated carbon, pollutant qualities, and ambient circumstances. Mechanisms include hydrophobic interactions and chemical affinities. Various isotherms (e.g., Langmuir, Freundlich) and kinetic models (e.g., pseudo-first-order, intraparticle diffusion) are used to effectively characterize the process (Jeirani et al., 2017). Table 6.3 depicts the removal of emerging contaminants by bioadsorbents.

Pharmaceutical wastewater is a significant environmental issue. Drugs in natural streams can affect both humans and ecosystems. Adsorption is the most used approach due to its low cost and ease of operation. Single treatment procedures cannot completely eliminate pharmaceutical contaminants. Both conventional and modern procedures have advantages and disadvantages. Wastewater characteristics and biosorbent types influence removal efficiency. Biosorption is the process of removing pollutants from water using biological materials. It is emerging as an alternative to conventional procedures. Recent research reveals that biosorbents can successfully remove hormones and antibiotics from wastewater. This method provides a sustainable solution for treating pharmaceutical effluent (Sahay et al., 2023). Figure 6.1 shows the removal of emerging contaminants using advanced adsorbents.

Emerging contaminants of concern include pharmaceuticals, personal care items, flame retardants, and endocrine disruptors. Their removal from water is critical for protecting health and ecosystems. The activated carbon, carbonaceous resin, and two high-silica zeolites was used for emerging contaminants elimination. Experiments were conducted using 28 emerging contaminants at low con centrations. Activated carbon was the most efficient adsorbent. Activated carbon, at doses below 10 mg/L, was most effective, achieving 2-log removal of nearly all tested ECCs. The carbonaceous resin performed less well due to a lack of

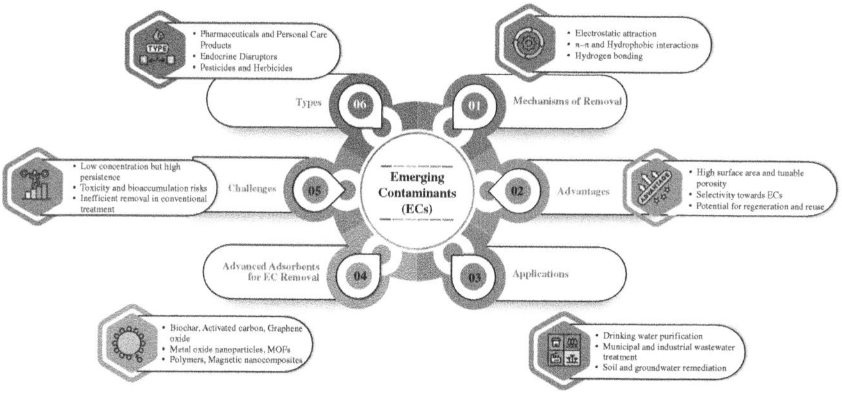

Figure 6.1 Advanced adsorbents for the removal of Emerging Contaminants (ECs).

Table 6.3 Removal of emerging contaminants by bioadsorbents.

Bioadsorbent	Emerging contaminants removed	Removal efficiency (%)	Optimum pH	Optimum dosage	Optimum contact time	Adsorption capacity (mg/g)	References
Banana peel	Estrone (EE2), and (SMX)	80	6	500.0 mg in 25.0 mL	90 min	0.0387	de Sousa et al. (2019)
Banana peel	17α-ethinylestradiol	80	6	500.0 mg in 25.0 mL	90 min	0.0420	de Sousa et al. (2019)
Banana peel	Sulfamethoxazole	70	6	500.0 mg in 25.0 mL	90 min	0.0413	de Sousa et al. (2019)
Activated charcoal	Primidone	86.11	7–9	5 mg/L	30 min	0.258	Appa et al. (2019)
Activated carbon of vegetable origin	Carbamazepine	90	6.15	100 mg/L	72 hr	242	Delgado et al. (2019)
Activated carbon of vegetable origin	Sildenafil citrate	90	5.97	100 mg/L	72 hr	395	Delgado et al. (2019)
Reduced graphene oxide aerogel	Diclofenac	96	6	0.25 g/L	120 min	596.71	Hiew et al. (2019)
Granular-activated carbon	Ciprofloxacin	—	7	0.025 g/10 mL	90 min	8.87	Rao et al. (2021)

Granular-activated carbon	Acetaminophen	—	7	0.025 g/10 mL	90 min	9.17	Rao et al. (2021)
Granular-activated carbon	Caffeine	—	7	0.025 g/10 mL	90 min	9.46	Rao et al. (2021)
Granular-activated carbon	Benzophenone	—	7	0.025 g/10 mL	90 min	15.47	Rao et al. (2021)
Granular-activated carbon	Irgasan	—	7	0.025 g/10 mL	90 min	22.38	Rao et al. (2021)
CS-MgO/Zeolite	Chlorpyrifos	80.9	7	0.4 g/L	40 min	83.3	Hamad and Mahran (2024)
Modified nanozeolite	Phytoestrogens daidzein	98	4	1 g/L	180 min	40.74	Goyal et al. (2018)
Modified nanozeolite	coumestrol	99	4	1 g/L	180 min	42.87	Goyal et al. (2018)

pore capacity in the appropriate size range. Only a few emerging contaminants with the same pore size were eliminated by zeolites, but the majority were not. According to the study, adsorbents having a wide range of pore diameters and a high pore volume are most effective at removing emerging contaminants (Rossner et al., 2009).

Bio-sorbents are gaining popularity as environmentally friendly materials for eliminating emerging pollutants (EPs). They are environmentally beneficial alternatives to conventional adsorbents, as they are derived from renewable and low-cost sources such as biochar and agricultural waste. Their removal efficiency is linked to a variety of methods, including physical adsorption (van der Waals and electrostatic forces), chemical interactions via functional groups, and pore formation in porous materials. Surface chemistry determines selectivity, allowing for the removal of certain pollutants. Modifications, hybridization, and enhanced production processes can help to improve performance even further. Many bio-sorbents can be regenerated and reused, saving waste and cost. While scaling up and maintaining long-term stability are still problems, these materials show great promise for sustainable water treatment solutions (Jatoi et al., 2023).

Adsorption capacity is determined by the chemical composition of ECs as well as adsorbent surface properties. The selection of adsorbents is critical in filtration and biofiltration systems. Proper selection improves removal efficiency and increases system longevity (Almeida-Naranjo et al., 2023). Valorization of agricultural and industrial residues offers a long-term solution to water contamination and waste creation. Many residues, whether raw or modified, have demonstrated high efficacy in eliminating EPs. To avoid harsh chemicals, green, low-cost modification approaches such as enzyme-assisted activation or dopants produced from waste are required. Hybrid materials, such as biochar-nanoparticle composites and residue-based aerogels, should be investigated for durability and reuse. Dual-function adsorbents that also promote catalysis, nutrient recovery, or microorganisms have the potential to increase their uses, particularly in low-resource situations (Almeida-Naranjo et al., 2025).

6.4 Gas and Air Purification

Industrial flue gas treatment has become a global priority as emissions rise. Fossil fuel combustion emits hazardous gasses and particles into the atmosphere. Power generation is a significant source. Pollutants include carbon dioxide, sulfur oxides, nitrogen oxides, volatile organics, particulate matter, and trace metals. Controlling these requires efficient separation technologies. The choice of treatment is determined not only by process knowledge but also by pollutant characteristics. Different technologies have different costs, efficiency, and environmental impact. Flue gas treatment remains critical to reduce industrial air pollution (Elehinafe et al., 2024).

6.4.1 VOCs Removal

Adsorption is an economical way to remove VOCs (Mobasser et al., 2022). In this investigation, microbial biosorbents were made from lyophilized fungal and bacterial cells that had been aminomethylated. The fungus biosorbent demonstrated the best performance. It has adsorption capabilities of 620 mg/g for ethyl acetate, 454 mg/g for α-pinene, and 374 mg/g for n-hexane. These values exceeded those of other biosorbents. The substance had a surface area of 20 m^2/g and was primarily made of hydrocarbons and polysaccharides. VOC adsorption followed the Langmuir isotherm and the pseudo-first-order model, showing physical adsorption. The fungal biosorbent remained effective after five reuse cycles, demonstrating excellent regeneration capabilities. This highlights the possibility for long-term VOC control (Cheng et al., 2020). Figure 6.2 depicts the VOCs removal by adsorption.

Activated carbons were made by chemically activating vegetable-tanned leather wastes (shavings, trimmings, and buffing dust) with KOH, NaOH, and K_2CO_3. KOH activation produced the best results, with a BET surface area of 2719 m^2/g and a micropore volume of 0.829 cm^3/g. However, K_2CO_3 activation resulted in activated carbons with excellent textural characteristics and adsorption capacity at a reduced cost. The toluene adsorption values (27 mg/g at 5 ppm, 288 mg/g at 50 ppm, and 700 mg/g at 100 ppm) were greater than those

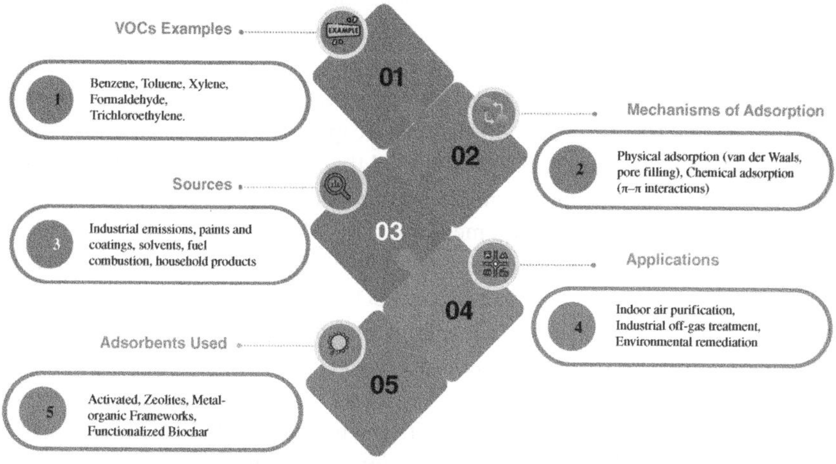

Figure 6.2 VOCs removal by adsorption.

of numerous commercial activated carbons. This indicates the long-term application of leather waste to control VOCs (Gil et al., 2014). A cobalt-adeninate metal-organic framework (bio-MOF-11) was created and used for VOC adsorption for the first time. Four VOCs were tested: methanol, acetone, benzene, and toluene. The saturation adsorption capacity ranged from 0.73 to 3.57 mmol/g, in the order of toluene, benzene, acetone, and methanol. Adsorption dropped by 7.3–35.6% as the temperature rose from 288 to 308 K, with acetone being the most affected due to its low boiling point and polarity. The results showed that adsorption was impacted by molecular size and polarity, as well as nitrogen and amino groups in the framework. Bio-MOF-11 has potential for VOC elimination (Chen et al., 2020).

6.4.2 CO_2 Removal

Three activated carbons made from phenol–formaldehyde resins (No1KCla-600, No1KClb-1000, and No2OS-1000) were investigated for their ability to extract CO_2. The experiments employed CO_2/H_2 and CO_2/CH_4 gas mixes under fixed-bed dynamic settings. Resin-based carbons beat commercial carbons for CO_2 adsorption, breakthrough time, and efficiency. They also demonstrated great selectivity, full regeneration, and steady cycling. The highest breakthrough capacities were 2.11 and 2.03 mmol/g for CO_2/H_2 and CO_2/CH_4, respectively. The No2OS-1000 adsorbent, consisting of resin and olive stones, has the highest volumetric capacity and CO_2/CH_4 selectivity, making it ideal for biogas purification (Gil et al., 2015). The carbons were generated using biochar derived from the rapid pyrolysis of whitewood. Three activation procedures were used: steam, CO_2, and potassium hydroxide. Adsorption studies were conducted in a fixed-bed reactor at 25–65 °C with CO_2 concentrations of 10–30 mol%. Adsorption with N_2 and CO_2 was used to investigate surface area, porosity, and chemistry. KOH-activated carbon demonstrated the best performance. It possessed a surface area of 1400 m^2/g, a micropore volume of 0.62 cm^3/g, and an adsorption capacity of 1.8 mol/kg at 25 °C and 30 mol% CO_2. Shahkarami et al. (2015) found that both KOH- and CO_2-activated carbons retained steady capacity after 50 cycles of low-temperature regeneration. Figure 6.3 shows the CO_2 removal methods.

Using biowaste as a CO_2 adsorbent is a low-cost and sustainable solution. It promotes the circular economy and reduces CO_2 emissions. Adsorbents that work well have a large surface area and a lot of micropores. Activation and surface modification are critical stages toward improving adsorption performance. Chemical activation, particularly with potassium hydroxide, had the best results. Multistep procedures can increase efficiency, but they are more complicated, expensive, and environmentally harmful. High adsorption effectiveness between 0 and 25 °C is promising since it eliminates the need to modify exhaust gas temperatures.

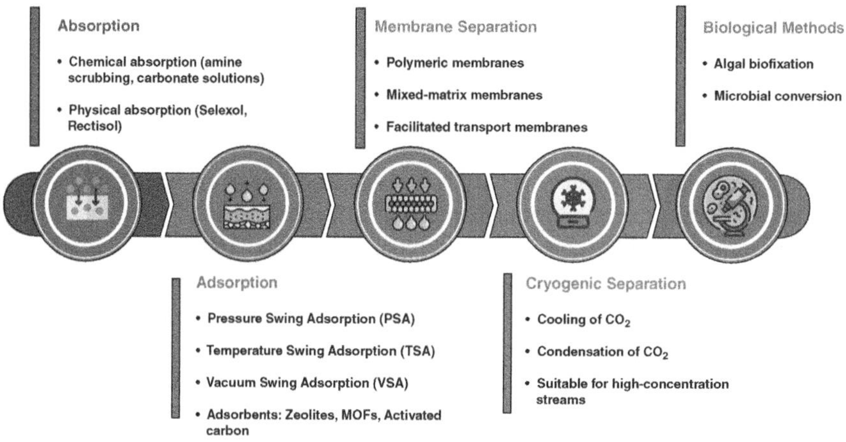

Figure 6.3 CO_2 removal methods.

Regenerability and a lengthy cycle life also save expenses. However, large-scale production and regeneration technologies for modified biochar are limited, needing future research (Sołtysik et al., 2022).

The Spanish food industry's byproducts, pine cone leaves (PCL) and pine kernel shells (PKS), were used to create activated carbons (ACs) for CO_2 collection. The goal was to increase narrow microporosity to boost adsorption capacity. Experiments with $CO_2/H_2O/N_2$ mixtures at 50 °C demonstrated excellent separation. In humid flue gases, both CO_2 and H_2O were adsorbed, but CO_2 adsorption predominate in the initial phases. Under humid conditions, the total adsorption was around 2 mmol/g. Cyclic adsorption–desorption tests demonstrated stability and regeneration potential. The AC achieved 1.08 mmol/g at 50 °C with 30 vol% CO_2 in dry flue gas. Kinetic investigation revealed rapid CO_2 uptake in cement flue gas conditions (Querejeta et al., 2024).

A facile approach was established to manufacture activated microporous carbons from Arundo donax by heating solid KOH to 600 °C for 2 hours. The surface area and pore structure were adjusted by varying the KOH-to-biomass ratio. Carbons manufactured in a single stage have superior textural qualities to those produced in two steps. The best sample, KLB2, had a surface area of 1122 m^2/g and a micropore volume of 0.50 cm^3/g. The CO_2 adsorption capacity was 6.3 mmol/g at 273 K and 1 pressure, and increased to 15.4 mmol/g at 30 bar. Singh et al. (2017) found that stable carbons from waste biomass can be used as cost-effective adsorbents for CO_2 capture both before and after combustion.

Sargassum horneri, the marine garbage that generates golden tide, can be converted into porous bioadsorbents. A one-step catalytic oxidative pyrolysis using

K_2FeO_4 was used to generate carbon materials for CO_2 capture. The surface area and pore volume rose until the K_2FeO_4 ratio reached 1 : 4, after which they declined. 25%-ASHC was the most effective sample for CO_2 adsorption due to its large surface area. Kinetic study revealed that the double exponential model fitted best, validating both physical and chemical adsorption. The adsorbent also showed good stability over time. High porosity and richer nitrogen and oxygen groups were shown to enhance performance (Ying et al., 2024).

Coffee ground residue can be used as bioadsorbents for CO_2 capture in energy and heavy industries. Carbonization was used to generate the material, which was then activated with KOH. Its characteristics were investigated with LECO, TG, SEM, BET, and FT-IR techniques. Adsorption studies with CO_2, O_2, and N_2 showed strong CO_2 selectivity. Adsorption capacity varies with CO_2 concentration and temperature, as demonstrated by tests. The bioadsorbent remained stable and could be renewed using multistep adsorption–desorption cycles. Vacuum swing adsorption (VSA) experiments showed good performance at various pressures. Sołtysik et al. (2025) found that coffee-based bioadsorbents are effective and affordable for CO_2 capture applications.

6.5 Soil Remediation

Toxic metal buildup in soil and crops poses a significant threat to food safety. Low-cost bioadsorbents derived from agricultural waste are one possible answer. Coffee bean waste (CBW), which is produced in large quantities worldwide, was investigated as a nickel (Ni) adsorbent in soil. In this investigation, CBW (0–5%) was combined with soil in containers containing coriander plants. Samples were tested at intervals of up to 40 days. The results revealed that CBW rapidly lowered nickel levels, with considerable adsorption happening in the early days. Adsorption efficiency was determined by contact time, concentration, pH, and dosage. The best results were seen with 3% CBW after 30 days. This highlights CBW's significant potential for soil detoxification (Tajik et al., 2020).

Municipal biowaste (MBW) can be converted into valuable goods for sustainable supply chains. The biosurfactants (BPS) formed from MBW has been used to remediate soil. Three BPS types were created using food digestate, gardening compost, and a compost-digestate-sludge mixture. Their molecular weight varied greatly, and they contained a variety of functional groups. BPS solutions were used on metal-polluted soil in Italy. CVDF BPS demonstrated the best extraction effectiveness, eliminating up to 98% of heavy metals while surpassing traditional extractants. A two-step remediation approach was proposed: first, soil washed with BPS, then acidified and membrane filtered. The results demonstrated exceptional efficiency and sustainability. MBW-derived BPS provide a cost-effective route from waste to resource (Tabasso et al., 2020).

Apple pomace (AP), an agricultural waste from the food industry, is used as a bio-sorbent to remove heavy metals from rice fields. Experiments were carried out in Babol paddies, Mazandaran province, Iran, with fresh AP obtained from companies in Tehran and Saveh. Rice samples were tested for nickel, cadmium, and lead using established wet digestion procedures and flame emission spectrophotometry. Statistical study (ANOVA, t-test) revealed significant reductions, particularly in lead concentrations. Untreated rice contained 2.71 mg/kg DW Pb, whereas AP-treated paddies had just 0.01 mg/kg DW Pb when administered 10 days before transplantation. The results show that AP can effectively remediate polluted rice soils and minimize hazardous metal uptake (Gholizadeh and Ziarati, 2016).

6.6 Conclusion

Bioadsorbents have shown strong potential as sustainable materials for environmental remediation. Their natural origin, low cost, and availability make them attractive alternatives to conventional adsorbents. In water and wastewater treatment, bioadsorbents effectively remove heavy metals, dyes, and emerging contaminants. Their diverse surface chemistry and functional groups enhance adsorption capacity and selectivity. For air and gas purification, bioadsorbents play a vital role in capturing VOCs and CO_2. These applications support cleaner air and help in climate change mitigation. In soil remediation, bioadsorbents immobilize toxic metals and improve nutrient retention, which supports sustainable agriculture. Their multifunctional role highlights their adaptability across different environmental systems. Despite these benefits, challenges remain in large-scale application. Variability in raw materials, regeneration efficiency, and disposal of spent adsorbents require further study. Modification and functionalization strategies can enhance stability and reusability. Integration with advanced treatment processes may also improve overall efficiency. Overall, bioadsorbents represent a green and versatile solution for pollution control. They offer pathways toward resource recovery and circular economy practices. Continued research, scale-up studies, and industry collaboration will help transform laboratory findings into real-world technologies.

References

Abate, G. Y., Alene, A. N., Habte, A. T., & Getahun, D. M. (2020). Adsorptive removal of malachite green dye from aqueous solution onto activated carbon of Catha edulis stem as a low cost bio-adsorbent. Environmental Systems Research, 9(1), 29.

Abdelfattah, I., Ismail, A. A., Al Sayed, F., Almedolab, A., & Aboelghait, K. M. (2016). Biosorption of heavy metals ions in real industrial wastewater using peanut husk as efficient and cost effective adsorbent. Environmental Nanotechnology, Monitoring & Management, 6, 176–183.

Ahmed, R. S., Abuarab, M. E., Baiomy, M. A., & Ibrahim, M. M. (2024). Heavy metals removal from industrial wastewater using bio-adsorbent materials based on agricultural solid wastes through batch and continuous flow mechanisms. Journal of Water Process Engineering, 57, 104665.

Akköz, Y., Coşkun, R., & Delibaş, A. (2019). Preparation and characterization of sulphonated bio-adsorbent from waste hawthorn kernel for dye (MB) removal. Journal of Molecular Liquids, 287, 110988.

Alardhi, S. M., Salih, H. G., Ali, N. S., Khalbas, A. H., Salih, I. K., Saady, N. M. C., Zendehboudi, S., Albayati, T. M., & Harharah, H. N. (2023). Olive stone as an eco-friendly bio-adsorbent for elimination of methylene blue dye from industrial wastewater. Scientific Reports, 13(1), 21063.

Ali, N. S., Jabbar, N. M., Alardhi, S. M., Majdi, H. S., & Albayati, T. M. (2022). Adsorption of methyl violet dye onto a prepared bio-adsorbent from date seeds: isotherm, kinetics, and thermodynamic studies. Heliyon, 8(8), e10276.

Almeida-Naranjo, C. E., Guerrero, V. H., & Villamar-Ayala, C. A. (2023). Emerging contaminants and their removal from aqueous media using conventional/non-conventional adsorbents: a glance at the relationship between materials, processes, and technologies. Water, 15(8), 1626.

Almeida-Naranjo, C. E., Tejedor, J., Villamar-Ayala, C. A., & Vizuete, G. (2025). Transforming waste into solutions: raw and modified bioadsorbents for emerging contaminant removal. Journal of Environmental Chemical Engineering, 13, 116720.

Ameri, A., Tamjidi, S., Dehghankhalili, F., Farhadi, A., & Saati, M. A. (2020). Application of algae as low cost and effective bio-adsorbent for removal of heavy metals from wastewater: a review study. Environmental Technology Reviews, 9(1), 85–110.

Amode, J. O., Santos, J. H., Md. Alam, Z., Mirza, A. H., & Mei, C. C. (2016). Adsorption of methylene blue from aqueous solution using untreated and treated (Metroxylon spp.) waste adsorbent: equilibrium and kinetics studies. International Journal of Industrial Chemistry, 7(3), 333–345.

Amuda, O. S., Giwa, A., & Bello, I. A. (2007). Removal of heavy metal from industrial wastewater using modified activated coconut shell carbon. Biochemical Engineering Journal, 36(2), 174–181.

Appa, R., Mhaisalkar, V. A., Naoghare, P. K., & Lataye, D. H. (2019). Adsorption of an emerging contaminant (primidone) onto activated carbon: kinetic, equilibrium, thermodynamic, and optimization studies. Environmental Monitoring and Assessment, 191(4), 215.

Babapoor, A., Rafiei, O., Mousavi, Y., Azizi, M. M., Paar, M., & Nuri, A. (2022). Comparison and optimization of operational parameters in removal of heavy metal ions from aqueous solutions by low-cost adsorbents. International Journal of Chemical Engineering, 2022(1), 3282448.

Baby, R., Hussein, M. Z., Zainal, Z., & Abdullah, A. H. (2023). Preparation of functionalized palm kernel shell bio-adsorbent for the treatment of heavy metal-contaminated water. Journal of Hazardous Materials Advances, 10, 100253.

Chatterjee, S., Kumar, A., Basu, S., & Dutta, S. (2012). Application of response surface methodology for methylene blue dye removal from aqueous solution using low cost adsorbent. Chemical Engineering Journal, 181, 289–299.

Chen, R., Yao, Z., Han, N., Ma, X., Li, L., Liu, S., Sun, H., & Wang, S. (2020). Insights into the adsorption of VOCs on a cobalt-adeninate metal–organic framework (Bio-MOF-11). ACS Omega, 5(25), 15402–15408.

Cheng, Z., Feng, K. E., Su, Y., Ye, J., Chen, D., Zhang, S., Zhang, X., & Dionysiou, D. D. (2020). Novel biosorbents synthesized from fungal and bacterial biomass and their applications in the adsorption of volatile organic compounds. Bioresource Technology, 300, 122705.

de Farias Silva, C. E., da Gama, B. M. V., da Silva Gonçalves, A. H., Medeiros, J. A., & de Souza Abud, A. K. (2020). Basic-dye adsorption in albedo residue: effect of pH, contact time, temperature, dye concentration, biomass dosage, rotation and ionic strength. Journal of King Saud University-Engineering Sciences, 32(6), 351–359.

de Sousa, P. A. R., Furtado, L. T., Neto, J. L. L., de Oliveira, F. M., Siqueira, J. G. M., Silva, L. F., & Coelho, L. M. (2019). Evaluation of the adsorption capacity of banana peel in the removal of emerging contaminants present in aqueous media—study based on factorial design. Brazilian Journal of Analytical Chemistry, 6(22), 14–28.

Delgado, N., Capparelli, A., Navarro, A., & Marino, D. (2019). Pharmaceutical emerging pollutants removal from water using powdered activated carbon: study of kinetics and adsorption equilibrium. Journal of Environmental Management, 236, 301–308.

Elehinafe, F. B., Aondoakaa, E. A., Akinyemi, A. F., Agboola, O., & Okedere, O. B. (2024). Separation processes for the treatment of industrial flue gases–effective methods for global industrial air pollution control. Heliyon, 10(11), e3242.

Elhady, S., Bassyouni, M., Elshikhiby, M. Z., Saleh, M. Y., & Elzahar, M. H. (2024). Optimization of anionic dye removal using cross-linked chitosan composite as eco-friendly bio-adsorbent. Applied Water Science, 14(7), 159.

Fathy, N. A., El-Shafey, O. I., & Khalil, L. B. (2013). Effectiveness of alkali-acid treatment in enhancement the adsorption capacity for rice straw: the removal of methylene blue dye. International Scholarly Research Notices, 2013(1), 208087.

Ghanbari, T., Abnisa, F., & Daud, W. M. A. W. (2020). A review on production of metal organic frameworks (MOF) for CO_2 adsorption. Science of The Total Environment, 707, 135090.

Gholizadeh, E., & Ziarati, P. (2016). Remediation of contaminated rice farmlands soil and Oryza sativa rice product by apple pomace as adsorbent. Biosciences Biotechnology Research Asia, 13(4), 2245.

Gil, E. R., Ruiz, B., Lozano, M. S., Martín, M. J., & Fuente, E. (2014). VOCs removal by adsorption onto activated carbons from biocollagenic wastes of vegetable tanning. Chemical Engineering Journal, 245, 80–88.

Gil, M. V., Álvarez-Gutiérrez, N., Martínez, M., Rubiera, F., Pevida, C., & Morán, A. (2015). Carbon adsorbents for CO_2 capture from bio-hydrogen and biogas streams: breakthrough adsorption study. Chemical Engineering Journal, 269, 148–158.

Gomes, C. S., Piccin, J. S., & Gutterres, M. (2016). Optimizing adsorption parameters in tannery-dye-containing effluent treatment with leather shaving waste. Process Safety and Environmental Protection, 99, 98–106.

Goyal, N., Bulasara, V. K., & Barman, S. (2018). Removal of emerging contaminants daidzein and coumestrol from water by nanozeolite beta modified with tetrasubstituted ammonium cation. Journal of Hazardous Materials, 344, 417–430.

Gupta, T. B., & Lataye, D. H. (2018). Adsorption of indigo carmine and methylene blue dye: Taguchi's design of experiment to optimize removal efficiency. Sādhanā, 43(10), 170.

Hamad, M. T., & Mahran, B. N. (2024). Biosynthesized CS-MgO/zeolite hybrid material: an efficient adsorbent for chlorpyrifos removal-kinetic studies and response surface methodology. Emerging Contaminants, 10(3), 100324.

Hiew, B. Y. Z., Lee, L. Y., Lai, K. C., Gan, S., Thangalazhy-Gopakumar, S., Pan, G. T., & Yang, T. C. K. (2019). Adsorptive decontamination of diclofenac by three-dimensional graphene-based adsorbent: response surface methodology, adsorption equilibrium, kinetic and thermodynamic studies. Environmental Research, 168, 241–253.

Hossain, N., Bhuiyan, M. A., Pramanik, B. K., Nizamuddin, S., & Griffin, G. (2020). Waste materials for wastewater treatment and waste adsorbents for biofuel and cement supplement applications: a critical review. Journal of Cleaner Production, 255, 120261.

Hummadi, K. K., Zhu, L., & He, S. (2023). Bio-adsorption of heavy metals from aqueous solution using the ZnO-modified date pits. Scientific Reports, 13(1), 22779.

Ighalo, J. O., & Adeniyi, A. G. (2020). Adsorption of pollutants by plant bark derived adsorbents: an empirical review. Journal of Water Process Engineering, 35, 101228.

Jagadeesh, N., & Sundaram, B. (2023). Adsorption of pollutants from wastewater by biochar: a review. Journal of Hazardous Materials Advances, 9, 100226.

Jatoi, A. S., Nguyen, H. M., Ahmed, J., & Jeyapaul, A. (2023). Bio-sorbents derived from agricultural biomass for the removal of emerging pollutants and its adsorption mechanisms. Journal of the Iranian Chemical Society, 20(10), 2457–2470.

Jeirani, Z., Niu, C. H., & Soltan, J. (2017). Adsorption of emerging pollutants on activated carbon. Reviews in Chemical Engineering, 33(5), 491–522.

Jiang, H., Yang, Y., Lin, Z., Zhao, B., Wang, J., Xie, J., & Zhang, A. (2020). Preparation of a novel bio-adsorbent of sodium alginate grafted polyacrylamide/graphene oxide hydrogel for the adsorption of heavy metal ion. Science of The Total Environment, 744, 140653.

Kalantari, K., Ahmad, M. B., Masoumi, H. R. F., Shameli, K., Basri, M., & Khandanlou, R. (2015). Rapid and high capacity adsorption of heavy metals by Fe_3O_4/montmorillonite nanocomposite using response surface methodology: preparation, characterization, optimization, equilibrium isotherms, and adsorption kinetics study. Journal of the Taiwan institute of Chemical Engineers, 49, 192–198.

Karimi, F., Ayati, A., Tanhaei, B., Sanati, A. L., Afshar, S., Kardan, A., Dabirifar, Z., & Karaman, C. (2022). Removal of metal ions using a new magnetic chitosan nano-bio-adsorbent; A powerful approach in water treatment. Environmental Research, 203, 111753.

Khadir, A., Negarestani, M., kheradmand, A., Azad, A., & Sillanpää, M. (2021). The utilization of biomaterials for water purification: dyes, heavy metals, and pharmaceuticals. In: Novel Materials for Dye-containing Wastewater Treatment (pp. 27–58). Singapore: Springer.

Khan, N. A., Hasan, Z., & Jhung, S. H. (2013). Adsorptive removal of hazardous materials using metal-organic frameworks (MOFs): a review. Journal of hazardous materials, 244, 444–456.

Khattri, S. D., & Singh, M. K. (2000). Colour removal from synthetic dye wastewater using a bioadsorbent. Water, Air, and Soil Pollution, 120(3), 283–294.

Kılıç, M., Kırbıyık, Ç., Çepelioğullar, Ö., & Pütün, A. E. (2013). Adsorption of heavy metal ions from aqueous solutions by bio-char, a by-product of pyrolysis. Applied Surface Science, 283, 856–862.

Kim, D., Hwang, S. J., Kim, Y., Jeong, C. H., Hong, Y. P., & Ryoo, K. S. (2019). Removal of heavy metals from water using chicken egg shell powder as a bio-adsorbent. Bulletin of the Korean Chemical Society, 40(12), 1156–1161.

Kolya, H., & Kang, C. W. (2023). Bio-based polymeric flocculants and adsorbents for wastewater treatment. Sustainability, 15(12), 9844.

Kumar, A., Jeyabalan, J., Patra, C. C., & Narayanasamy, S. (2023). Fabrication of a novel bio-polymer adsorbent with high adsorptive capacity towards organic dyes. Industrial Crops and Products, 203, 117166.

Lee, S. Y., & Choi, H. J. (2018). Persimmon leaf bio-waste for adsorptive removal of heavy metals from aqueous solution. Journal of Environmental Management, 209, 382–392.

Li, W., Mu, B., & Yang, Y. (2019). Feasibility of industrial-scale treatment of dye wastewater via bio-adsorption technology. Bioresource Technology, 277, 157–170.

Li, X., Zhang, L., Yang, Z., Wang, P., Yan, Y., & Ran, J. (2020). Adsorption materials for volatile organic compounds (VOCs) and the key factors for VOCs adsorption process: a review. Separation and Purification Technology, 235, 116213.

Liu, L., Gao, Z. Y., Su, X. P., Chen, X., Jiang, L., & Yao, J. M. (2015). Adsorption removal of dyes from single and binary solutions using a cellulose-based bioadsorbent. ACS Sustainable Chemistry & Engineering, 3(3), 432–442.

Matouq, M., Jildeh, N., Qtaishat, M., Hindiyeh, M., & Al Syouf, M. Q. (2015). The adsorption kinetics and modeling for heavy metals removal from wastewater by Moringa pods. Journal of Environmental Chemical Engineering, 3(2), 775–784.

Mobasser, S., Wager, Y., & Dittrich, T. M. (2022). Indoor air purification of volatile organic compounds (VOCs) using activated carbon, zeolite, and organosilica sorbents. Industrial & Engineering Chemistry Research, 61(20), 6791–6801.

Mohrazi, A., & Ghasemi-Fasaei, R. (2023). Removal of methylene blue dye from aqueous solution using an efficient chitosan-pectin bio-adsorbent: kinetics and isotherm studies. Environmental Monitoring and Assessment, 195(2), 339.

Moussavi, G., & Khosravi, R. (2011). The removal of cationic dyes from aqueous solutions by adsorption onto pistachio hull waste. Chemical Engineering Research and Design, 89(10), 2182–2189.

Nady, D. S., Abdel-Halim, S., Hegazy, M. E. F., El-Desouky, M. A., & Hanna, D. H. (2024). Use of Carica papaya waste as bio-adsorbent for sewage wastewater treatment. Biomass Conversion and Biorefinery, 15, 1–18.

Naushad, M., Alqadami, A. A., AlOthman, Z. A., Alsohaimi, I. H., Algamdi, M. S., & Aldawsari, A. M. (2019). Adsorption kinetics, isotherm and reusability studies for the removal of cationic dye from aqueous medium using arginine modified activated carbon. Journal of Molecular Liquids, 293, 111442.

Nejadshafiee, V., & Islami, M. R. (2019). Adsorption capacity of heavy metal ions using sultone-modified magnetic activated carbon as a bio-adsorbent. Materials Science and Engineering: C, 101, 42–52.

Ponnusamy, S. K., & Subramaniam, R. (2013). Process optimization studies of Congo red dye adsorption onto cashew nut shell using response surface methodology. International Journal of Industrial Chemistry, 4(1), 17.

Querejeta, N., Gil, M. V., Rubiera, F., Pevida, C., Wawrzyńczak, D., Panowski, M., & Majchrzak-Kucęba, I. (2024). Bio-engineering of carbon adsorbents to capture CO_2 from industrial sources: the cement case. Separation and Purification Technology, 330, 125407.

Raghuvanshi, S. P., Singh, R., Kaushik, C. P., & Raghav, A. K. (2005). Removal of textile basic dye from aqueous solutions using sawdust as bio-adsorbent. International Journal of Environmental Studies, 62(3), 329–339.

Rao, A., Kumar, A., Dhodapkar, R., & Pal, S. (2021). Adsorption of five emerging contaminants on activated carbon from aqueous medium: kinetic characteristics and computational modeling for plausible mechanism. Environmental Science and Pollution Research, 28(17), 21347–21358.

Reddy, P. M. K., Verma, P., & Subrahmanyam, C. (2016). Bio-waste derived adsorbent material for methylene blue adsorption. Journal of the Taiwan Institute of Chemical Engineers, 58, 500–508.

Rossner, A., Snyder, S. A., & Knappe, D. R. (2009). Removal of emerging contaminants of concern by alternative adsorbents. Water Research, 43(15), 3787–3796.

Sahay, P., Mohite, D., Arya, S., Dalmia, K., Khan, Z., & Kumar, A. (2023). Removal of the emergent pollutants (hormones and antibiotics) from wastewater using different kinds of biosorbent—a review. Emergent Materials, 6(2), 373–404.

Saheed, I. O., Da Oh, W., & Suah, F. B. M. (2021). Chitosan modifications for adsorption of pollutants–A review. Journal of hazardous materials, 408, 124889.

Shahkarami, S., Azargohar, R., Dalai, A. K., & Soltan, J. (2015). Breakthrough CO_2 adsorption in bio-based activated carbons. Journal of Environmental Sciences, 34, 68–76.

Singh, G., Kim, I. Y., Lakhi, K. S., Srivastava, P., Naidu, R., & Vinu, A. (2017). Single step synthesis of activated bio-carbons with a high surface area and their excellent CO_2 adsorption capacity. Carbon, 116, 448–455.

Sołtysik, M., Majchrzak-Kucęba, I., & Wawrzyńczak, D. (2022). Bio-waste as a substitute for the production of carbon dioxide adsorbents: a review. Energies, 15(19), 6914.

Sołtysik, M., Majchrzak-Kucęba, I., & Wawrzyńczak, D. (2025). A coffee-based bioadsorbent for CO_2 capture from flue gas using VSA: TG-vacuum tests. Energies, 18(15), 3965.

Tabasso, S., Ginepro, M., Tomasso, L., Montoneri, E., Nisticò, R., & Francavilla, M. (2020). Integrated biochemical and chemical processing of municipal bio-waste to obtain bio based products for multiple uses. The case of soil remediation. Journal of Cleaner Production, 245, 119191.

Tajik, S., Ziarati, P., & Cruz-Rodriguez, L. (2020). Coffee waste as novel bio-adsorbent: detoxification of nickel from contaminated soil and *Coriandrum sativum*. Methods, 38(41), 2693–2504.

Tamjidi, S., & Ameri, A. (2020). A review of the application of sea material shells as low cost and effective bio-adsorbent for removal of heavy metals from wastewater. Environmental Science and Pollution Research, 27(25), 31105–31119.

Tee, G. T., Gok, X. Y., & Yong, W. F. (2022). Adsorption of pollutants in wastewater via biosorbents, nanoparticles and magnetic biosorbents: a review. Environmental Research, 212, 113248.

Van Nguyen, T. T., Phan, A. N., Nguyen, T. A., Nguyen, T. K., Nguyen, S. T., Pugazhendhi, A., & Phuong, H. H. K. (2022). Valorization of agriculture waste biomass as biochar: as first-rate biosorbent for remediation of contaminated soil. Chemosphere, 307, 135834.

Wang, F., Pan, Y., Cai, P., Guo, T., & Xiao, H. (2017). Single and binary adsorption of heavy metal ions from aqueous solutions using sugarcane cellulose-based adsorbent. Bioresource Technology, 241, 482–490.

Xiang, Z., Tang, N., Jin, X., & Gao, W. (2022). Fabrications and applications of hemicellulose-based bio-adsorbents. Carbohydrate Polymers, 278, 118945.

Ying, H., Jia, C., Zeng, G., & Ai, N. (2024). Turn waste golden tide into treasure: bio-adsorbent synthesis for CO_2 capture with K_2FeO_4 as catalytic oxidative activator. Molecules, 29(6), 1345.

Younas, F., Mustafa, A., Farooqi, Z. U. R., Wang, X., Younas, S., Mohy-Ud-Din, W., Ashir Hameed, M., Mohsin Abrar, M., Maitlo, A. A., Noreen, S., & Hussain, M. M. (2021). Current and emerging adsorbent technologies for wastewater treatment: trends, limitations, and environmental implications. Water, 13(2), 215.

Zhou, R., Zhou, R., Zhang, X., Tu, S., Yin, Y., Yang, S., & Ye, L. (2016). An efficient bio-adsorbent for the removal of dye: adsorption studies and cold atmospheric plasma regeneration. Journal of the Taiwan Institute of Chemical Engineers, 68, 372–378.

Ziarati, P., Mostafidi, M., Shirkhan, F., & Zahedi, M. T. (2018). Analysis of removal methods of toxic heavy metals using bio-absorbs. Technogenic and Ecological Safety, 4, 62–76

Ziarati, P., Zahirnejad, M., & Asgar Pahanh, J. (2017). The efficiency of bio-adsorption of heavy metals from pharmaceutical effluent by *Rumex crispus* L. Seed. Journal of Pharmaceutical & Health Sciences, 5(3), 231–243.

Zulys, A., Yulia, F., Muhadzib, N., & Nasruddin. (2020). Biological metal–organic frameworks (Bio-MOFs) for CO_2 capture. Industrial & Engineering Chemistry Research, 60(1), 37–51.

7

Bioadsorbents in Dye Removal

7.1 Introduction

The discharge of dye-laden effluents into the environment has become a major ecological concern, contaminating soil, surface water, and groundwater. Among industrial dyes, it is widely used in the textile industry and is of special concern because of its poisonous, persistent, and nonbiodegradable properties. Conventional treatment processes frequently fail to entirely remove such dyes, causing additional contamination. Activated carbon has proven to be quite effective in dye adsorption; nevertheless, its commercial use is limited due to high production costs. This has prompted the investigation of biowaste-derived adsorbents as sustainable, low-cost, and renewable alternatives. Agricultural, animal, and wood-based wastes have high sorption effectiveness, making them attractive candidates for dye remediation (Hamad and Idrus, 2022).

Textile dyes are broadly classed based on their chemical structure—such as azo, nitro, indigo, anthraquinone, phthalein, triphenyl methyl, and nitrated dyes—or by their industrial usage (El Harfi and El Harfi, 2017). Reactive dyes dominate textile manufacturing because of their broad color spectrum, great fastness, and adaptability. However, reactive dyes frequently face constraints such as charge accumulation on synthetic fibers, whereas dispersion dyes offer benefits by avoiding such concerns and rectifying problems such as the Barré effect in polyester and nylon. Acid dyes, applied under acidic circumstances, are typically used for nylon, whereas direct dyes, while simple to apply and accessible in a wide range of colors, frequently exhibit poor fastness. These various dye classes pose significant obstacles to wastewater cleanup (Ismail and Sakai, 2022). Textile and related industry effluents contain considerable levels of such dyes, which are very soluble and can spread widely, posing dangers to crops, aquatic systems, and human health (Adeyemo et al., 2017).

Adsorption is widely acknowledged as a simple, efficient, and effective approach for removing persistent dyes from wastewater (Yagub et al., 2014). Commercial activated carbon is widely used because of its high porosity and huge surface area, but its high production and regeneration costs prevent large-scale utilization (Tan et al. 2015). This has increased interest in low-cost, sustainable bio-sorbents. Biological adsorption has several advantages, including adaptability to changing pH and temperature, as well as selective binding of pollutants via functional groups in proteins, uronic acids, and polysaccharides (Agarwala and Mulky, 2023). Physical factors such as van der Waals interactions, hydrogen bonding, and π–π stacking drive adsorption, making it a cost-effective and environmentally acceptable wastewater cleanup method.

Several operational and physical parameters influence liquid-phase adsorption efficiency in wastewater treatment (Rápó and Tonk, 2021). Adsorbent–adsorbate interactions, surface chemistry, pore structure, particle size, and the presence of competing ions in the solution all have an effect on sorption performance. Environmental factors such as pH, temperature, and contact time are also important, and the molecular weight, structure, size, and polarity of dyes influence their adsorption behavior. In batch systems, equilibrium is reached when no more adsorption occurs, which is sometimes assisted by agitation to reduce mass transfer resistance. However, excessive mixing might be harmful to biosorbents. Thus, adjusting these parameters is critical for increasing removal efficiency while preserving the adsorbent's structural integrity.

Adsorption isotherms are critical for understanding the interaction between adsorbent and adsorbate, providing information about adsorption capacity and surface behavior (Yagub et al., 2014). Depending on the system, adsorption can be monolayer or multilayer, with the Langmuir and Freundlich models being the most commonly used to represent equilibrium data. Furthermore, adsorption kinetics are important in calculating dye adsorption rates and adsorbate residence times at the solid–liquid interface. Kinetic studies can assist in determining whether the process is controlled by a chemical reaction, diffusion, or mass transfer, directing the selection of optimal operating parameters for large-scale applications. Several models have been used to investigate the kinetics of anionic and cationic dye adsorption.

Exhausted activated carbon can be regenerated to restore its porous structure and adsorption capacity, with the process selected based on whether adsorption is physical or chemical. Thermal and wet air oxidation, as well as chemical and solvent regeneration, are common techniques. Thermal regeneration is versatile but energy-intensive, and it can damage pore structure, whereas moist air oxidation and chemical treatments are somewhat efficient but have downsides such as surface oxidation and reagent expense. Solvent regeneration, which includes supercritical fluids, is an alternative method for desorbing certain contaminants. Recent research has also investigated the recovery of bio-based adsorbents such as banana peel, coco peat, and rice husk ash, revealing their reusability and potential for integration into sustainable wastewater treatment (Ani et al., 2020).

Furthermore, bioadsorbents derived from renewable resources offer an eco-friendly alternative to conventional activated carbon, with the added advantage of regeneration and reuse potential. The novelty of this chapter lies in providing a comprehensive overview that links dye classification, adsorption mechanisms, influencing factors, isotherms, kinetics, and real-case applications, highlighting biowaste-derived adsorbents as next-generation sustainable solutions for dye remediation.

7.2 Types of Dyes and Their Impact

Dyes are a major class of synthetic organic chemicals used in textiles, leather, paper, cosmetics, and plastics. Their widespread usage and inappropriate disposal make them significant industrial pollutants, with textile processing playing a major role. Increasing worldwide competitiveness, along with stringent environmental restrictions, has prompted enterprises to strike a compromise between product quality and environmentally sustainable operations (Pereira and Alves, 2011). Table 7.1 gives the physicochemical properties of common dyes. However, many dyes are extremely stable, resistant to biodegradation, and can have hazardous, mutagenic, or carcinogenic consequences. The presence of color in effluents inhibits light penetration, impacting photosynthesis and aquatic biodiversity. While physicochemical and biological treatment approaches exist, none of them provide total removal or water reuse, emphasizing the importance of understanding dye types, structures, and their effects on the environment. Figure 7.1 depicts the classification of dyes used in industries.

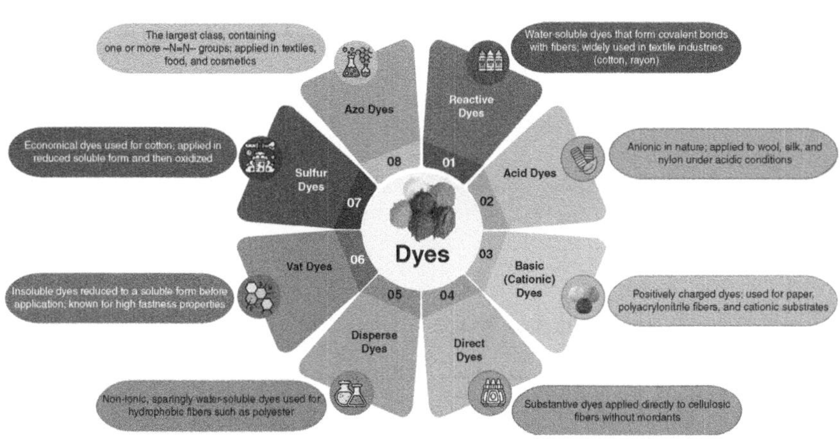

Figure 7.1 Classification of dyes used in industries.

Table 7.1 Physicochemical properties of common dyes.

Dye name	Chemical formula	Molecular weight (g/mol)	Density	λmax (nm)	pKa	References
Acid blue 25	$C_{20}H_{13}N_2NaO_5S$	416.38	1.491 g/cm^3	603	—	Pérez-Jasso et al. (2025)
Basic blue 3	$C_{20}H_{26}ClN_3O$	359.89	—	655	—	Pérez-Jasso et al. (2025)
Congo red	$C_{32}H_{22}N_6Na_2O_6S_2$	696.66	0.995	497	4.1 and 3.0	Yaneva and Georgieva et al. (2012)
Crystal violet	$C_{25}H_{30}ClN_3$	407.99	1.190 g/cm^3	590	5.31 and 8.64	Carmen and Daniela (2012); Parisi (2020)
Indigo carmine	$C_{16}H_8N_2Na_2O_8S_2$	466.35	0.71 g/cm^3	609	11.0	Gamoudi and Srasra (2019)
Methyl orange	$C_{14}H_{14}N_3NaO_3S$	327.34	1.28 g/cm^3	465	3.47	Trabelsi et al. (2013)
Methylene blue	$C_{16}H_{18}ClN_3S$	319.85	—	660	3.8	Carmen and Daniela (2012); Khan et al. (2022)
Phenol red	$C_{19}H_{14}O_5S$	354.37	1.5 g/cm^3	430	8.0	Gamoudi and Srasra (2019)
Reactive blue 4	$C_{23}H_{14}C_{12}N_6O_8S_2$	637.43	1.5315 g/cm^3	600	1.22	Pérez-Jasso et al. (2025)
Reactive orange	$C_{20}H_{17}N_3Na_2O_{11}S_3$	617.54	2.07 g/cm^3	495	3.75	Carmen and Daniela (2012); Ekman et al. (2023)
Rhodamine B	$C_{28}H_{31}ClN_2O_3$	479.2	0.79 g/cm^3	550	3.7	Carmen and Daniela (2012); Wierzbicka et al. (2022)
Tartrazine	$C_{16}H_9N_4Na_3O_9S_2$	534.37	—	425	—	Roy et al. (2023)

7.2.1 Reactive Dyes

Reactive dyes are among the most popular synthetic dyes because of their vibrant colors, adaptability, and good fastness on cotton, wool, silk, and regenerated cellulose fibers. They have reactive groups that can establish covalent connections with nucleophilic sites in fibers, resulting in better wash fastness (Lewis, 2014). Unlike traditional dyes, their fixation forms long-lasting connections, preventing color loss during laundering. Although regulatory restrictions and reduced production in Europe hindered the development of novel chemistries, reactive dyes continue to play an essential role in textile dyeing and printing. They are also used on polyamide fibers as eco-friendly alternatives to chromium dyes, with some varieties minimizing fiber degradation at high temperatures, confirming their industrial and environmental importance. Once adsorbed onto the fiber surface, reactive dyes are fixed through nucleophilic substitution or addition processes between the reactive groups and cellulose's hydroxyl groups. This covalent link formation separates reactive dyes from other dye classes and improves wash fastness. However, the fixing process is incomplete, and a large portion of the dye remains in hydrolyzed or unfixed form, resulting in severe dye losses during textile processing. These unfixed dye molecules are released into effluents, causing high color intensity, recalcitrant organics, and toxicity in aquatic systems (Asgher, 2012). As a result, despite their benefits in textile coloration, reactive dyes are recognized as ecologically persistent contaminants requiring effective treatment solutions.

Reactive dyes are an important family of synthetic dyes because they may establish stable covalent connections with fibers, resulting in exceptional wash fastness and brilliant colors on cellulose and polyamide substrates. They are commonly used in the textile industry to create vibrant and long-lasting colors. However, there are limitations to generating deep, complete hues on cotton and nylon because the sulfonate groups of reactive dyes produce significant negative charges, resulting in the repulsion of further anionic dye molecules and hence constrained buildup. Furthermore, long-liquor dyeing with reactive colors necessitates significant quantities of inorganic electrolytes, which raises environmental problems. As a result, research is focused on developing low-salt or salt-free dyeing processes that improve color-fixing efficiency while limiting environmental impact (Lewis, 2011).

7.2.2 Azo Dyes

Azo dyes are the most numerous and flexible type of synthetic dye, distinguished by the presence of one or more azo linkages (—N=N—), which provide vibrant colors and structural diversity. They account for more than half of global dye output, with uses in textiles, medicines, food, cosmetics, polymers, and inks (Chung, 2016). Despite their stability and resistance to biodegradation, azo dyes are persistent environmental contaminants, with ~10% failing to bind fibers during dyeing and entering wastewater streams. Some azo dyes and their metabolites, including benzidine derivatives, have been associated with mutagenic, genotoxic, and carcinogenic effects, raising serious concerns about human and environmental health. As a result, their extensive use necessitates strict monitoring and safer alternatives (Chung, 2016).

Azo dye toxicity is principally caused by their metabolic conversion into aromatic amines, which can then be oxidized to produce reactive electrophilic species capable of covalent DNA binding (Brown and De Vito, 1993). Toxicity can occur through three major pathways: (1) azo bond reduction by intestinal anaerobic bacteria or hepatic reductases, resulting in carcinogenic amines such as benzidines; (2) direct oxidation of free aromatic amine groups without azo reduction; and (3) oxidation of the azo linkage itself to form highly reactive diazonium salts. Because these pathways are compound-specific, the carcinogenicity of azo dyes cannot be reliably anticipated. However, structural guidelines and short-term mutagenicity assays are useful tools for risk assessment and guiding the development of safer azo dye alternatives.

7.2.3 Acid and Basic Dyes

In biomedical applications, dyes have traditionally been classed as "acid" or "basic," despite the fact that this terminology does not correspond to modern scientific definitions of acids and bases. Acids and bases, according to the Brönsted–Lowry and Lewis theories, are determined by proton transfer or electron pair interactions, regardless of charge. In histochemistry, dyes having anionic groups (e.g., $-SO_3^-$ or $-COO^-$) are called "acid" dyes, while cationic dyes are called "basic." However, chemically, these are anionic bases and Lewis acids, respectively. The inclusion of ammonium groups does not confer basic qualities; rather, it changes dye color by preventing lone electron pairs. To ensure precision, anionic and cationic dyes are preferable over acid and basic dyes (Puchtler et al., 1985).

Basic dyes are cationic colorants with positively charged nitrogen or sulfur cores. They provide vivid colors even at low doses (<1 ppm). Their high tinctorial value makes them the most visible family of soluble dyes, yet many of them are potential or proven carcinogens, posing serious health concerns (Leodopoulos et al., 2015). Chemically diverse, they are primarily produced from substituted aromatic groups, specifically triarylmethane derivatives. Cationic dye complex pigments are formed by precipitating water-soluble basic dyes with massive inorganic counterions, such heteropolyacids or copper ferrocyanide salts. These pigments provide beautiful red, violet, blue, and green hues with great transparency, making them useful in printing inks for packaging and advertising; however, their fastness remains moderate (Christie and Abel, 2021). Acid dyes, which have acidic functional groups in their molecular structure, are generally used on non-cellulosic fibers such as proteins and polyamides under acidic circumstances. Their chromophores are often based on sulfonated azo, anthraquinone, nitrodiphenylamine, triphenylmethane, or xanthene derivatives. The contact takes place via electrostatic attraction, with cationized fibers binding to the anionic groups of acid dyes (Haque et al., 2022).

7.2.4 Environmental Effects

Synthetic dyes are typically nonbiodegradable due to their complicated chemical structures, resulting in long-term persistence in the environment, sometimes for more than 50 years (Ardila-Leal et al., 2021). Most are recalcitrant, carcinogenic, and poisonous, with their effects generally amplified throughout trophic levels. Persistence is directly related to chemical reactivity; aromatic molecules containing numerous substitutions, particularly halogens, are more resistant to degradation. Among dye classes, azo, anthraquinone, and triarylmethane are the most common and environmentally harmful. Notably, 60–70% of azo dyes are poisonous, carcinogenic, and resistant to traditional treatments. Their toxicity is mostly due to the reductive cleavage of the azo link, which produces aromatic amines that undergo metabolic oxidation to electrophilic intermediates capable of covalently binding DNA, producing mutagenicity and cancer.

Untreated azo dyes provide a significant environmental problem because they degrade into aromatic amines, which are known carcinogens that damage both ecosystems and human health (Alzain et al., 2023). These byproducts are produced when the azo link is cleaved, and they persist in water bodies, increasing biochemical and chemical oxygen demand while harming aquatic ecosystems, plants, and crops. Growing data emphasizes the necessity of effective azo dye degradation solutions for reducing their environmental and toxicological effects. Untreated synthetic dyes discharged into aquatic systems not only cause coloration and restrict light penetration into the photic zone, but they also have carcinogenic, mutagenic, and teratogenic effects on plants, animals, and people (Lin et al., 2023; Islam et al., 2023). Their release has a substantial impact on water quality by increasing biochemical oxygen demand (BOD) and chemical oxygen demand (COD), total dissolved solids (TDS), total suspended solids (TSS), and causing pH imbalance. Conventional chemical treatments, such as coagulation, are commonly used but necessitate careful sludge control. Alternatively, biological techniques based on enzymes, microbes, and plants are both ecologically beneficial and energy efficient, however, they are frequently constrained by longer reaction durations and substrate selectivity. As a result, eco-friendly and sustainable wastewater treatment technologies are urgently required to reduce environmental and health threats.

Synthetic dyes' refractory nature allows them to bioaccumulate in soil and sediments, eventually poisoning public water supplies (Mehra et al., 2021). Azo dyes frequently break down into poisonous aromatic amines, and their structural complexity upsets natural balance. Textile dyes commonly contain heavy metal cations such as nickel, cobalt, copper, and chromium, which accumulate in aquatic species and biomagnify down the food chain, posing a threat to human health. Intermediates such as benzidine and 2-naphthylamine are recognized bladder carcinogens. Prolonged exposure to dye effluents has been associated with dermatitis, central nervous system problems, and enzymatic activity inhibition caused by cofactor displacement. These findings emphasize the serious environmental and health risks connected with textile dye contamination.

7.3 Adsorption Mechanisms for Dyes

Dye adsorption on solid surfaces happens primarily through two mechanisms: physisorption and chemisorption. Physisorption is characterized by weaker, reversible interactions such as electrostatic forces, hydrogen bonding, and π–π stacking, which frequently result in multilayer development. Chemisorption, on the other hand, is based on stronger covalent bonds and usually results in monolayer adsorption, with dye molecules spreading into the adsorbent's interior. Under favorable conditions, both processes can occur at the same time, driven by decreases in free energy and entropy. Table 7.2 gives the summary of adsorption mechanisms and their key interactions. Adsorption efficiency is heavily controlled by pH, temperature, initial dye concentration, sorbent dosage, surface structure, and morphology. Adsorption isotherm models are often used to represent the equilibrium connection between adsorbate concentration and uptake capacity, shedding light on dye-adsorbent interactions (Bushra et al., 2021). Figure 7.2 represents the schematic representation of adsorption mechanisms.

7.3.1 Electrostatic Interaction

Electrostatic interactions, defined as the attraction of oppositely charged molecules, are critical in dye adsorption. Their strength is dependent on surface charge intensity and distance; hence, solution pH is an important consideration. When the pH varies from zero, adsorbents like hydrogels can acquire positive or negative charges, allowing them to interact with oppositely charged dyes or ions. Carboxymethyl cellulose/chitosan hydrogels effectively adsorb methylene blue (MB) at alkaline pH due to deprotonated $-COO^-$ groups, while protonated

Table 7.2 Summary of adsorption mechanisms and their key interactions.

Adsorption mechanism	Key interactions	Adsorbent	Dye	References
Van der Waals force	Attraction between charged dye molecules and oppositely charged adsorbent surface	Mulberry branches	Acid blue 93	Liu et al. (2015)
Hydrogen bonding interactions	Hydrogen bonding between hydroxyl groups of adsorbent and dye molecules	Mulberry branches	Methylene blue	Liu et al. (2015)
Van der Waals force	Attraction between charged dye molecules and oppositely charged adsorbent surface	Mulberry branches	Methylene blue	Liu et al. (2015)
Hydrogen bonding interactions	Hydrogen bonding between hydroxyl groups of adsorbent and dye molecules	Mulberry branches	Acid blue 93	Liu et al. (2015)
Electrostatic attraction	Attraction between negatively charged carboxyl/phenolic groups and cationic dye	Orange peel, golden shower pod, coconut shell	Methylene green 5	Tran et al. (2017)
Hydrogen bonding (Dipole–dipole)	Interaction between hydroxyl groups and N/O atoms of dye	Golden shower pod, coconut shell	Methylene green 5	Tran et al. (2017)
Hydrogen bonding	Interaction between hydroxyl groups of adsorbent and aromatic rings of dye	Golden shower pod, coconut shell	Methylene green 5	Tran et al. (2017)
n–π interactions	Carbonyl oxygens with aromatic rings	Golden shower pod, coconut shell	Methylene green 5	Tran et al. (2017)
Acidic functional group binding	Carboxylic groups forming negatively charged COO^- binding dye molecules	Golden shower pod, coconut shell, orange peel	Methylene green 5	Tran et al. (2017)
Weak electrostatic attraction (pH dependent)	Ionized oxygen groups binding dye molecules when pH > pKa	Orange peel	Methylene green 5	Tran et al. (2017)

(Continued)

Table 7.2 *(Continued)*

Adsorption mechanism	Key interactions	Adsorbent	Dye	References
Structural contribution	Hemicellulose, cellulose, lignin functional groups supporting dye binding	Golden shower pod, coconut shell, orange peel	Methylene green 5	Tran et al. (2017)
Ion exchange	Exchange of H^+ from $-COOH$, $-SO_3H$, $-OH$ groups with MB cations	Sulfonated hawthorn kernel (SHK)	Methylene blue	Akköz et al. (2019)
Electrostatic interaction	Attraction between negatively charged acidic groups and cationic MB molecules	SHK	Methylene blue	Akköz et al. (2019)
π–π interactions	Stacking between aromatic rings of MB and SHK surface groups	SHK	Methylene blue	Akköz et al. (2019)
Electrostatic interaction (pH-dependent)	Attraction between negatively charged $-COO^-$ groups (deprotonated carboxyls) and cationic MB molecules	Oxidized chitosans	Methylene blue	León et al. (2018)
Ion exchange	Release of H^+ from $-COOH$ groups during MB binding	Oxidized chitosans	Methylene blue	León et al. (2018)
Electrostatic interaction	Attraction between negatively charged $-SO_3^-$ groups of CR and positively charged ZnO/CTs surface ($pH < pHpzc = 8.73$)	ZnO/CTs	Congo red	Mchich et al. (2025)
Hydrogen bonding	Bonding between hydroxyl groups of ZnO/CTs and amine ($-NH$) groups of CR; also between CO_3^{2-} groups and N–H groups	ZnO/CTs	Congo red	Mchich et al. (2025)
Lewis acid–base interaction	Interaction of surface Ca^{2+} ions (Lewis acid sites) with nitrogen atoms of CR (Lewis base sites)	ZnO/CTs	Congo red	Mchich et al. (2025)

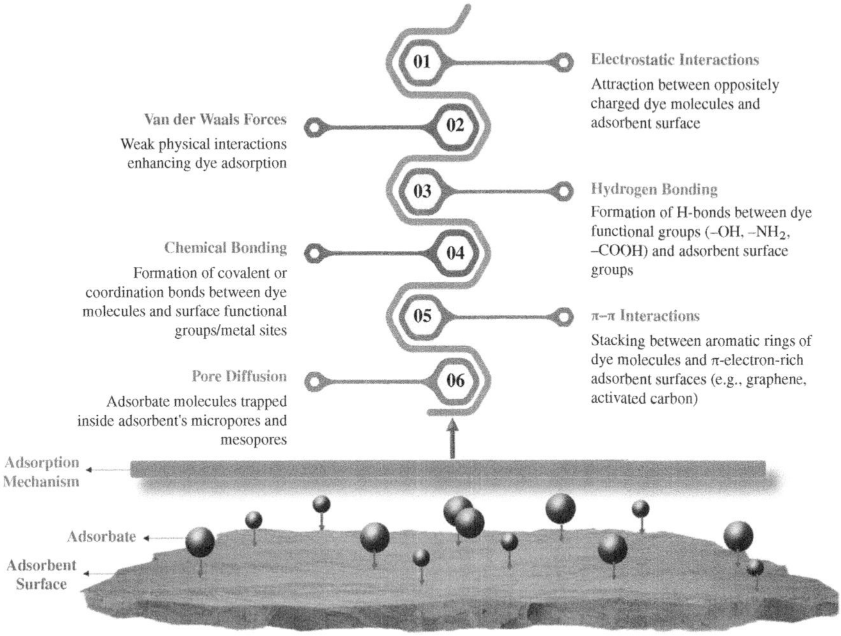

Figure 7.2 Schematic representation of adsorption mechanisms.

$-NH_4^+$ groups at acidic pH favor acid orange II removal. As pH increased, sodium alginate/hectorite hydrogels removed more Cu^{2+}, indicating less competition with H^+ and better ionization of functional groups (Zhu et al., 2023). Electrostatic interaction, or the attraction of oppositely charged adsorbents and adsorbates, is a prominent process in dye adsorption that is heavily impacted by solution pH. Hydrogen bonding, which involves fluorine, oxygen, nitrogen, and hydrogen atoms, plays an important role in adsorption processes. Furthermore, acid–base interactions can occur when metal doping raises surface acidity, which promotes dye uptake. Metals can also form metal-oxygen bonds on carbon surfaces, changing surface electronegativity and facilitating adsorption. Metal-doped porous carbon materials are extremely excellent adsorbents for organic dyes due to their synergistic effects (Xiao et al., 2021).

7.3.2 Hydrogen Bonding

Traditionally, hydrogen bonding was thought to be a connection between proton donors such as F—H, O—H, or N—H and electronegative acceptors such as O or N. This description has recently been broadened to encompass directed noncovalent interactions, including divalent sulfur, halogens, and various electron donors. The geometry and strength of such interactions depend greatly on donor ability: strong donors lead to shorter, collinear connections, while weaker donors produce longer, less directed interactions. Gospodinova and Tomšík (2015) found that hydrogen bonding and π–π stacking determine molecular order in functional organic materials, including organic semiconductors. Hydrogen bonds are commonly formed between electronegative atoms like fluorine, nitrogen, or oxygen and electropositive hydrogen atoms. In dye adsorption, the presence of different functional groups in both adsorbents and dye molecules increases the likelihood of hydrogen bond formation. Although hydrogen bonding is commonly regarded as the principal driving factor, clear proof of this interaction is sparse. Electrostatic interactions can also play an important role, as seen by the adsorbent's surface charge and the charge of dye molecules. Furthermore, pore size and dye molecular size influence adsorption efficiency. More extensive research is needed to clarify these pathways (Shi et al., 2022).

7.3.3 π–π Interaction

Organic photocatalysts are gaining popularity due to their structural versatility and tunable optoelectronic properties. Noncovalent interactions, such as π–π interactions, have a significant impact on the electronic structure, charge migration, and interfacial reactions of materials. In aromatic systems, orbital overlap determines the strength of interactions, which can range from 1 to 50 kJ/mol. Intermolecular distance and orientation have a significant impact on contact strength, particularly at 3.3 Å. π–π stacking has four major modes: T-shaped, edge-to-face, offset face-to-face, and face-to-face. These interactions improve photocatalyst stability, adsorption, and catalytic efficiency, making them necessary for long-term photocatalytic performance (Jing et al., 2025).

7.4 Parameters Affecting Adsorption

Many factors influence organic dye adsorption on porous carbon-based materials. The amount of doped metal, solution pH, and temperature are very critical for metal-doped porous carbons, since they have a considerable impact on adsorption capacity and mechanism (Xiao et al., 2021). Table 7.3 presents the parameters affecting adsorption and their influence on efficiency. Furthermore, initial dye concentration, pH, and temperature have all been shown to play important roles in dye adsorption processes (Yagub et al., 2014). Understanding and optimizing these factors is critical to improving adsorption efficiency and enabling the development of industrial-scale dye removal methods. Figure 7.3 shows the various factors affecting the dye adsorption efficiency.

Table 7.3 Parameters affecting adsorption and their influence on efficiency.

Parameter	Range/condition studied	Influence on efficiency	Optimum condition	Dye	Adsorbent	References
Initial dye concentration	50–1,000 mg/L	Higher concentration increases driving force but may saturate active sites	100 mg/L	Methylene blue	Albedo residue	de Farias Silva et al. (2020)
Temperature	30–60 °C	Influences adsorption capacity and kinetics	30 °C	Methylene blue	Albedo residue	de Farias Silva et al. (2020)
Time	0–60 min	Rapid adsorption initially, then equilibrium	60 min	Methylene blue	Albedo residue	de Farias Silva et al. (2020)
Adsorbent Dosage	0.1–5 g in 50 mL	Removal efficiency increased from 53.35 to 62.35% due to more active sites and surface	0.5 g in 50 mL	Methyl violet	Date seed bioadsorbent	Ali et al. (2022)
Initial Concentration	20, 40, 60, 80 mg/L	Removal efficiency decreased (89.9–57.6%) at higher concentration due to site saturation	20 mg/L	Methyl violet	Date seed bioadsorbent	Ali et al. (2022)
Contact Time	60–120 min	Rapid uptake within 120 min, then equilibrium; no further increase after 2 h	120 min (2 h)	Methyl violet	Date seed bioadsorbent	Ali et al. (2022)
Temperature	25 °C–45 °C	Removal efficiency increased from 89.9 to 93.4% with rise in temperature due to enhanced diffusion and pore expansion	45 °C	Methyl violet	Date seed bioadsorbent	Ali et al. (2022)
Contact Time	0–120 min	Adsorption increased with time until equilibrium; plateau at 60 min (CCC) and 90 min (SCC)	60 min	Methylene blue	Corn cobs	Reddy et al. (2016)

(Continued)

Table 7.3 *(Continued)*

Parameter	Range/condition studied	Influence on efficiency	Optimum condition	Dye	Adsorbent	References
Adsorbent Mass	100–200 mg	Increasing mass increased % removal but decreased unit adsorption capacity due to unoccupied sites	100 mg	Methylene blue	Corn cobs	Reddy et al. (2016)
Initial dye concentration	100–200 mg/L	Higher MB concentration increased unit adsorption capacity but reduced % removal	100 mg/L	Methylene blue	Corn cobs	Reddy et al. (2016)
Solution pH	2–12	Removal efficiency increased with pH due to adsorbent surface deprotonation; at acidic pH, repulsion reduced uptake	10	Malachite green	*Catha edulis*	Abate et al. (2020)
Adsorbent Dose	0.2–1.4 g	Removal % increased for higher mass caused aggregation, reducing uptake	0.5 g	Malachite green	*Catha edulis*	Abate et al. (2020)
Initial Concentration	10–50 mg/L	% Removal decreased with higher concentration while adsorption capacity (qe) increased	10 mg/L	Malachite green	*Catha edulis*	Abate et al. (2020)
Contact Time	10–80 min	Rapid initial adsorption, equilibrium at 60 min; efficiency rose from (10 to 60 min)	60 min	Malachite green	*Catha edulis*	Abate et al. (2020)

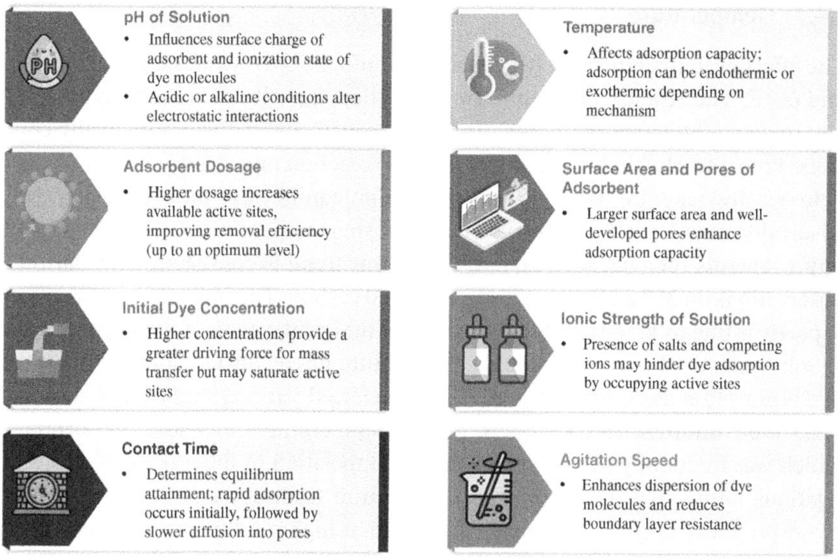

Figure 7.3 Factors affecting the dye adsorption efficiency.

7.4.1 pH

Solution pH is important in dye adsorption because it influences both adsorbent surface charge and solute ionization state. pH changes can either accelerate or decrease dye uptake, depending on how surface functional groups interact with dye molecules (Foo and Hameed, 2010). Hydrogen and hydroxyl ions have a considerable effect on the dissociation of active functional groups, changing adsorption kinetics and equilibrium. During activation, oxygen-containing groups such as carbonyl and carboxyl are introduced, acting as electron acceptors and enhancing electrostatic interaction with dyes. At low pH levels, extra H^+ ions compete with cationic dyes, limiting adsorption effectiveness. At increased pH, OH^- ions improve binding of cationic dyes by reducing competition and promoting electrostatic interactions. As a result, acidic conditions favor anion adsorption, while alkaline conditions promote cation adsorption, emphasizing the substantial pH dependence of dye adsorption processes. The adsorption of acid yellow 42 (AY) dye on calcined layered double hydroxide (CLDH) is greatly controlled by solution pH. At pH 4.0, adsorption efficiency was lowered due to CLDH's higher solubility in acidic circumstances. At pH 12.0, adsorption capacity decreased because CLDH's surface charge became extremely negative, resulting in electrostatic repulsion with the anionic dye. The best effectiveness was seen at neutral pH (7.0), when the CLDH surface had a positive zeta potential, promoting significant electrostatic interaction with negatively charged dye molecules (dos Santos et al., 2017).

7.4.2 Temperature

The effect of temperature on RB adsorption onto MOMD was tested at 22, 35, 50, and 60 °C. The results reveal that the adsorption capacity increased from 125 to 160 mg/g as the temperature rose from 22 to 60 °C, indicating a kinetically controlled process. Higher temperatures increase reaction rates while decreasing particle density, resulting in voids and shortening equilibrium time. Similarly, MB adsorption onto MOMD improved with increasing temperature, with considerable improvements seen between 50 and 60 °C. This trend is consistent with Barton's observations on MB adsorption using activated carbon. The increase in adsorption capacity is due to kinetic factors, specifically the greater monomer concentration in solution (Al-Ghouti et al., 2005). Temperature has a considerable influence on dye biosorption. Smoczyński et al. (2020) reviewed various studies and summarized their findings. In 21 studies, greater temperatures increased biosorption, which was frequently tied to the chemical composition of the process. Ten investigations found that increasing the temperature from 20 to 30–40 °C increased biosorption capacity, but efficiency decreased at higher temperatures. According to some reports, increasing the temperature from 10 to 20 °C resulted in a three- to fourfold increase. However, 13 investigations found that greater temperatures inhibited biosorption, indicating a physical process. These opposite findings demonstrate the complex function of temperature, as well as the importance of taking into account both chemical and physical parameters in biosorption research.

7.4.3 Initial Dye Concentration

The adsorption of Direct Red 80 (DR 80) dye on almond shells was examined using contact time and starting concentration. Mixed almond shells were employed as an adsorbent. The dye concentration varied from 50 to 150 mg/L. Adsorption was initially rapid, but eventually slowed until equilibrium was attained after 300 minutes. At equilibrium, dye removal was around 94%. After 300 minutes, adsorption remained steady for 1500 minutes. Increasing dye concentration decreased removal efficiency. Removal was 94% at 50 mg/L, but reduced to 83.9% at 150 mg/L (Ardejani et al., 2008). The effect of initial dye concentration on SOA adsorption rate is shown as discrete points, with model predictions represented as solid lines. Experiments with starting drim yellow-K4G concentrations ranging from 50 to 380 mg/L demonstrated a remarkable correlation between experimental and theoretical data. For each experiment, the adsorption rate reduced over time until a plateau was established due to the decreasing concentration driving force. Furthermore, at higher dye concentrations, the initial adsorption rate increased because mass transfer resistance decreased with increasing driving power (Al-Qodah, 2000).

7.4.4 Contact Time

To some extent, increasing contact duration enhances dye removal rate. Once dye molecules have occupied all of the accessible adsorption sites on the adsorbent surface, any additional contact time has no significant impact on adsorption. At this point, the amount of dye desorbing from the adsorbent is in dynamic equilibrium with the amount being adsorbed. The time necessary to establish this balance is known as the equilibrium time, and the dye uptake represents the adsorbent's maximal adsorption capacity under the particular operating conditions (Razi et al., 2017). The adsorption behavior of dyes onto clinoptilolite was investigated under controlled conditions, using a dye concentration of 4 mg/L, a pH of 7.0, and a solid concentration of 40 mg/L (Qiu et al., 2009). The results showed that adsorption increased drastically with increasing contact duration, particularly during the first 360 minutes. Following this interval, the adsorption rate slowed as the system approached equilibrium. Beyond 360 minutes, there was no significant increase in dye uptake, indicating that the majority of adsorption sites were already occupied. This illustrates that adsorption time is essential initially, but equilibrium is reached after a few hours of contact.

7.4.5 Adsorbent Dosage

Dye removal by fly ash increases with contact time until equilibrium is reached. The greatest color removal is highly dependent on the adsorbent dosage. For MB, equilibrium was attained after 40 minutes with 4 g and 60 minutes with 8 g of adsorbent, resulting in 30 and 89.5% elimination, respectively. A parallel trend was observed for Congo red (CR)dye. The adsorption curves were smooth and continuous, indicating monolayer adsorption at the adsorbent surface. The maximum color removal rate was 98.5% with 2 g of activated carbon and 96% with 12 g of fly ash. These findings suggest that fly ash, a less expensive alternative to activated carbon, could be used successfully (Rao and Rao, 2006). To investigate the effect of adsorbent dose on basic dye adsorption, a series of experiments were carried out with various adsorbent masses at a fixed beginning dye concentration of 50 ppm. The results show that increasing the adsorbent dosage improved the removal effectiveness of MB and crystal violet. With higher adsorbent dosages, there is more surface area and active sites available for adsorption, which can explain this result. However, the amount of dye adsorbed per unit mass of adsorbent reduced as dosage rose, most likely due to particle aggregation and site overlap, which lower adsorption efficiency per gram of adsorbent (Akbal, 2005).

7.5 Removal of Dyes Using Bioadsorbent

Wastewater from agricultural, domestic, and industrial operations frequently contains dangerous contaminants such as heavy metals, pesticides, phenols, and dyes, which, if left untreated, destroy ecosystems and threaten human health. Dyes are especially dangerous due to their toxicity, persistence, and carcinogenic or mutagenic properties. MB can cause serious health concerns such as nausea, respiratory problems, and skin irritation, whereas rhodamine B (RB) causes gastrointestinal and reproductive problems. Although conventional treatment procedures such as coagulation, filtering, and oxidation exist, adsorption stands out as the most cost-effective and efficient. Bioadsorbents, particularly marine algae like Sargassum muticum and Codium decorticatum, have recently acquired popularity due to their availability, eco-friendliness, and high dye removal capacity (Hamad and Idrus, 2022).

Polymeric bioadsorbents have grown in popularity due to their high efficiency, selectivity, and environmentally friendly nature in wastewater treatment (Cestari et al., 2004). Among them, graphene oxide (GO) integrated semi-interpenetrating network (semi-IPN) hydrogels, made utilizing chitosan (CS), acrylamide (AAm), and itaconic acid (IA), show exceptional performance. GO increases the hydrogel's temperature stability, swelling, and mechanical strength. The ideal circumstances (0.025 g hydrogel, 0.5 wt% GO, pH 8) resulted in a 99.8% elimination of MB with a maximum adsorption capacity of 247.47 mg/g. The process followed pseudo-second-order kinetics and the Langmuir isotherm, and thermodynamics confirmed that it was spontaneous and endothermic. Even after seven cycles, the hydrogel preserved over 96% MB removal efficiency, demonstrating its reusability and structural stability (Tamer et al., 2022).

In this study, a hemp hurd nanocomposite (HHNC) adsorbent was created by inserting silver nanoparticles in a simple and environmentally friendly way. The material was tested for the simultaneous removal of brilliant green (BG, cationic) and CR, anionic dyes, as well as antibacterial activity against wastewater pathogens. Adsorption followed the Langmuir isotherm and pseudo-second-order kinetic model, with capacities of around 256 mg/g for BG and ~455 mg/g for CR. At pH 6.5, HHNC removed 99.4% BG and 88.6% CR. A thermodynamic investigation demonstrated that the adsorption occurred spontaneously and endothermically. After six cycles, the adsorbent retained over 82% efficiency and showed antibacterial activity with MIC 12.7 μg/ml against *E. coli and S. aureus* (Mandal et al., 2022).

The elimination of RB is critical due to its negative impact on aquatic and soil systems. This study looked into the effectiveness of *Calophyllum inophyllum* seed char as a bioadsorbent. Slow pyrolysis at 823 K produced mesoporous char with a BET surface area of 155.39 m^2/g and a pore volume of 0.628 cc/g.

pH, dose, dye concentration, stirring speed, contact time, and temperature all had an effect on adsorption efficacy. The optimal conditions for 95.5% RB elimination were pH 2, 100 rpm, 25 minutes, 308 K, and 1.2 g/L of dose. The Langmuir isotherm showed a maximal capacity of 169.5 mg/g, with pseudo-second-order kinetics. Thermodynamic investigation revealed spontaneous, endothermic adsorption, and reusability was demonstrated for three cycles (Behera et al., 2024).

A mesoporous Ag metal-organic framework (Ag-MOF) was created as an environmentally friendly adsorbent for dye removal from industrial effluent. Analysis with SEM, XPS, XRD, FT-IR, pHpzc, and BET revealed a high surface area of 676 m^2/g. Adsorption experiments found that malachite green (MG) uptake followed the Langmuir isotherm and pseudo-second-order kinetics, with a maximum adsorption capacity of 809.71 mg/g (2.25 mmol/g). Electrostatic, hydrogen bonding, pore filling, and π–π interactions all played a role in adsorption. Thermodynamic study revealed a spontaneous, endothermic reaction. The Box–Behnken design validated the importance of pH, dose, temperature, and time. The Ag-MOF remained highly efficient after five reuse cycles, demonstrating its promise for wastewater treatment (Aljohani et al., 2023).

The adsorption of MG dye over Arjun saal powder was studied with various parameters. The results showed that adsorption effectiveness increased with adsorbent dose, temperature, and contact time, while higher starting dye concentration decreased removal efficiency. SEM investigation indicated a multilayer porous structure, while elemental analysis confirmed carbon and oxygen as the main components. FTIR spectra revealed functional groups such as C—H and N—H stretching, which contributed to dye retention. The optimal conditions for dye removal were 3 g adsorbent, 120 minutes of contact time, 10 mg/L dye, pH 10, and 40°C. Kinetic results supported the pseudo-first-order model, but the Freundlich isotherm best described adsorption. Mandale et al. (2024) found that Arjun saal powder was a cost-effective way to remove MG.

Zeolite, clay, fly ash, and activated carbon have all been investigated in recent studies as adsorption materials for wastewater treatment purposes. Bioadsorbents are particularly attractive due to their low cost, availability, and biodegradability. Hemicelluloses, with their branching structure and abundance of hydroxyl groups, are suited for converting into efficient bioadsorbents. By grafting functional groups like —OH, —COOH, and —NH, hemicellulose-based adsorbents obtain more active binding sites, increasing adsorption efficiency. Hemicellulose hydrogels, particularly those based on xylan, have been shown to interact significantly with dyes, heavy metals, and water. Xylan is abundant in agricultural waste and easy to extract, making it ideal for large-scale uses. Furthermore, magnetic hemicellulose-based hydrogels are easily separated and reused, indicating a sustainable future trend (Xiang et al., 2022).

7.6 Regeneration and Reusability

Industrialization and increasing urbanization have exacerbated anthropogenic pollution, harming ecosystems and human health. Aquatic systems are particularly sensitive to dye-rich effluents and organic contaminants. Adsorption has evolved as a simple and effective method of dye removal; nevertheless, it only transfers pollutants to adsorbents, resulting in secondary sludge pollution. Thus, for long-term treatment, saturated adsorbents must be regenerated and reused. Low-cost adsorbents (LCAs) generated from agricultural wastes, industrial byproducts, natural materials, and marine biomass were studied for numerous adsorption–desorption cycles. Various eluents, including acids, bases, salts, alcohols, and chelating agents, can effectively regenerate. Agricultural waste (39.56%) had the highest regeneration and reusability, followed by sea materials, natural sources, and industrial residues (El Messaoudi et al., 2024).

Regeneration is the recycling or recovery of spent adsorbents by cost-effective and technically possible methods, which is critical for long-term pollution control. Clays, in particular, require regeneration to preserve their ability to adsorb organic contaminants. Several techniques have been used, including thermal, steam, pressure swing, vacuum, microwave, ultrasound, chemical, oxidative, ozone, and bioregeneration. Chemical regeneration is a popular method that involves desorption with solvents or the breakdown of adsorbed species with oxidants. Supercritical fluid extraction (SFE), which uses fluids above their critical temperature and pressure, has emerged as a possible alternative to solvent extraction or incineration due to its effective desorption of pollutants from adsorbent matrices. The efficiency is determined by pH, oxidation rates, and degradation mechanisms (Shahadat and Isamil, 2018).

Cold atmospheric plasma (CAP) regeneration of exhausted adsorbents provides a long-term and efficient solution. In this investigation, MB-saturated CA-WT adsorbent (2 g) was treated in a plasma reactor with 20 ml of distilled water. The reactor used a glass chamber with a microplasma jet array (6 × 6 units) powered by an alternating current supply. Tungsten wires placed into hollow fibers served as high-voltage electrodes, using O_2 as the supply gas at a flow rate of 2.0 SLM. Plasma regeneration studies were carried out for 45 minutes at 4.5 kV, equivalent to 35 W discharge power. The results demonstrated excellent adsorbent desorption and regeneration, indicating CAP as a promising regeneration approach (Zhou et al., 2016).

The regeneration and reusable properties of alginate/bentonite (A-B) composite beads for color removal were explored. MB desorption from A-B 1/1 was carried out with distilled water at pH 2, whereas CR desorption from A-B 1/3 was investigated at pH 10. After each adsorption cycle, beads were washed, dried, and treated with 100 mL of desorbing solution under the same adsorption conditions (pHMB = 6.5, pHCR = 7.2, T = 24 °C, agitation = 200 rpm). The desorbed samples were tested to evaluate the dye removal efficiency in each cycle. Regeneration tests demonstrated that the beads retained stable adsorption capabilities for up to six cycles for MB and five cycles for CR, demonstrating that this low-cost adsorbent technology is effectively reusable (Oussalah et al., 2019).

Gupta et al. (2014) explored the regeneration and reuse of ABAC saturated with RB dye. The exhausted adsorbent was dispersed in deionized water and microwaved at 600 W for 15 minutes. After cooling, the regenerated ABAC was filtered, dried, and employed in adsorption experiments to test its efficiency. Reusability tests revealed effective performance for 10 cycles, after which dye removal declined to 70% and regeneration efficiency to 84%.

7.7 Real Case Studies

Ahmed et al. (2020) compared the efficacy of waste shell powder (WSP) and moringa seed powder (MSP) in eliminating color, heavy metals, and resistant bacteria from industrial effluent. Industrial wastewater (IWW) samples from the tannery and textile sectors contained dyes, heavy metals, ~7.0 log CFU/ml aerobic bacteria, and 3.0–4.0 log CFU/ml pathogens such as *Escherichia coli, Pseudomonas aeruginosa, Citrobacter freundii, Serratia liquefaciens, and Bacillus cereus*. Several isolates were resistant to several metals (Cr^{6+}, Cd^{2+}, Pb^{2+}) and antibiotics. The combination of MSP (0.8 g/100 ml) and WSP (0.2 g/100 ml) effectively discolored wastewater in 4 hours while also eliminating heavy metals and pathogens in 24 hours. Individual treatments were found to be less effective, validating MSP-WSP's more effective combination.

Aljohani et al. (2023) created an environmentally friendly inorganic–organic silver metal-organic framework (Ag-MOF) for extracting MG dye from industrial effluent. Analysis using SEM, XPS, XRD, FT-IR, BET, and pHpzc revealed a high surface area of 676 m^2/g. Adsorption studies revealed that isotherms fit the Langmuir model and kinetics followed the pseudo-second-order equation. Electrostatic attraction, hydrogen bonding, ion exchange, pore filling, and π–π interactions contributed to a maximum adsorption capacity of 2.25 mmol/g (809.71 mg/g). The response surface methodology's (RSM) Box–Behnken design was used to optimize process parameters. Thermodynamic tests demonstrated both spontaneous and endothermic adsorption. Over five cycles, the Ag-MOF displayed outstanding regeneration and reusability, indicating that it is a viable mesoporous adsorbent for dye removal and industrial wastewater treatment.

Alardhi et al. (2023) studied raw olive stone (OS) as an environmentally favorable and low-cost adsorbent for the removal of MB from industrial effluent. Analysis with SEM, FTIR, and BET revealed high porosity and a surface area of 0.9 m^2/g. Langmuir and Temkin isotherms best described the adsorption equilibrium, which had a maximum monolayer capacity of 44.5 mg/g. Kinetic studies revealed that the pseudo-second-order and Elovich models accurately described adsorption kinetics. Using 0.2 g OS for 90 minutes, batch studies removed 93.65% of the dye, while five subsequent regeneration cycles maintained 91–78% efficiency. Thermodynamic studies demonstrated that the reaction was spontaneous and endothermic. Thus, OS provides a sustainable, effective, and economically viable alternative for treating MB-contaminated wastewater.

Kumari et al. (2022) investigated the potential of walnut (*Juglans regia*) shell biomass as a sustainable and low-cost adsorbent for removing MB from aqueous solutions. The ideal parameters for maximum dye removal (97.70%) were discovered using RSM using a face-centered central composite design. These included MB concentration of 30 mg/L, 1.5 g of biomass, pH 6, contact period of 60 minutes, and temperature of 25 °C. SEM examination confirmed particle adsorption on the biomass surface, while FTIR spectra identified functional groups involved in dye binding. The study found that *Juglans regia* shell is a biodegradable, eco-friendly, cost-effective, and efficient biosorbent appropriate for industrial wastewater treatment.

7.8 Conclusion

This chapter reviewed dyes and their environmental impact. Reactive, azo, basic, and acid dyes are widely used in industries. They pollute water and harm aquatic life and humans. Adsorption is an effective and simple method to remove dyes. Electrostatic interactions, hydrogen bonding, and π–π interactions are key mechanisms. Factors such as pH, temperature, initial dye concentration, contact time, and adsorbent dosage affect adsorption efficiency. Bioadsorbents from agricultural waste, seaweeds, and other natural sources are low-cost, biodegradable, and readily available. They show high adsorption capacity and selectivity. Regeneration and reuse make the process sustainable. Multiple adsorption–desorption cycles are possible with minimal loss in efficiency. Case studies show bioadsorbents perform as well as synthetic materials. This method reduces secondary pollution and saves resources. Future research should focus on novel bioadsorbents, surface modifications, and process optimization. Large-scale applications can provide sustainable wastewater treatment. Overall, bioadsorption is a promising and environmentally friendly approach for industrial dye removal.

References

Abate, G. Y., Alene, A. N., Habte, A. T., & Getahun, D. M. (2020). Adsorptive removal of malachite green dye from aqueous solution onto activated carbon of Catha edulis stem as a low cost bio-adsorbent. Environmental Systems Research, 9(1), 29.

Adeyemo, A. A., Adeoye, I. O., & Bello, O. S. (2017). Adsorption of dyes using different types of clay: a review. Applied Water Science, 7(2), 543–568.

Agarwala, R., & Mulky, L. (2023). Adsorption of dyes from wastewater: a comprehensive review. ChemBioEng Reviews, 10(3), 326–335.

Ahmed, S., Aktar, S., Zaman, S., Jahan, R. A., & Bari, M. L. (2020). Use of natural bio-sorbent in removing dye, heavy metal and antibiotic-resistant bacteria from industrial wastewater. Applied Water Science, 10(5), 1–10.

Akbal, F. (2005). Adsorption of basic dyes from aqueous solution onto pumice powder. Journal of Colloid and Interface Science, 286(2), 455–458.

Akköz, Y., Coşkun, R., & Delibaş, A. (2019). Preparation and characterization of sulphonated bio-adsorbent from waste hawthorn kernel for dye (MB) removal. Journal of Molecular Liquids, 287, 110988.

Alardhi, S. M., Salih, H. G., Ali, N. S., Khalbas, A. H., Salih, I. K., Saady, N. M. C., Zendehboudi, S., Albayati, T.M., & Harharah, H. N. (2023). Olive stone as an eco-friendly bio-adsorbent for elimination of methylene blue dye from industrial wastewater. Scientific Reports, 13(1), 21063.

Al-Ghouti, M., Khraisheh, M. A. M., Ahmad, M. N. M., & Allen, S. (2005). Thermodynamic behaviour and the effect of temperature on the removal of dyes from aqueous solution using modified diatomite: a kinetic study. Journal of Colloid and Interface Science, 287(1), 6–13.

Ali, N. S., Jabbar, N. M., Alardhi, S. M., Majdi, H. S., & Albayati, T. M. (2022). Adsorption of methyl violet dye onto a prepared bio-adsorbent from date seeds: isotherm, kinetics, and thermodynamic studies. Heliyon, 8(8), e10276.

Aljohani, M. M., Al-Qahtani, S. D., Alshareef, M., El-Desouky, M. G., El-Bindary, A. A., El-Metwaly, N. M., & El-Bindary, M. A. (2023). Highly efficient adsorption and removal bio-staining dye from industrial wastewater onto mesoporous Ag-MOFs. Process Safety and Environmental Protection, 172, 395–407.

Al-Qodah, Z. (2000). Adsorption of dyes using shale oil ash. Water Research, 34(17), 4295–4303.

Alzain, H., Kalimugogo, V., Hussein, K., & Karkadan, M. (2023). A review of environmental impact of azo dyes. International Journal Of Engineering Research and General Science, 10(6), 673–689.

Ani, J. U., Akpomie, K. G., Okoro, U. C., Aneke, L. E., Onukwuli, O. D., & Ujam, O. T. (2020). Potentials of activated carbon produced from biomass materials for sequestration of dyes, heavy metals, and crude oil components from aqueous environment. Applied Water Science, 10(2), 69.

Ardejani, F. D., Badii, K. H., Limaee, N. Y., Shafaei, S. Z., & Mirhabibi, A. R. (2008). Adsorption of direct red 80 dye from aqueous solution onto almond shells: effect of pH, initial concentration and shell type. Journal of Hazardous Materials, 151(2-3), 730–737.

Ardila-Leal, L. D., Poutou-Piñales, R. A., Pedroza-Rodríguez, A. M., & Quevedo-Hidalgo, B. E. (2021). A brief history of colour, the environmental impact of synthetic dyes and removal by using laccases. Molecules, 26(13), 3813.

Asgher, M. (2012). Biosorption of reactive dyes: a review. Water, Air, & Soil Pollution, 223(5), 2417–2435.

Behera, A. K., Shadangi, K. P., & Sarangi, P. K. (2024). Efficient removal of Rhodamine B dye using biochar as an adsorbent: study the performance, kinetics, thermodynamics, adsorption isotherms and its reusability. Chemosphere, 354, 141702.

Brown, M. A., & De Vito, S. C. (1993). Predicting azo dye toxicity. Critical Reviews in Environmental Science and Technology, 23(3), 249–324.

Bushra, R., Mohamad, S., Alias, Y., Jin, Y., & Ahmad, M. (2021). Current approaches and methodologies to explore the perceptive adsorption mechanism of dyes on low-cost agricultural waste: a review. Microporous and Mesoporous Materials, 319, 111040.

Carmen, Z., & Daniela, S. (2012). Textile organic dyes-characteristics, polluting effects and separation/elimination procedures from industrial effluents-a critical overview (Vol. 3, pp. 55–86). Rijeka: IntechOpen.

Cestari, A. R., Vieira, E. F., Dos Santos, A. G., Mota, J. A., & De Almeida, V. P. (2004). Adsorption of anionic dyes on chitosan beads. 1. The influence of the chemical structures of dyes and temperature on the adsorption kinetics. Journal of Colloid and Interface Science, 280(2), 380–386.

Christie, R., & Abel, A. (2021). Cationic (basic) dye complex pigments. Physical Sciences Reviews, 6(10), 557–567.

Chung, K. T. (2016). Azo dyes and human health: a review. Journal of Environmental Science and Health, Part C, 34(4), 233–261.

de Farias Silva, C. E., da Gama, B. M. V., da Silva Gonçalves, A. H., Medeiros, J. A., & de Souza Abud, A. K. (2020). Basic-dye adsorption in albedo residue: effect of pH, contact time, temperature, dye concentration, biomass dosage, rotation and ionic strength. Journal of King Saud University-Engineering Sciences, 32(6), 351–359.

dos Santos, R. M. M., Gonçalves, R. G. L., Constantino, V. R. L., Santilli, C. V., Borges, P. D., Tronto, J., & Pinto, F. G. (2017). Adsorption of acid yellow 42 dye on calcined layered double hydroxide: effect of time, concentration, pH and temperature. Applied Clay Science, 140, 132–139.

Ekman, S., Dos Reis, G. S., Laisné, E., Thivet, J., Grimm, A., Lima, E. C., Naushad, M., & Dotto, G. L. (2023). Synthesis, characterization, and adsorption properties of

nitrogen-doped nanoporous biochar: efficient removal of reactive orange 16 dye and colorful effluents. Nanomaterials, 13(14), 2045.
El Harfi, S., & El Harfi, A. (2017). Classifications, properties and applications of textile dyes: a review. Applied Journal of Environmental Engineering Science, 3(3), 00000J-Envir.
El Messaoudi, N., El Khomri, M., El Mouden, A., Bouich, A., Jada, A., Lacherai, A., Iqbal, H. M. N., Mulla, S.I., Kumar, V., & Américo-Pinheiro, J. H. P. (2024). Regeneration and reusability of non-conventional low-cost adsorbents to remove dyes from wastewaters in multiple consecutive adsorption–desorption cycles: a review. Biomass Conversion and Biorefinery, 14(11), 11739–11756.
Foo, K. Y., & Hameed, B. H. (2010). An overview of dye removal via activated carbon adsorption process. Desalination and Water Treatment, 19(1–3), 255–274.
Gamoudi, S., & Srasra, E. J. J. O. M. S. (2019). Adsorption of organic dyes by $HDPy^+$-modified clay: effect of molecular structure on the adsorption. Journal of Molecular Structure, 1193, 522–531.
Gospodinova, N., & Tomšík, E. (2015). Hydrogen-bonding versus π–π stacking in the design of organic semiconductors: from dyes to oligomers. Progress in Polymer Science, 43, 33–47.
Gupta, V. K., Bhushan, R., Nayak, A., Singh, P., & Bhushan, B. (2014). Biosorption and reuse potential of a blue green alga for the removal of hazardous reactive dyes from aqueous solutions. Bioremediation Journal, 18(3), 179–191.
Hamad, H. N., & Idrus, S. (2022). Recent developments in the application of bio-waste-derived adsorbents for the removal of methylene blue from wastewater: a review. Polymers, 14(4), 783.
Haque, A. N. M. A., Sultana, N., Sayem, A. S. M., & Smriti, S. A. (2022). Sustainable adsorbents from plant-derived agricultural wastes for anionic dye removal: a review. Sustainability, 14(17), 11098.
Islam, T., Repon, M. R., Islam, T., Sarwar, Z., & Rahman, M. M. (2023). Impact of textile dyes on health and ecosystem: a review of structure, causes, and potential solutions. Environmental Science and Pollution Research, 30(4), 9207–9242.
Ismail, G. A., & Sakai, H. (2022). Review on effect of different type of dyes on advanced oxidation processes (AOPs) for textile color removal. Chemosphere, 291, 132906.
Jing, L., Li, P., Li, Z., Ma, D., & Hu, J. (2025). Influence of π–π interactions on organic photocatalytic materials and their performance. Chemical Society Reviews, 54, 2054–2090.
Khan, I., Saeed, K., Zekker, I., Zhang, B., Hendi, A. H., Ahmad, A., Ahmad, S., Zada, N., Ahmad, H., Shah, L. A., Shah, T., & Khan, I. (2022). Review on methylene blue: its properties, uses, toxicity and photodegradation. Water, 14(2), 242.
Kumari, S., Rajput, V. D., Minkina, T., Rajput, P., Sharma, P., Verma, A. K., Agarwal, S., & Garg, M. C. (2022). Application of RSM for bioremoval of methylene blue dye from

industrial wastewater onto sustainable walnut shell (Juglans regia) biomass. Water, 14(22), 3651.
Leodopoulos, C., Doulia, D., & Gimouhopoulos, K. (2015). Adsorption of cationic dyes onto bentonite. Separation & Purification Reviews, 44(1), 74–107.
León, O., Muñoz-Bonilla, A., Soto, D., Pérez, D., Rangel, M., Colina, M., & Fernández-García, M. (2018). Removal of anionic and cationic dyes with bioadsorbent oxidized chitosans. Carbohydrate Polymers, 194, 375–383.
Lewis, D. M. (2011). The chemistry of reactive dyes and their application processes. In: Handbook of Textile and Industrial Dyeing (pp. 303–364). Woodhead Publishing.
Lewis, D. M. (2014). Developments in the chemistry of reactive dyes and their application processes. Coloration Technology, 130(6), 382–412.
Lin, J., Ye, W., Xie, M., Seo, D. H., Luo, J., Wan, Y., & Van der Bruggen, B. (2023). Environmental impacts and remediation of dye-containing wastewater. Nature Reviews Earth & Environment, 4(11), 785–803.
Liu, L., Gao, Z. Y., Su, X. P., Chen, X., Jiang, L., & Yao, J. M. (2015). Adsorption removal of dyes from single and binary solutions using a cellulose-based bioadsorbent. ACS Sustainable Chemistry & Engineering, 3(3), 432–442.
Mandal, S., Alankar, T., Hughes, R., Marpu, S. B., Omary, M. A., & Shi, S. Q. (2022). Removal of hazardous dyes and waterborne pathogens using a nanoengineered bioadsorbent from hemp–fabrication, characterization and performance investigation. Surfaces and Interfaces, 29, 101797.
Mandale, P., Kulkarni, K., Jadhav, K., Kulkarni, A., & Mahajan, R. (2024). Adsorption of malachite green dye from aqueous solution on bioadsorbent as low-cost adsorbent. Materials Today: Proceedings.
Mchich, Z., Kjidaa, B., Bouzid, T., Saffaj, N., El Haddad, M., & Mamouni, R. (2025). Engineering of highly bio-adsorbent Zinc Oxide/Cellana Tramoserica seashells for efficient removal of anionic dye: BBD optimization, density functional theory study, and adsorption mechanism. Separation and Purification Technology, 363, 132015.
Mehra, S., Singh, M., & Chadha, P. (2021). Adverse impact of textile dyes on the aquatic environment as well as on human beings. Toxicology International, 28(2), 165.
Oussalah, A., Boukerroui, A., Aichour, A., & Djellouli, B. (2019). Cationic and anionic dyes removal by low-cost hybrid alginate/natural bentonite composite beads: adsorption and reusability studies. International Journal of Biological Macromolecules, 124, 854–862.
Parisi, F. (2020). Adsorption and separation of crystal violet, cerium (III) and lead (II) by means of a multi-step strategy based on k10-montmorillonite. Minerals, 10(5), 466.

Pereira, L., & Alves, M. (2011). Dyes—environmental impact and remediation. In: Environmental Protection Strategies for Sustainable Development (pp. 111–162). Dordrecht: Springer Netherlands.

Pérez-Jasso, A. N., Aguayo-Villarreal, I. A., Rojas-Mayorga, C. K., Muñiz-Valencia, R., Mendoza-Castillo, D. I., & Bonilla-Petriciolet, A. (2025). Comparative study of the physicochemical properties of metal organic frameworks in the adsorption of textile dyes. International Journal of Environmental Science and Technology, 22(10), 9611–9626.

Puchtler, H., Meloan, S. N., & Spencer, M. (1985). Current chemical concepts of acids and bases and their application to anionic ("acid") and cationic ("basic") dyes. Histochemistry, 82(4), 301–306.

Qiu, M., Qian, C., Xu, J., Wu, J., & Wang, G. (2009). Studies on the adsorption of dyes into clinoptilolite. Desalination, 243(1–3), 286–292.

Rao, V. B., & Rao, S. R. M. (2006). Adsorption studies on treatment of textile dyeing industrial effluent by flyash. Chemical Engineering Journal, 116(1), 77–84.

Rápó, E., & Tonk, S. (2021). Factors affecting synthetic dye adsorption; desorption studies: a review of results from the last five years (2017–2021). Molecules, 26(17), 5419.

Razi, M. A. M., Hishammudin, M. N. A. M., & Hamdan, R. (2017). Factor affecting textile dye removal using adsorbent from activated carbon: a review. In: MATEC Web of Conferences (Vol. 103, p. 06015). EDP Sciences.

Reddy, P. M. K., Verma, P., & Subrahmanyam, C. (2016). Bio-waste derived adsorbent material for methylene blue adsorption. Journal of the Taiwan Institute of Chemical Engineers, 58, 500–508.

Roy, D., Majumder, S., Mallick, K., Roy, N., Saha, B., Saha, S., & Sinha, B. (2023). Exploration of solvation consequences of ionic liquids prevalent in the aqueous media of food additive azo dye tartrazine by physicochemical and computational studies. Journal of Chemical & Engineering Data, 69(1), 38–58.

Shahadat, M., & Isamil, S. (2018). Regeneration performance of clay-based adsorbents for the removal of industrial dyes: a review. RSC Advances, 8(43), 24571–24587.

Shi, Y., Chang, Q., Zhang, T., Song, G., Sun, Y., & Ding, G. (2022). A review on selective dye adsorption by different mechanisms. Journal of Environmental Chemical Engineering, 10(6), 108639.

Smoczyński, L., Pierożyński, B., & Mikołajczyk, T. (2020). The effect of temperature on the biosorption of dyes from aqueous solutions. Processes, 8(6), 636.

Tamer, Y., Koşucu, A., & Berber, H. (2022). Graphene oxide incorporated chitosan/acrylamide/itaconic acid semi-interpenetrating network hydrogel bio-adsorbents for highly efficient and selective removal of cationic dyes. International Journal of Biological Macromolecules, 219, 273–289.

Tan, K. B., Vakili, M., Horri, B. A., Poh, P. E., Abdullah, A. Z., & Salamatinia, B. (2015). Adsorption of dyes by nanomaterials: recent developments and adsorption mechanisms. Separation and Purification Technology, 150, 229–242.
Trabelsi, H., Khadhraoui, M., Hentati, O., & Ksibi, M. (2013). Titanium dioxide mediated photo-degradation of methyl orange by ultraviolet light. Toxicological & Environmental Chemistry, 95(4), 543–558.
Tran, H. N., You, S. J., Nguyen, T. V., & Chao, H. P. (2017). Insight into the adsorption mechanism of cationic dye onto biosorbents derived from agricultural wastes. Chemical Engineering Communications, 204(9), 1020–1036.
Wierzbicka, E., Kuśmierek, K., Świątkowski, A., & Legocka, I. (2022). Efficient rhodamine B dye removal from water by acid-and organo-modified halloysites. Minerals, 12(3), 350.
Xiang, Z., Tang, N., Jin, X., & Gao, W. (2022). Fabrications and applications of hemicellulose-based bio-adsorbents. Carbohydrate Polymers, 278, 118945.
Xiao, W., Jiang, X., Liu, X., Zhou, W., Garba, Z. N., Lawan, I., Wang, L., & Yuan, Z. (2021). Adsorption of organic dyes from wastewater by metal-doped porous carbon materials. Journal of Cleaner Production, 284, 124773.
Yagub, M. T., Sen, T. K., Afroze, S., & Ang, H. M. (2014). Dye and its removal from aqueous solution by adsorption: a review. Advances in Colloid and Interface Science, 209, 172–184.
Yaneva, Z. L., & Georgieva, N. V. (2012). Insights into Congo Red adsorption on agro-industrial materials- spectral, equilibrium, kinetic, thermodynamic, dynamic and desorption studies. A review. International Review of Chemical Engineering, 4(2), 127–146.
Zhou, R., Zhou, R., Zhang, X., Tu, S., Yin, Y., Yang, S., & Ye, L. (2016). An efficient bio-adsorbent for the removal of dye: adsorption studies and cold atmospheric plasma regeneration. Journal of the Taiwan Institute of Chemical Engineers, 68, 372–378.
Zhu, H., Chen, S., & Luo, Y. (2023). Adsorption mechanisms of hydrogels for heavy metal and organic dyes removal: a short review. Journal of Agriculture and Food Research, 12, 100552.

8

Bioadsorbents in Heavy Metal Removal

8.1 Introduction

Water is necessary for life, but rising industrialization, urbanization, climate change, and pollution endanger its availability and purity. Industries consume massive amounts of water and frequently discharge effluent into natural bodies, causing significant damage to aquatic ecosystems. Industrial effluents generally contain toxic heavy metals as well as hazardous compounds that must be effectively treated. Biological treatments are effective for removing organic waste; however, heavy metals remain since they are nonbiodegradable. These toxins represent substantial health risks: arsenic (As) exposure can lead to cancer and organ damage, whilst mercury (Hg) can impair neurological, cardiovascular, and immunological systems (Hussain et al., 2024). As a result, eliminating heavy metals from wastewater is an important step toward sustainable and safe water treatment.

Heavy metals are among the most persistent and harmful pollutants found in industrial effluents, which frequently infiltrate municipal wastewater via industrial emissions or household chemicals. lead (Pb), copper (Cu), nickel (Ni), cadmium (Cd), zinc (Zn), mercury (Hg), arsenic (As), and chromium (Cr) are examples of common pollutants that represent significant environmental and health hazards due to their toxicity and bioaccumulation potential (Al-Enezi et al., 2004). Industries that emit Cd, Cr, Cu, Ni, As, Pb, and Zn are especially harmful since these metals are highly soluble in aquatic systems and easily adsorbed by organisms, resulting in buildup in the food chain and major health problems when concentrations exceed acceptable concentrations. Conventional treatment processes such as precipitation, ion exchange, and electrochemical approaches are extensively utilized, but they have limitations such as partial removal, high costs, and harmful sludge production (Zhao et al., 2016).

Conventional adsorbents such as activated carbon, clays, and metal oxides have been widely used, but their low adsorption capacity, lack of functional tunability, and poor recyclability limit their usage on a broad scale. To overcome these difficulties, new adsorbents such as nanomaterials are gaining attention because of their high surface area, numerous functional groups, reusability, and decreased sludge production (Singh et al., 2021). Biomass-derived adsorbents are also a low-cost and environmentally beneficial alternative, with chemical modification increasing their efficiency. Various adsorbents, both native and modified, have been investigated for the removal of metallic trace elements from wastewater, including heavy metals and metalloids. These include activated carbon, zeolites, clays, nano-sized metal oxides, animal feces, agricultural residues, industrial byproducts, and innovative designed adsorbents. Biosorbents are particularly appealing for their inexpensive cost, excellent metal-binding affinity, and widespread availability. They are typically formed from nonliving biomass (e.g., bark, crab shells, shrimp shells, fish scales), algal biomass, and biochar, or microbial biomass (e.g., bacteria, fungi, yeast). Plant fibers, which are mostly constituted of cellulose, hemicellulose, lignin, and pectin, are also efficient bioadsorbents because their carboxyl, hydroxyl, phenolic, and carbonyl functional groups interact effectively with metal ions through ion exchange and complexation (Raji et al., 2023).

In recent decades, significant progress has been achieved in improving the efficiency and cost-effectiveness of heavy metal and metalloid (e.g., arsenic) removal from water using adsorption. However, there has been little attention paid to the recycling of wasted adsorbents and the recovery of metals from regeneration agents. Acids, alkalis, and chelating agents such as EDTA have been used for regeneration, but the results are usually limited to a few adsorption–desorption cycles. Only a few research studies have addressed metal recovery from saturated adsorbents or desorbing solvents. Despite their importance for sustainable wastewater treatment, the management and safe disposal of spent adsorbents has received little attention (Lata et al., 2015).

Heavy metal pollution in water is a growing global concern because these contaminants are toxic, nonbiodegradable, and accumulate in the food chain. Conventional treatments often face challenges such as high cost, incomplete removal, and secondary waste. Adsorption offers a simple, low-cost, and efficient alternative for metal removal. The novelty of this chapter is that it brings together insights on adsorption mechanisms, metal-specific behavior, bioadsorbents, regeneration methods, and real wastewater applications. By focusing on both removal and regeneration, it highlights sustainable solutions that can move from lab studies to real-world wastewater treatment.

8.2 Common Heavy Metals in Wastewater

Heavy metals are commonly found in wastewater from industrial activities such as electroplating, battery manufacturing, mining, tanning, textile processing, smelting, petrochemicals, and paper making. These contaminants are extremely

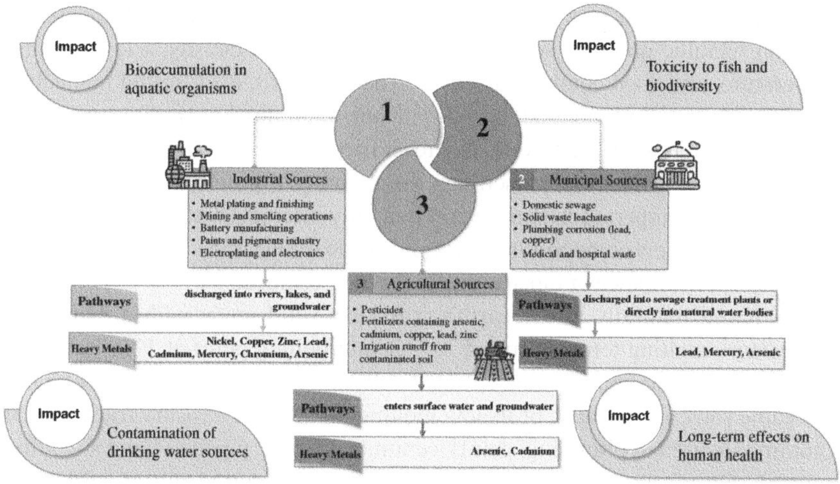

Figure 8.1 Schematic representation of heavy metal contamination in wastewater.

concerning because heavy metals are nonbiodegradable and many are carcinogenic, posing major threats to ecosystems and human health. Even in trace concentrations, metals like Pb, Zn, Hg, Ni, Cd, Cu, Cr, and As can be dangerous. Additional metals typically found in wastewater include silver, iron, manganese, molybdenum, boron, calcium, antimony, and cobalt. Figure 8.1 depicts the schematic representation of heavy metal contamination in wastewater. Their persistence and toxicity underscore the critical need for effective removal measures to protect water resources, reduce health risks, and promote environmental sustainability (Qasem et al., 2021).

8.2.1 Lead

Pb is a hazardous heavy metal that has serious public health consequences, including neurological, reproductive, and hypertensive effects. It is considered as a significant environmental threat, with widespread pollution recorded worldwide (Zhang et al., 2019). Pb has been utilized for centuries due to its unique qualities such as softness, malleability, low melting point, and corrosion resistance, making it useful in a variety of sectors. It is found in almost every area of the environment and biological systems, owing to its long history of industrial and home use. Pb is produced from both mined ores (primary sources) and recycled products such as batteries and pipes (secondary sources), which contributes to its long-term presence (Flora et al., 2006). Although elemental lead exists naturally, it is uncommon; instead, lead is typically found in the form of Pb(II) coupled with sulfur or oxygen in a variety of minerals with varied compositions. It has common metallic qualities such as metallic luster, high density, a low melting temperature, and good conductivity, as well as being soft, ductile, and malleable. When freshly cut, lead seems shiny, but in damp air, it quickly produces a surface coating of Pb(II) oxide. This oxide may react further with carbon dioxide to form Pb(II) carbonate. Under normal settings, these surface films function as protective layers. At high temperatures, however, Pb rapidly interacts with sulfur and halogens to produce new compounds (Casas and Sordo, 2006). Table 8.1 represents the adsorption performance of bioadsorbents for Pb^{2+} removal.

Pb is found naturally in the earth's crust as a solid metal with a density of 11.34 g/cm^3, melting temperature of 327°C, and boiling point of 1755°C. It dissolves in nitric acid and hot concentrated sulfuric acid, but not in water at 25°C or organic solvents. Because of its low melting point and corrosion resistance, it is commonly used in plumbing, paints, ceramics, solders, pipelines, and shielding materials. Currently, the battery industry accounts for around 80% of worldwide Pb consumption, primarily for automobile batteries, with China leading production. Pb remains in water, air, and soil, causing significant health concerns. The WHO limits Pb in drinking water to 0.01 mg/L to protect human health (Omar et al., 2019).

Pb exposure has been linked to health problems similar to Cd, as it enters the human body by inhalation, ingestion, or skin absorption. It predominantly affects the respiratory and gastrointestinal tracts, neurological system, kidneys, and skin, causing inflammation, tooth loss, hypertension, cardiovascular disease, and kidney impairment (Bouida et al., 2022). Pb can substitute critical cations, including Ca^{2+}, Mg^{2+}, Fe^{2+}, and Na^{+}, altering vital biological processes such as protein folding, apoptosis, ion transport, enzyme control, and neurotransmitter release. Prolonged exposure causes bioaccumulation, with Pb retained in soft tissues and eventually in bones and teeth. It causes oxidative stress by producing reactive oxygen species (ROS), decreasing antioxidant defenses, and affecting glutathione function, resulting in cellular damage, hemolysis, and an increased risk of cardiovascular disease and cancer (Collin et al., 2022).

8.2.2 Cadmium

Cd is a naturally occurring heavy metal that causes worry due to its persistence, mobility, and toxicity. Industrial operations, mining, smelting, and the use of phosphate fertilizers all contribute significantly to soil, water, and air pollution. Atmospheric deposition from fuel burning and urbanization spreads Cd into the environment. $CdCl^{-}$ and $Cd(SO_4)_2^{2-}$ complexes can keep it in solution by regulating its mobility based on pH, redox state, and ionic strength. Cd also interacts with dissolved organic materials, which increases its persistence. Cd's high

Table 8.1 Adsorption performance of bioadsorbents for Pb^{2+} removal.

Bioadsorbent source	Modification/Treatment	Experimental conditions	Adsorption capacity (mg/g)	Adsorption isotherm	Adsorption kinetics	References
Almond shell	Dried at 105 °C, grinded, and sieved	Initial Concentration (IC)—100 ppm Time—45 min pH—6 Temp.—Room Temp. Adsorbent Dosage (AD)—0.5 g in 100 mL	48.14	Freundlich isotherm	Pseudo-second-order kinetic model	Kayranli et al. (2022)
Peanut shell	Dried at 105 °C, grinded, and sieved	IC—100 ppm Time—45 min pH—6 Temp.— Room Temp. AD—0.5 g in 100 mL	37.14	Freundlich isotherm	Pseudo-second-order kinetic model	Kayranli et al. (2022)
Pistachio shell	Dried at 105 °C, grinded, and sieved	IC—100 ppm Time—45 min pH—6 Temp.—Room Temp. AD—0.5 g in 100 mL	36.73	Freundlich isotherm	Pseudo-second-order kinetic model	Kayranli et al. (2022)
Lettuce leaves	Leaves heated at 343 K for 20 h, Grinded and sieved Crosslinking with alginate (ALG) and functionalization with diethylenetriamine-pentaacetic acid (DTPA)	IC—80 ppm Time—10 min pH—6 Temp.—303 K AD—1 g in 1000 mL	168.35	Langmuir isotherm	Pseudo-second-order kinetic model	Nguyen et al. (2025)

(Continued)

Table 8.1 *(Continued)*

Bioadsorbent source	Modification/Treatment	Experimental conditions	Adsorption capacity (mg/g)	Adsorption isotherm	Adsorption kinetics	References
Schleichera oleosa bark	Dried at 60 °C, grinded and sieved	IC—50 mg/L Time—20 min pH—6 Temp.—50 °C AD—1 g/L	69.44	Langmuir isotherm	Pseudo-second-order kinetic model and intraparticle diffusion	Khatoon et al. (2018)
Natural clay honeycomb monoliths	Calcination treatments at 400 °C for 4 h	IC—250 ppm Time—60 min pH—4.5 Temp.—Room temp. AD—3 g/L	50	Langmuir isotherm	Pseudo-second-order kinetic model	Ahrouch et al. (2019)
Azadirachta indica leaves	Dried at room temperature, grinded, and sieved	IC—50 mg/L Time—40 min pH—7 Temp.—298 K AD—0.60 g	39.7	Freundlich isotherm	Pseudo-second-order kinetic model	Elkhaleefa et al. (2021)
Juniperus procera leaves	Dried at 40 °C for 3 h, crushed and sieved Immersed in diethyl ether and left overnight Acid activation using sulfuric acid (1:1.8)	IC—50 mg/L pH—4.6 Time—100 min Temp.—40 °C AD—0.8 mg/L	30.3	Langmuir isotherm	Pseudo-second-order kinetic model	Ali et al. (2019)

Enterobacter chuandaensis DGI-2	Live biomass	IC—750 μg/mL pH—6.5 Time—60 min Temp.— 30 ± 2°C AD—0.5 g/L	102.95	Freundlich isotherm	Pseudo-second-order kinetic model	Pal et al. (2025)
ZnO biochar nanocomposite	Dried and grounded	IC—100 mg/L pH—5.5 Time—120 min, AD—0.05 g	79.302	Langmuir isotherm	Pseudo-second-order kinetic model	Mudassar et al. (2025)
Rice husk	Dried at room temperature Alkali Treatment—1 M NaOH (1:10) for 24 hours at room temperature Dried at 100°C Carbonizes 400°C for 2 h (Nitrogen atmosphere)	IC—150 mg/L AD—5 g/L	1666.67	Langmuir and Freundlich isotherm	Pseudo-second-order kinetic model	Taha et al. (2014)
Pressmud (Sugar Industry byproduct)	Dried and sieved	IC—500 mg/L pH—7 Time—4 h Temp.—37°C AD—8 g/L	43.7	Langmuir isotherm	Pseudo-first-order kinetic model	Haider et al. (2025)
Phytolacca americana leaves	Dried at 60°C for 72h Acid Treatment—HNO_3 solution (0.1 mol/L), Washed and dried	IC—30 mg/L pH—6 Time—120 min Temp.—318 K AD—2 gL	14.51	Freundlich isotherm	Pseudo-second-order kinetic model	Wang et al. (2015)

(Continued)

Table 8.1 *(Continued)*

Bioadsorbent source	Modification/Treatment	Experimental conditions	Adsorption capacity (mg/g)	Adsorption isotherm	Adsorption kinetics	References
Spent tea grains biomass (STG)	Heated at 80°C for 24 h Dried and sieved	IC—5 mg/L pH—5 Time—90 min Temp.—Room Temp. AD—0.5 g/100 mL	24.272	Langmuir isotherm	Pseudo-second-order kinetic model	Chauhan and Dikshit (2023)
Momordica charantia leaves	Sun dried, grinded and sieved	IC—100 mg/L pH—5 Time—240 min AD—0.1 g/L	48.56	Freundlich isotherm	Pseudo-second-order kinetic model	
Bagasse biochar	Pyrolyzed at 300±10°C for 2.5 h Washed and dried	IC—50 mg/L pH—5 Time—140 min Temp.—Room Temp. AD—5 g/L	12.74	Langmuir isotherm	Pseudo-second-order kinetic model	Poonam et al. (2018)

mobility and bioaccumulation potential offer significant threats to ecosystems and human health, emphasizing the necessity for strict monitoring and preventive actions (Ankush et al., 2024).

Cd poisoning of groundwater is mostly caused by human activities such as mining, smelting, agriculture, sewage irrigation, and traffic. Nonferrous mining and smelting around the Xiangjiang River have been recognized as significant Cd sources in southern China, particularly in Hunan province. Acid mine drainage, wastewater, and toxic slag all contribute to soil and water contamination. Agricultural operations, particularly rice production, worsen Cd contamination by irrigating with contaminated water, using organic manure, and applying phosphate fertilizers, which lower soil pH and increase Cd mobility. Furthermore, inefficient e-waste recycling in developing countries such as China has increased Cd deposition in soils and sediments. These operations pose significant concerns to food safety and groundwater quality (Hoareau et al., 2022). Table 8.2 summarizes the adsorption performance of bioadsorbents for Cd^{2+} removal.

Cd is a very poisonous transition metal and an environmental contaminant resulting from agricultural and industrial activities. Human exposure is primarily through contaminated food, drink, inhalation, and cigarette smoking. Cd accumulates in living organisms and has a half-life of 25–30 years. It has been associated with breast, lung, prostate, kidney, pancreatic, and nasopharyngeal malignancies, as well as osteoporosis risk factors. The liver and kidneys are especially sensitive because Cd induces metallothionein production, which binds harmful ions. Cd causes oxidative stress and mitochondrial dysfunction, decreasing ATP and raising ROS. It also causes epigenetic alterations, such as DNA methylation and histone modifications, which modify chromatin and contribute to illness development (Genchi et al., 2020).

8.2.3 Chromium

Cr is the 21st most prevalent element in the Earth's crust, with typical values of around 40 mg/kg in US soils. It occurs primarily in three oxidation states: Cr(0), Cr(III), and Cr(VI), with trivalent Cr being the most common natural form, often coupled with iron or other oxides. Although only ~15% of mined Cr is employed for chemical synthesis, its unique chemistry supports numerous applications. Stainless steel is corrosion resistant due to protective Cr oxides, while chrome plating, wood preservation with CCA, and leather tanning all rely on Cr chemistry. In natural and biological systems, Cr(III) oxide is the most stable, with low solubility, reactivity, mobility, and toxicity, making it less dangerous to the environment than Cr(VI) (Barnhart, 1997). Cr has three common forms: elemental chromium (Cr^0), trivalent chromium (Cr^{3+}), and hexavalent chromium (Cr^{6+}). Trivalent chromium is naturally abundant, while Cr^0 and Cr^{6+} are primarily

Table 8.2 Adsorption performance of bioadsorbents for Cd^{2+} removal.

Bioadsorbent source	Modification/ Treatment	Experimental conditions (pH, contact time, temp.)	Adsorption capacity (mg/g)	Adsorption isotherm	Adsorption kinetics	Regeneration/ reusability	References
Surface-engineered *Escherichia coli*	Expressing cyanobacterial MT	IC—110 mg/L Time—120 min pH—8 Temp.— 40 °C AD—1 mg/mL	31.5	Freundlich isotherm	Pseudo-second-order kinetic model	68.6% Cd desorbed with EDTA; 56.5% with $CaCl_2$	Tafakori et al. (2017)
Rice husk biochar	Pyrolyzed at 500–600 °C Surface modified—$KMnO_4$ (0.01 M)	IC—50 mg/L pH—6 Time—120 min Temp.—25–30 °C, AD—5 g/L	15.3	Langmuir isotherm	Pseudo-second-order kinetic model	—	Gautam et al. (2023)
Red mud	Dried at 105 °C for 48 h MnO_2-modified	IC—50 mg/L pH—6 Time—180 min Temp.—Room Temp. AD—1 g/L	103.59	Langmuir isotherm	Pseudo-first-order kinetic model	Good; ~65% capacity retained after 5 cycles	Pang et al. (2022)
Pumpkin peel	Base-treated	IC—220 mg/L pH—5.6 Time—24 h Temp.—20 ± 2 °C AD—5 g/L	29	Freundlich	Pseudo-second-order kinetic model	Acid washing; effective regeneration	Allah et al. (2023)

Potato peel	Base-treated	IC—220 mg/L pH—5.6 Time—24 h Temp.—20 ± 2°C AD—5 g/L	29	Freundlich	Pseudo-second-order kinetic model	Acid washing; effective regeneration	Allah et al. (2023)
Activated charcoal (AC) and bone charcoal (BC)	Grinded, sieved, and combined in equal parts	IC—110 mg/L Time—120 min pH—6 Temp.—40°C AD—5 mg/mL	7	Langmuir isotherm	Pseudo-second-order kinetic model	Desorption of 94% for Cd (II), Reuse: 9–10 cycles	Bhran et al. (2025)
Khat (*Catha edulis*) stem	Dried at 105°C for 24 h Acid impregnated with 0.1 M H_2SO_4 (1:4 w/v) for 24 h. Neutralized by 2 % $NaHCO_3$ for 24 h Activated at 600°C for 1 h	IC—20 mg/L pH—5 Time—30 min Temp.—25°C AD—0.125 g	25.55	Langmuir isotherm	Pseudo-second-order kinetic model	Desorption—1 M HNO3 for 1 h Reuse: upto 3 cycles (63.7%)	Gelaw et al. (2024)
Capparis decidua branches (CDB)	Powdered bioadsorbent	IC—249 mg/L pH—6 Time—60 min Temp.—25°C AD—0.1 g/100 mL	248.62	Freundlich	Pseudo-second order	Acid washing regeneration; 6 reuse cycles with minor capacity loss	Bilal et al. (2021)

(Continued)

Table 8.2 *(Continued)*

Bioadsorbent source	Modification/ Treatment	Experimental conditions (pH, contact time, temp.)	Adsorption capacity (mg/g)	Adsorption isotherm	Adsorption kinetics	Regeneration/ reusability	References
Ziziphus mauritiana leaves (ZML)	Powdered bioadsorbent	IC—249 mg/L pH—6 Time—60 min Temp.—25 °C AD—0.1 g/100 mL	235.65	Langmuir	Pseudo-second order	Acid washing regeneration; 6 reuse cycles with minor capacity loss	Bilal et al. (2021)
Orange peel	Dried at 80 °C for 12 h, grinded, and seived	IC—240 mg/L pH—5.5 Time—120 min Temp.—45 °C AD—0.04 g/L	27.916	Langmuir isotherm	Pseudo-first-order kinetic model	—	Akinhanmi et al. (2020)
Luffa peels (base-treated)	Soaked in 0.4 M NaOH 24 h, dried at 50 °C	IC—250 ppm pH—8 Time—60 min Temp.—20 ± 2 °C, AD—0.1 g/10 mL	36.9	Freundlich isotherm	Pseudo-second-order kinetic model	Regeneration in fixed bed with 0.4 M HNO_3; recovery efficiency >90% for 3 cycles	Feng and Wang (2025)
Luffa peels (acid-treated)	Soaked in 0.4 M HNO_3 24 h, dried at 50 °C	IC—250 ppm pH—8 Time—60 min Temp.—20 ± 2 °C, AD—0.1 g/10 mL	30.9	Freundlich isotherm	Pseudo-second-order kinetic model	Regeneration in fixed bed with 0.4 M HNO_3; recovery efficiency >87% for 3 cycles	Feng and Wang (2025)

Chamomile flowers (base-treated)	Soaked in 0.4 M NaOH 24 h, dried at 50 °C	IC—250 ppm pH—8 Time—60 min Temp.—20 ± 2 °C, AD—0.1 g/10 mL	42.8	Freundlich isotherm	Pseudo-second-order kinetic model	Regeneration in fixed bed with 0.4 M HNO_3; recovery efficiency >90% for 3 cycles	Feng and Wang (2025)
Chamomile flowers (acid-treated)	Soaked in 0.4 M HNO_3 24 h, dried at 50 °C	IC—250 ppm pH—8 Time—60 min Temp.—20 ± 2 °C, AD—0.1 g/10 mL	34.0	Freundlich isotherm	Pseudo-second-order kinetic model	Regeneration in fixed bed with 0.4 M HNO_3; recovery efficiency >90% for 3 cycles	Feng and Wang (2025)
Tectona grandis leaves (TGLB)	Powdered leaves activated with 0.1 N HNO_3 (24 h soaking), dried at 373 K	IC—10 mg/L pH—5 Time—45 min Temp.— 298 ± 1.5 K AD—5 g/L	5.1	Langmuir isotherm	Pseudo-second-order kinetic model	Desorption with 0.1 M HCl; up to 3 regeneration cycles with decreasing desorption efficiency: 93.21% (1st cycle) to 65.42% (4th cycle)	Negi (2024)
Eupatorium adenophorum leaves (AEA)	Powdered leaf biomass protonated with 0.1 N HNO_3 (24 h), dried at 60°C	IC—10–50 mg/L pH—7 Time—90 min Temp.—298 K AD—2 g/L	45.45	Langmuir isotherm	Pseudo-second-order kinetic model	Desorption with 0.1 M HNO_3; reuse studied (details in supplementary)	Joshi et al. (2022)
Acer oblongum leaves (AAO)	Powdered leaf biomass protonated with 0.1 N HNO_3 (24 h), dried at 60°C	IC—10–50 mg/L pH—7 Time—90 min Temp.—298 K AD—2 g/L	44.64	Langmuir isotherm	Pseudo-second-order kinetic model	Desorption with 0.1 M HNO_3; reuse studied (details in supplementary)	Joshi et al. (2022)

derived via industrial processes. Chromium is commonly used in electroplating, corrosion prevention, and leather tanning. It binds proteins in animal skin, resulting in durable, water- and bacteria-resistant leather. Cr^0 is mostly utilized in the production of stainless steel and alloys, but it also has use in anti-corrosive coatings, textile dyes, and wood preservation. Drinking water contamination with Cr^{6+} is a significant health hazard, particularly in industries such as stainless steel, dye, and wood (Kaprara et al., 2015). Wastewater irrigation also increases Cr levels in soil and plants (El-Kady and Abdel-Wahhab, 2018). Table 8.3 gives the adsorption performance of bioadsorbents for chromium (Cr^{3+}/Cr^{6+}) removal.

Weathering of ultramafic rocks is a significant source of increased hexavalent Cr(VI) in soils, sediments, and groundwater. Ultramafic rocks and their related serpentine soils are abundant and possess high Cr levels, averaging 2200 mg/kg in rocks and 2650 mg/kg in soils. Groundwater in these places contains Cr(VI) values ranging from 0.2 to 180 μg/L, typically surpassing drinking water restrictions (50 μg/L Cr tot, 5 μg/L in Italy, and 10 μg/L in California). Cr is released when Cr(III) dissolves from mineral hosts, sorbs to high-valence Mn oxides, and oxidizes to mobile Cr(VI) in alkaline circumstances. This geogenic process emphasizes the environmental danger of natural Cr(VI) contamination in populated ultramafic regions (Chrysochoou et al., 2016).

Cr's oxidation states range from −IV to +VI, resulting in compounds having square planar, tetrahedral, and octahedral shapes. Cr contamination is mostly caused by industrial sources such as ore refining, chemical and refractory processes, cement factories, brake linings, catalytic converters, leather tanning, and chrome pigments. Hexavalent Cr(VI) is almost 100 times more hazardous than trivalent chromium due to its high solubility, mobility, and reduction potential. The respiratory tract is the primary target after inhalation, resulting in septum perforation, ulceration, bronchitis, impaired lung function, pneumonia, and nasal irritation. Prolonged exposure can also harm the liver, kidneys, gastrointestinal, immunological, and hematological systems. Dermal contact to Cr(VI) induces dermatitis, irritation, and skin ulceration, making it a serious health concern (Saha et al., 2011).

8.2.4 Arsenic

As is a controlled hazardous element and a persistent pollutant that causes environmental, agricultural, and health problems globally. Atomic spectrometry is commonly used to analyze As, with detection limits as low as 1 μg/L, making it suitable for environmental monitoring. Concentrations vary greatly: up to 1660 ng/m^3 in urban air, 2000 μg/L in groundwater, 4600 mg/kg in soils, and 2500 mg/kg in sediment. Natural deposits, fossil fuel burning, pesticides, and industrial effluents are also significant sources. Many reported levels surpass the recommended

Table 8.3 Adsorption performance of bioadsorbents for chromium (Cr^{3+}/Cr^{6+}) removal.

Bioadsorbent source	Modification/ Treatment	Experimental conditions (pH, contact time, temp.)	Adsorption capacity (mg/g)	Adsorption isotherm	Adsorption kinetics	Regeneration/ Reusability	References
Portieria hornemannii biomass	None (dried, ground algal biomass)	IC—300 mg/L pH—2 Time—200 min, Temp.—40 °C Adsorbent dose—6 g/L	18.28	Freundlich isotherm	Pseudo-second-order kinetic model	Regeneration with 0.1 M EDTA; stable efficiency up to 6 cycles	Sekaran and Thiagarajan (2025)
EDTA modified graphene oxide/ chitosan composite (EDTA/GO/CS)	EDTA modification of GO/CS composite	C0 = 100 mg/L; adsorbent dose = 1 g/L; T = 25 °C; pH = 5.0	86.17	Freundlich isotherm	Pseudo-second-order kinetic model	Adsorption capacity lowered by 5% after 7 cycles	Liu et al. (2023)
Arecanut leaf sheath (CALS)	Acid modification with concentrated H_2SO_4	pH—2 Time—150 min Temp.—298 K	109.89	Langmuir isotherm	Pseudo-second-order kinetic model	Capacity decreased by 15.67% after 4 adsorption/ desorption cycles; reusable at least 3 times; desorption with 1.0 M NaOH; stable adsorbent structure	Pant et al. (2022)
Walnut shell powder (native)	None	pH—2 Time—40 min Temp.—298 K	64.82	Langmuir isotherm	Pseudo-second-order kinetic model	Effective up to 10 cycles with slight reduction in % removal	Garg et al. (2023)

(Continued)

Table 8.3 *(Continued)*

Bioadsorbent source	Modification/ Treatment	Experimental conditions (pH, contact time, temp.)	Adsorption capacity (mg/g)	Adsorption isotherm	Adsorption kinetics	Regeneration/ Reusability	References
Walnut shell powder (alkali treated, AWP)	Sodium hydroxide treatment	pH—2 Time—40 min Temp.—298 K	69.56	Langmuir isotherm	Pseudo-second-order kinetic model	Effective up to 10 cycles with slight reduction in % removal	Garg et al. (2023)
Walnut shell powder (citric acid treated, CWP)	Citric acid treatment	pH—2 Time—40 min Temp.—298 K	75.26	Langmuir isotherm	Pseudo-second-order kinetic model	Effective up to 10 cycles though with slightly more reduction in % removal compared to native and AWP	Garg et al. (2023)
Eggshell	None (washed, dried, ground)	Initial Cr (III)—3.21 g/L pH—3.81 Time—14 h Temp.—25 °C Adsorbent dose—20 g/L	200.25 ± 3.63	Langmuir isotherm	Pseudo-second-order kinetic model	—	Elabbas et al. (2016)
Powdered marble	Raw, powder	Initial Cr (III)—3.21 g/L pH—3.81 Time—30 min Temp.—25 °C Adsorbent dose—12 g/L	434.82 ± 2.51	Langmuir isotherm	Pseudo-second-order kinetic model	—	Elabbas et al. (2016)

Aspergillus niger	Dried	pH ~3.0–4.6, contact time = 48–120 min, Temp.—27–36 °C	11.79	Temkin and Freundlich isotherm	Pseudo-second-order kinetic model	—	Islam et al. (2023)
Scrap tire-derived activated carbon (STAC) doped with nano ZnO (nZnO-STAC)	$ZnCl_2$ chemical activation + wetness impregnation with ZnO nanoparticles + calcination at 250 °C	pH—3 Time—120 min Temp.—298 K	152.2	Langmuir isotherm	Pseudo-second-order kinetic model	—	Chigova and Mudono (2022) [this paper]
Activated carbon from water hyacinth (*Eichhornia crassipes*)	Chemical activation with 37% phosphoric acid soaking for 24 h + thermal activation at 500 °C for 15 min	pH—2 Time—120 min Temp. ~25 °C	6.03	Langmuir isotherm	Pseudo-second-order model	Tested 7 regeneration cycles with 1 M NaOH desorption; removal efficiency decreased from 93.25 (1st cycle) to 21.35% (7th cycle)	Worku et al. (2023)
Cranberry kernel shell (CKS)	Washed with water and 0.1 M HCl; dried, crushed, sieved	pH—2 Time—60 min, Temp.—25 ± 1 °C	6.81	Langmuir isotherm	Pseudo-second-order kinetic model	Not reported	Parlayici and Pehlivan (2019)
Rosehip seed shell (RSS)	Washed with water and 0.1 M HCl; dried, crushed, sieved	pH—2 Time—60 min, Temp.—25 ± 1 °C	15.17	Langmuir isotherm	Pseudo-second-order kinetic model	Not reported	Parlayici and Pehlivan (2019)

(Continued)

Table 8.3 *(Continued)*

Bioadsorbent source	Modification/ Treatment	Experimental conditions (pH, contact time, temp.)	Adsorption capacity (mg/g)	Adsorption isotherm	Adsorption kinetics	Regeneration/ Reusability	References
Banana peel (BP)	Washed with water and 0.1 M HCl; dried, crushed, sieved	pH—2 Time—60 min Temp.—25 ± 1 °C	10.42	Langmuir isotherm	Pseudo-second-order kinetic model	Not reported	Parlayici and Pehlivan (2019)
NaOH-pretreated *Humicola phialophoroides* fungus biomass	NaOH pretreatment	pH >8 Time—90 min, varied temperature	NA	Langmuir isotherm	Pseudo-second-order kinetic model	Desorption with 0.1 M HNO3 showed up to ~54% desorption	Netpae et al. (2018)
Tamarind fruit shell (PSh) biosorbent	Raw biosorbent	pH—2 Time—45 min Temp.—25–45 °C	2.48–3.49	Langmuir isotherm	Pseudo-second-order kinetic model	Regeneration with NaOH: <5% loss in Cr (VI) removal efficiency up to 3 cycles	Rzig et al. (2021)

limits, emphasizing As's serious concern. However, many elements of its chemistry, mobility, and elimination are still unclear (Patel et al., 2023).

As is a highly poisonous chemical that poses significant concerns to big populations. Historically, it was utilized in skin infections, cosmetics, and even as a poison for murder. Groundwater contamination in South Asia was initially documented in the mid-1990s, and substantial research has since been conducted. The use of contaminated groundwater for drinking and irrigation has had serious consequences for humans, cattle, and crops, often resulting in death. According to reports, As-contaminated groundwater affects more than 200 million people in 70 nations. As poisoning is expected to affect 1.2 million people in Bangladesh each year. This broad contamination emphasizes the urgent need for appropriate remediation solutions (Shrivastava et al., 2015).

8.2.5 Mercury

Mercury (Hg) exists in various chemical forms, each with unique features that influence its behavior in combustion emissions and atmospheric movement. Coal combustion emits around 20–50% elemental Hg and 50–80% divalent Hg(II), primarily as $HgCl_2$. Waste incinerators emit around 10–20% Hg and 75–85% Hg(II). The distribution is determined by contaminants such as particulate carbon and hydrochloric acid. Activated carbon injection, sodium sulfide injection, and flue gas desulfurization using wet lime/limestone are examples of Hg control technologies. While Hg(II) is water-soluble and accumulates near sources, Hg has a high vapor pressure, allowing for long-range air transport. Hg eventually oxidizes to Hg(II), which is then eliminated by deposition mechanisms (Carpi, 1997).

Hg is abundant in the environment, naturally occurring in volcanic regions, and employed in science, industry, and agriculture. Because of its toxicity, many applications have been replaced; however, coal burning and metal smelting continue to emit considerable amounts of Hg. Because of its volatility, Hg flows globally, contaminating even pristine areas with no local sources. The chemical form determines toxicity and mobility. In flooded soils and sediments, microorganisms convert inorganic Hg into methylmercury (MeHg), a bioaccumulative neurotoxin that enters the food chain through fish and rice intake, endangering human health. Elemental mercury vapor can injure the neurological system, while inorganic Hg primarily affects the kidneys (Beckers and Rinklebe, 2017).

Hg is one of the most hazardous heavy metals, found throughout the lithosphere, hydrosphere, atmosphere, and biosphere. It exists primarily in three oxidation states and undergoes intricate changes that enable ongoing cycling in the environment. In soils, sediments, and water, inorganic forms like $HgCl_2$, $Hg(OH)_2$, and HgS dominate, whereas organic forms like CH_3HgCl and CH_3HgOH appear in lower levels. Natural processes, industrial activities, and the re-emission of

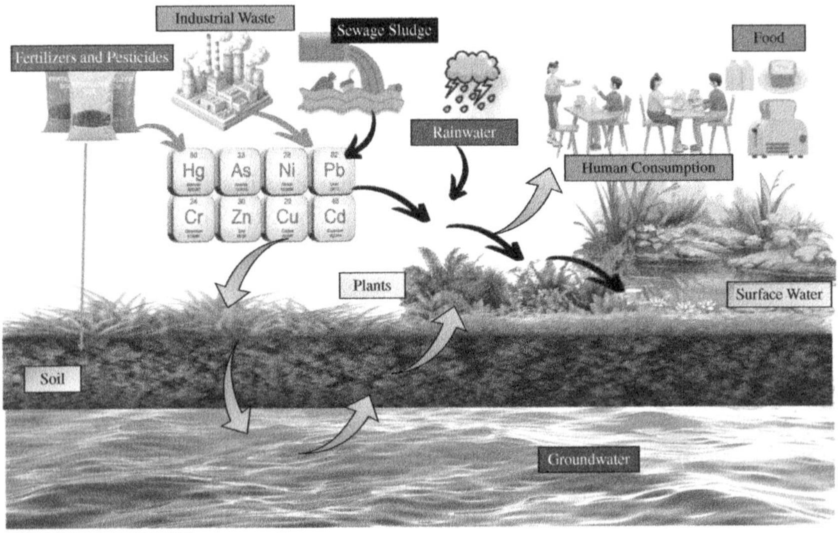

Figure 8.2 Environmental cycle of heavy metals.

previously deposited Hg are all major sources of Hg emissions. Human-related causes include industrial discharges, mining, agriculture, combustion, and urban garbage. Since the industrial revolution, atmospheric Hg has increased roughly 10-fold, making deposition a significant source of environmental contamination (Zhang and Wong, 2007). Figure 8.2 shows the environmental cycle of heavy metals.

8.3 Mechanisms of Heavy Metal Adsorption

Metal adsorption on biochar happens through both physical and chemical mechanisms. Physical adsorption is mostly determined by biochar's surface area and pore structure, notably the micropores generated during pyrolysis at 500–600 °C, where van der Waals forces are dominant. However, functional groups on the biochar surface have an effect on chemical adsorption. The primary mechanisms are ion exchange, precipitation, complexation, redox reactions, and electrostatic interactions. Ion exchange involves the reversible replacement of metal ions with surface cations. For example, Pb(II) removal using biochars generated from sawdust, canola straw, manure pellets, and wheat straw involved precipitation, ion exchange, and complexation, with quantifiable cation exchange capabilities for each biochar type (Awang et al., 2023).

Physisorption is frequently seen as a precursor to chemisorption, driven primarily by van der Waals forces and electrostatic interactions. These small pressures produce little adsorption heat, making the process rapid and partially reversible. The strength of physisorption is proportional to adsorbent surface area, surface charge, and atomic distance, with greater values boosting interactions. However, these are weaker than the chemical bonds generated during chemisorption. Chemisorption occurs when an adsorbate and an adsorbent create covalent bonds, which frequently result in the formation of chelates or metal complexes. Its rate is dependent on activation energy, which may delay equilibrium achievement. Metal ion adsorption can also occur through ion exchange, often including functional groups such as —OH, —COOH, $-SO_3H$, and $-PO_3H_2$, through proton-metal or metal–metal ion interchange (Kaur et al., 2022).

Polysaccharides in microbial cell walls play an important role in metal ion exchange, with functional groups such as carboxyl, phosphate, sulfate, and amine facilitating binding via electrostatic attraction, ion exchange, or covalent bonding. Fungi and algae commonly adsorb Pb(II) and Cd(II) by ion exchange, which is regulated by pH and surface charge availability. Pretreatment or surface modification, including nanotechnology, improves adsorption. In addition to ion exchange, complexation with groups like —COOH, $-NH_2$, and —SH increases metal binding. Microbes release organic acids that aid in solubilization (Priya et al., 2022). Metals can also precipitate on cell walls or within cells, either metabolically or chemically, offering a defensive mechanism against toxicity. Key biosorption mechanisms include ion exchange, complexation, and precipitation (Figure 8.3).

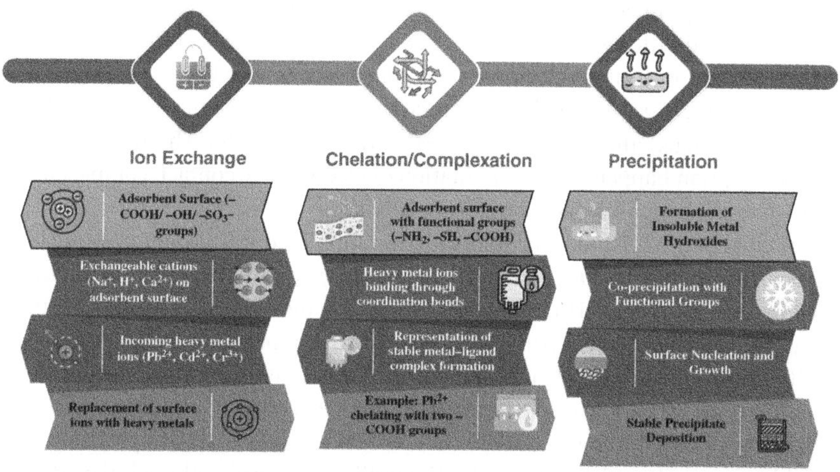

Figure 8.3 Mechanisms of heavy metal adsorption.

8.4 Removal of Heavy Metals Using Bioadsorbent

Chemical precipitation, ion exchange, electrochemical treatment, filtration, membrane processes, adsorption on activated carbon, and evaporation are some of the traditional heavy metal removal procedures for water. These approaches have disadvantages, including high cost, sludge production, and inefficiency at low metal concentrations (1–100 mg/L). For example, ion exchange and membrane technologies are costly for large-scale applications, but precipitation and electrochemical procedures generate enormous amounts of sludge that are difficult to manage. In recent years, biotechnology-based techniques have received more attention. Biosorption, which uses natural materials including bacteria, fungi, yeast, and algae, has shown significant promise to successfully remove metals from ppm to ppb levels, offering a low-cost and eco-friendly option (Wang and Chen, 2009). Agricultural byproduct has been identified as a viable sorbent for heavy metal and dye removal. It is increasingly being examined as a less expensive alternative to traditional procedures. Studies have shown that it may be used to remove metals such as As(V), Au, Cr(IV), Cu, Pb, Fe, Mn, Zn, Cu, and Cd(II). Activated carbon generated from rice husk has gained a lot of interest due to its effectiveness and manufacturing methods (Chuah et al., 2005).

8.4.1 Lead

Biosorbent produced from waste activated sludge was tested for Pb^{2+} and Zn^{2+} in aqueous single-metal systems. Optimal adsorption occurred at pH 6.0, a biosorbent-to-metal ion mass ratio of two, and temperatures below 55 °C. RSM was used to optimize biosorption, resulting in high determination coefficients ($R^2 > 0.99$) and little lack of fit ($p = 0.0838$ for Pb^{2+}, $p = 0.0782$ for Zn^{2+}). Atomic force microscopy verified electrostatic contact between metal ions and the biosorbent surface. X-ray photoelectron spectroscopy revealed complexation ($-NH_2$, $-CN$ with metals) and ion exchange ($-COOH$ with metals). The findings show that biosorption happens by a combination of electrostatic contact, complexation, and ion exchange pathways (Zhou et al., 2016).

The current work explores Pb removal from simulated wastewater utilizing low-cost bioadsorbents produced from mango seed cover with kernel (M) and jamun seed cover with kernel (JP). The response surface methodology (RSM) was used to optimize the adsorption parameters. The ideal parameters were determined to be 500 rpm, 60 mg dose, pH 6.5, and 120 minutes contact duration. Under these conditions, adsorption capacities reached 39.15 mg/g for M and 20.28 mg/g for JP, with maximum Pb removal efficiencies of 94.85 and 92.78%, respectively. The equilibrium results suited well to the Freundlich isotherm, although the adsorption kinetics followed a pseudo-first-order equation. Pal et al. (2022) found that agricultural waste-based products can serve as low-cost and sustainable adsorbents for wastewater treatment.

The biosorptive ability of three aquatic-based biosorbents—bivalve mollusk shells and two fish species' scales—was tested for Pb removal from aqueous solutions. A Box-Behnken design with response surface approach was used to investigate seven variables: contact time, temperature, beginning concentration, dosage, size, salinity, and pH. Dosage, beginning concentration, and pH were the most significant, while particle size had no discernible effect. Initial concentration was found to be the most important factor. SEM-EDX and XRF investigations confirmed that Pb ions bind to biosorbent surfaces. *Rutilus kutum* scales were found to have the highest sorption capabilities, followed by *Oncorhynchus mykiss* scales and *Cerastoderma glaucum* shells. According to Rezaei et al. (2022), the Langmuir model fits fish scales while the Freundlich model fits mollusk shells.

Papaya wood was tested as a biosorbent for removing Cd from aqueous solutions. A 60-minute interaction between 10 mg/L Cd^{2+} solution and 5 g/L papaya wood resulted in 94.9% elimination. Sorption was most effective at pH 5. The adsorption rose with a lower biomass-to-solution ratio but dropped as biomass quantity increased. At low Cd^{2+} concentrations, biosorption increased rapidly before plateauing at higher levels. Equilibrium results suited the Langmuir isotherm model ($R^2 = 0.99$), whereas Freundlich provided a worse fit. Kinetics used a second-order model with $R^2 = 0.99$. Loaded biomass was entirely regenerated with 0.1 N HCl, and Cd removal was efficient for five sorption–desorption cycles, confirming reusability (Saeed et al., 2005).

8.4.2 Cadmium

Papaya wood was tested as a biosorbent for removing Cd from aqueous solutions. A 60-minute interaction between 10 mg/L Cd^{2+} solution and 5 g/L papaya wood resulted in 94.9% elimination. Sorption was most effective at pH 5. The adsorption rose with a lower biomass-to-solution ratio but dropped as biomass quantity increased. At low Cd^{2+} concentrations, biosorption increased rapidly before plateauing at higher levels. Equilibrium results suited the Langmuir isotherm model ($R^2 = 0.99$), whereas Freundlich provided a worse fit. Kinetics used a second-order model with $R^2 = 0.99$. Loaded biomass was entirely regenerated with 0.1 N HCl, and Cd removal was efficient for five sorption–desorption cycles, confirming reusability (Saeed et al., 2005).

The viability of removing Cd(II) from aqueous solutions utilizing agricultural leftovers such as sugarcane bagasse (SCB), maize corncob (MCC), and Jatropha oil cake (JOC) has been investigated. Batch studies were carried out to investigate the effects of initial metal concentration, pH (2–7), adsorbent dose (250–2000 mg), and contact time (60 minutes). Maximum Cd removal was seen at pH 6 with a contact period of 60 minutes and an adsorbent dose of 20 g/L, attaining 99.5, 99, and 85% removal for JOC, MCC, and SCB, respectively, with JOC > MCC > SCB in order of efficiency. Langmuir and Freundlich isotherms suggested that Cd(II) formed monolayers on the adsorbents. FT-IR research revealed the presence of metal-binding functional groups. Garg et al. (2008) found that JOC has the potential to be a cost-effective biosorbent. Olive stone and SCB byproducts were examined as environmentally friendly biosorbents for removing Cd(II) from aqueous solutions. The biosorbents were evaluated utilizing FTIR, BET, SEM, and zeta potential measurements. The influence of pH, starting Cd(II) concentration, contact time, and temperature on adsorption was investigated. Both biosorbents demonstrated maximum Cd(II) uptake at pH 7 and 298 K, with equilibrium adsorption increasing with initial metal concentration, resulting in a maximum elimination of 96%. The adsorption behavior corresponded to the pseudo-second-order kinetic model, confirming chemisorption. According to Moubarik and Grimi (2015), low-cost agricultural byproducts such as olive stone and SCB may effectively and sustainably remove Cd from wastewater.

Marine chitin and kraft lignin were used to create novel chitin/lignin-based biosorbents, which were then studied for their physicochemical, morphological, and electrokinetic characteristics. The alteration of chitin with lignin occurred by hydrogen bond formation, resulting in useful materials for wastewater treatment. The biosorbents were highly efficient at removing Cd, achieving up to 98.4% adsorption. The work emphasizes the efficacy of these low-cost chitin/lignin composites as heavy metal adsorbents. Furthermore, this strategy provides an environmentally benign route to valorize the seafood sector's chitinous waste and lignin byproducts from the pulp and paper industry, solving both waste management and water pollution concerns (Wysokowski et al., 2014).

Fruit peel wastes, such as banana, durian, pummelo, and corn, were tested as biosorbents for removing Cd(II) from aqueous solutions at 25 °C. Banana peel had the highest removal effectiveness (73.15%), followed by durian (72.17%), pummelo (70.56%), and corn (51.22%). Cd uptake increased with pH, peaking at pH 5, whereas particle size had no influence. Equilibrium data supported the Langmuir isotherm, with maximal biosorption capacities of 20.88, 18.55, and 21.83 mg/g for banana, durian, and pummelo, respectively. FTIR research showed that Cd(II) adsorption involved carboxyl, hydroxyl, and amide groups on peel surfaces, indicating chemical interactions as the predominant mechanism (Saikaew and Kaewsarn, 2009). Agricultural wastes are a sustainable alternative to traditional adsorbents because of their local availability, low cost, abundance, and biodegradability. They are simple to process, apply, and recover without affecting the environment, in accordance with green chemistry and sustainable development concepts. These wastes are gradually replacing traditional adsorbents for Cd(II) removal and recovery. However, chemical alterations of agricultural wastes employing strong acids or alkalis might cause environmental damage, potentially negating the low-cost benefits if cleanup and chemical expenditures are considered (Kwikima et al., 2021).

8.4.3 Chromium

Heavy metal removal has traditionally relied on procedures such as ion exchange, precipitation, electroplating, and evaporation, but these methods have numerous limitations. Adsorption with low-cost biosorbents provides an eco-friendly option. This review assessed the efficacy of natural and agricultural wastes for removing Cr(VI) from contaminated water. Adsorption factors such as pH, contact time, adsorbent type, starting concentration, and temperature were investigated. Maximum Cr(VI) removal ranged from 50 to 100% at optimal pH 2–6 and contact period 30–180 minutes. Pseudo-second-order kinetics and Langmuir isotherms best described the endothermic and spontaneous adsorption process. Agricultural wastes can replace commercial activated carbon; however, better regeneration, reuse, and commercialization are needed (Bayuo, 2021). Native walnut shell powder (NWP) was chemically treated using alkali (AWP) and citric acid (CWP) to increase porosity, as confirmed by BET analysis. Batch adsorption studies improved Cr(VI) uptake at pH 2. Adsorption occurred after Langmuir isotherms, indicating monolayer formation, and pseudo-second-order kinetics, implying chemisorption. The maximum adsorption capacities were 75.26 mg/g (CWP), 69.56 mg/g (AWP), and 64.82 mg/g (NWP). Chemical alteration increased efficiency by 4.5–8.2%. The reaction was endothermic and spontaneous, highlighting walnut shell powder as a sustainable Cr(VI) adsorbent (Garg et al., 2023).

Neem leaves were found to remove the most Cr and reduce Cr(VI) to Cr(III). FTIR analysis identified functional groups involved in biosorption. Equilibrium data were fitted to Langmuir and Freundlich isotherms; Langmuir was more appropriate for neem leaves and water hyacinth, while Freundlich was best suited for sawdust, sludge, and byproducts. According to Aggarwal and Arora (2020), biosorbents are effective for treating industrial wastewater due to sorption-coupled reduction. The removal of Cr(III) utilizing waste fish scales as a bioadsorbent in batch adsorption tests with different adsorbent dosages, solution pHs, and contact times was studied. The maximum Cr(III) removal (99.75%) was reached at 0.8 g dose, pH 5, 90 minutes of contact time, and an initial concentration of 150 mg/L. Adsorption occurred via the Langmuir isotherm ($R^2 = 0.9998$, $qmax = 18.35$ mg/g, $RL = 0.000073$) and pseudo-first/second-order kinetics ($R^2 = 1$). Teshale et al. (2020) used SEM, XRF, FTIR, and thermal investigations to confirm surface characteristics and metal adsorption.

Lignocellulosic biomass-derived lignin has emerged as a low-cost, very effective biosorbent for heavy metal removal from wastewater. Lignin, extracted from a variety of sources, comprises functional groups such as hydroxyl, carbonyl, and methoxyl, which aid in adsorption. Chemical changes like sulfonation, carboxylation, and amination improve its heavy metal uptake. pH, temperature, adsorbent dose, contact time, and metal concentration all have an effect on adsorption performance, with an optimum pH ranging from 1 to 3. Adsorption is described using the Langmuir and Freundlich models. Lignin-based biosorbents effectively remove Cr(III) and Cr(VI), providing a cost-effective alternative to commercial adsorbents (Wang et al., 2019). However, further mechanistic studies are needed.

8.4.4 Arsenic

As contamination in water offers major health problems and requires appropriate cleanup procedures. Several biosorbents have been tested for their ability to remove the inorganic As species, arsenite [As(III)] and arsenate [As(V)]. The biosorption process is affected by pH, As concentration, competing ions, and As speciation. Kinetic studies reveal rapid adsorption, with over 80% of As biosorbed within the first hour, followed by a slower secondary phase. Adsorption follows a pseudo-second-order model, implying chemisorption, while the Langmuir isotherm indicates monolayer adsorption. According to Sahmoune (2016), As(III) has a greater removal efficiency compared to As(V). Shakoor et al. (2019) tested biosorbents such as eggshell, java plum seed, water chestnut shell, corn cob, tea waste, and pomegranate peel to remove arsenite [As(III)] and arsenate [As(V)] from water. At pH 7, eggshell and java plum seed removed the most As(III) (78–87%), while As(V) was removed by 71 and 67% at pH 4.1–5.3, respectively. After 2 hours, maximum sorption was observed, following the Langmuir isotherm and pseudo-second-order kinetics. Adsorption was facilitated by functional groups such as —OH, —COOH, $-NH_2$, and sulfur, resulting in up to 97% As recovery after four regeneration cycles.

Biosorbents such as eggshell, java plum seed, water chestnut shell, corn cob, tea waste, and pomegranate peel were tested for their ability to remove arsenite [As(III)] and arsenate [As(V)] from water. At pH 7, eggshell and java plum seed removed the most As(III) (78–87%), while at pH 4.1–5.3, As(V) removal was 71 and 67%, respectively. The maximum sorption occurred after 2 hours, following the Langmuir isotherm and pseudo-second-order kinetics. Adsorption was facilitated by functional groups such as —OH, —COOH, $-NH_2$, and sulfur, resulting in up to 97% As recovery after four regeneration cycles (Shakoor et al., 2019).

Giri et al. (2022) investigated the removal of As from water utilizing low-cost, solid waste-derived biosorbents. Cassia fistula pod biomass removed around 91% of As(III) at pH 6, 30 °C, and 60 mg dosage in 80 minutes, with adsorption fitting pseudo-first-order and Freundlich models. Agricultural byproducts such as corncobs and soybean hulls were tested on groundwater and lab-prepared As solutions, with removal rates of up to 98 and 71%, respectively, with greater efficiency at higher initial concentrations. The biosorbent type, As levels, and contact time all had an impact on removal efficiency variations. These investigations emphasize the potential of agricultural wastes as eco-friendly and effective biosorbents for As-contaminated water (Shabbir et al., 2023).

8.4.5 Mercury

Hg is a very harmful pollutant in water that must be removed at all levels owing to environmental and health risks. Agricultural and industrial wastes are plentiful, affordable, and can act as effective biosorbents because Hg(II) ions make linkages with functional groups on lignocellulosic biomass. In a comparative investigation, six biosorbents—banana and potato peels, eggshells, Eucalyptus globulus bark, water hyacinth, and coffee waste—were examined at an initial Hg(II) concentration of 50 $\mu g/m^3$. Banana peels had the highest removal effectiveness, removing over 90% of Hg(II) within 72 hours, with over 80% eliminated in the first 10 hours (Fabre et al., 2021).

Banana peels have been extensively researched as a low-cost biosorbent for removing Hg from aqueous solutions. Their adsorption performance is primarily determined by biosorbent dosage and contact time, with ionic strength having a minor effect. Banana peels effectively removed Hg(II) concentrations up to 50 $\mu g/dm^3$, meeting drinking water requirements. Kinetic investigations found that pseudo-second-order and Elovich models best described the sorption process, whereas equilibrium data suited well to the Freundlich isotherm ($R^2 = 0.991$), indicating multilayer adsorption with a Langmuir capacity of 0.75 mg/g (Fabre et al., 2020). Banana peels effectively removed $Hg(CN)_2$ from simulated gold mine effluents, attaining 94% removal in 90 minutes with monolayer adsorption defined by the Sips isotherm. However, excess free cyanide hindered efficiency (Aliprandini et al., 2023). These experiments show that banana peel can be used as an accessible and efficient biosorbent to detoxify Hg in water and mining effluents.

SCB-derived adsorbents treated with $ZnCl_2$ and H_3PO_4 show high effectiveness in removing Hg(II) from aqueous solutions. Batch adsorption investigations revealed more than 90% Hg uptake, with the process following pseudo-second-order kinetics. Adsorption occurs by complexation, electrostatic interactions, and π-electron interactions with functional groups on the adsorbent surface, including $-COO^-$, $-O^-$, C=C, and C=O. The maximum adsorption capacity exceeded 11.47 mg/g. These findings indicate that chemically activated SCB is a viable, low-cost, and ecologically acceptable solution for treating Hg-contaminated water (Giraldo et al., 2020). Key parameters such as solution pH, contact time, and temperature were investigated, and pH 4.0 resulted in the highest Hg removal (97.58%). The pseudo-second-order model best explained adsorption kinetics, although equilibrium data fitted both the Langmuir and Freundlich isotherms. The highest adsorption capacity was calculated to be 35.71 mg/g. Additional chemical treatments had no significant influence on Hg uptake, demonstrating SCB as an effective and low-cost biosorbent for Hg-contaminated water (Khoramzadeh et al., 2013).

8.5 Regeneration Techniques for Metal-loaded Adsorbents

Heavy metal removal from soil and wastewater has been extensively studied in recent years, with a focus on chemical adsorbents and biosorbents. However, little emphasis has been paid to recovering heavy metals from metal-loaded sorbents and then regenerating these sorbents for reuse. Only a few studies have focused on desorption processes and the efficacy of sorbent regeneration for further adsorption cycles, which are critical for long-term and cost-effective heavy metal remediation (Chatterjee and Abraham, 2019). Developing efficient recovery and reuse strategies could improve the environmental and economic viability of heavy metal removal technologies.

Adsorption regeneration and recovery are critical steps in wastewater treatment. Gupta and Ali discovered that Pb- and Cr-loaded columns may be regenerated with multiple eluents and reused for at least 10 cycles with no efficiency loss. This work successfully recovered bound metal ions from rice husk biomatrix with HCl or HNO_3. While regeneration is technically conceivable, it may not necessarily be required for rice husk, which is a cheap and abundant lignocellulosic material. According to Krishnani et al. (2008), rice husk may effectively remove heavy metals from wastewater at a reasonable cost through direct disposal after usage. Qiu et al. (2022) examined the effectiveness of regeneration solutions on activated carbon for five heavy metals: Pb^{2+}, Cu^{2+}, Zn^{2+}, Cd^{2+}, and Cr(VI). Deionized water had a poor desorption rate (<5%) due to physical forces like van der Waals interactions, while NaCl and NaOH showed marginally enhanced effects. HCl had the highest desorption efficiency, exceeding 85% for all metals. H^+'s rapid mobility, small ionic radius, and excellent ion-exchange capacity make it easy to replace adsorbed metal ions. NaOH showed limited desorption (~30%) due to decreased protonation and increased metal attraction. Overall, HCl is the most efficient regenerant for activated carbon.

After 10 cycles of adsorption–desorption, activated carbon treated with 6 M HCl lost only 13.3% of its Cu^{2+} adsorption capacity, with the majority of the loss occurring during the first cycle (Da'na and Awad, 2017). Carboxyl-functionalized adsorbents, such as MIL-121, use —COOH groups to bind metals, allowing for successful desorption even in mild acidic environments. MIL-121 exhibited 99% Cu^{2+}

desorption at pH 3.0, indicating good adsorption selectivity and simple regeneration, greatly exceeding other adsorbents needing stronger acids (Ji et al., 2022). Overall, controlling acid concentration and treatment time is critical for ensuring repeated adsorbent reuse without losing performance.

8.6 Conclusion

Heavy metal contamination in wastewater is a critical environmental and public health concern, as metals such as Pb, Cd, Cr, As, and Hg are toxic, nonbiodegradable, and persistent in ecosystems. This chapter highlights that biosorption using natural and low-cost bioadsorbents is a sustainable, efficient, and eco-friendly alternative to conventional treatment methods such as precipitation, ion exchange, and electroplating. Agricultural residues, industrial byproducts, and other biomass materials have demonstrated high adsorption efficiency due to functional groups like $-OH$, $-COOH$, $-NH_2$, and $-SH$. The adsorption performance of bioadsorbents depends on several factors, including solution pH, contact time, initial metal concentration, adsorbent dose, and temperature. Modification of biosorbents, either chemically or physically, can further improve their selectivity and capacity. The chapter also emphasizes the importance of regeneration techniques for metal-loaded biosorbents. Acidic solutions, particularly HCl, have shown effective desorption, allowing multiple reuse cycles while maintaining adsorption efficiency. Overall, the use of bioadsorbents provides a cost-effective, practical, and environmentally safe solution for heavy metal removal from wastewater. With careful optimization of adsorption parameters, structural modification of biosorbents, and regeneration strategies, these materials can play a significant role in mitigating industrial wastewater pollution.

References

Aggarwal, R., & Arora, G. (2020). Assessment of biosorbents for chromium removal from aqueous media. Materials Today: Proceedings, 28, 1540–1545.

Ahrouch, M., Gatica, J. M., Draoui, K., Bellido, D., & Vidal, H. (2019). Lead removal from aqueous solution by means of integral natural clays honeycomb monoliths. Journal of Hazardous Materials, 365, 519–530.

Akinhanmi, T. F., Ofudje, E. A., Adeogun, A. I., Aina, P., & Mayowa Joseph, I. (2020). Orange peel as low-cost adsorbent in the elimination of Cd(II) ions: kinetics, isotherm, thermodynamic and optimization evaluations. Bioresources and Bioprocessing, 7, 34.

Al-Enezi, G., Hamoda, M. F., & Fawzi, N. (2004). Ion exchange extraction of heavy metals from wastewater sludges. Journal of Environmental Science and Health. Part A, Toxic/Hazardous Substances & Environmental Engineering, 39(2), 455–464.

Ali, I. H., Al Mesfer, M. K., Khan, M. I., Danish, M., & Alghamdi, M. M. (2019). Exploring adsorption process of lead (II) and chromium -(VI) ions from aqueous solutions on acid activated carbon prepared from *Juniperus procera* leaves. Processes, 7(4), 217.

Aliprandini, P., Junior, A. B. B., Veiga, M. M., Marshall, B. G., Scarazzato, T., & Espinosa, D. C. R. (2023). Evaluation of biosorbents as an alternative for mercury cyanide removal from aqueous solution. Minerals Engineering, 204, 108431.

Allah, T. F., Jusuf, K., Shohatee, A. A., Feng, L., Trombetta, L., & Wang, E. (2023). Studies of adsorption of cadmium ions by biowaste adsorbent from aqueous solutions with ion-selective electrodes and ICP-OES. American Journal of Analytical Chemistry, 14, 274–286.

Ankush, Ritambhara, Lamba, S., Deepika, & Prakash, R. (2024). Cadmium in environment—an overview. In: Cadmium Toxicity in Water: Challenges and Solutions (pp. 3–20). Springer Nature.

Awang, N. A., Wan Salleh, W. N., Aziz, F., Yusof, N., & Ismail, A. F. (2023). A review on preparation, surface enhancement and adsorption mechanism of biochar-supported nano zero-valent iron adsorbent for hazardous heavy metals. Journal of Chemical Technology & Biotechnology, 98(1), 22–44.

Barnhart, J. (1997). Occurrences, uses, and properties of chromium. Regulatory Toxicology and Pharmacology, 26(1), S3–S7.

Bayuo, J. (2021). An extensive review on chromium (vi) removal using natural and agricultural wastes materials as alternative biosorbents. Journal of Environmental Health Science and Engineering, 19(1), 1193–1207.

Beckers, F., & Rinklebe, J. (2017). Cycling of mercury in the environment: sources, fate, and human health implications: a review. Critical Reviews in Environmental Science and Technology, 47(9), 693–794.

Bhran, A. A., Tadepalli, S., Murthy, K. S. R., & AlGhamdi, A. A. (2025). Biosorption and regeneration studies for Cu (II) and Cd (II) removal from industrial effluents using orange peel and composite adsorbents. Processes, 13(7), 1972.

Bilal, M., Ihsanullah, I., Shah, M. U. H., & Younas, M. (2021). Enhanced removal of cadmium from water using bio-sorbents synthesized from branches and leaves of *Capparis decidua* and *Ziziphus mauritiana*. Environmental Technology & Innovation, 24, 101922.

Bouida, L., Rafatullah, M., Kerrouche, A., Qutob, M., Alosaimi, A. M., Alorfi, H. S., & Hussein, M. A. (2022). A review on cadmium and lead contamination: sources, fate, mechanism, health effects and remediation methods. Water, 14(21), 3432.

Carpi, A. (1997). Mercury from combustion sources: a review of the chemical species emitted and their transport in the atmosphere. Water, Air, and Soil Pollution, 98(3), 241–254.

Casas, J. S., & Sordo, J. (2006). An overview of the historical importance, occurrence, isolation, properties and applications of lead. In: Lead (pp. 1–40). Elsevier Science BV.

Chatterjee, A., & Abraham, J. (2019). Desorption of heavy metals from metal loaded sorbents and e-wastes: a review. Biotechnology letters, 41(3), 319–333.

Chauhan, S. S., & Dikshit, P. K. S. (2023). Optimization of batch study parameters for the adsorption of lead (II) ions onto spent tea grains. Aqua – Water Infrastructure, Ecosystems and Society, 72(6), 996–1024.

Chigova, J. T., & Mudono, S. (2022). Adsorption of chromium (VI) using nano-ZnO doped scrap tire-derived activated carbon. Journal of Geoscience and Environment Protection, 10(9), 121–135.

Chrysochoou, M., Theologou, E., Bompoti, N., Dermatas, D., & Panagiotakis, I. (2016). Occurrence, origin and transformation processes of geogenic chromium in soils and sediments. Current Pollution Reports, 2(4), 224–235.

Chuah, T. G., Jumasiah, A., Azni, I., Katayon, S., & Choong, S. T. (2005). Rice husk as a potentially low-cost biosorbent for heavy metal and dye removal: an overview. Desalination, 175(3), 305–316.

Collin, M. S., Venkatraman, S. K., Vijayakumar, N., Kanimozhi, V., Arbaaz, S. M., Stacey, R. S., Anusha, J., Choudhary, R., Lvov, V., Tovar, G. I., Senatov, F., Koppala, S., & Swamiappan, S. (2022). Bioaccumulation of lead (Pb) and its effects on human: a review. Journal of Hazardous Materials Advances, 7, 100094.

Da'na, E., & Awad, A. (2017). Regeneration of spent activated carbon obtained from home filtration system and applying it for heavy metals adsorption. Journal of Environmental Chemical Engineering, 5(4), 3091–3099.

Elabbas, S., Mandi, L., Berrekhis, F., Pons, M. N., Leclerc, J. P., & Ouazzani, N. (2016). Removal of Cr(III) from chrome tanning wastewater by adsorption using two natural carbonaceous materials: eggshell and powdered marble. Journal of Environmental Management, 166, 589–595.

El-Kady, A. A., & Abdel-Wahhab, M. A. (2018). Occurrence of trace metals in foodstuffs and their health impact. Trends in Food Science & Technology, 75, 36–45.

Elkhaleefa, A., Ali, I. H., Brima, E. I., Shigidi, I., Elhag, A. B., & Karama, B. (2021). Evaluation of the adsorption efficiency on the removal of lead(II) ions from aqueous solutions using *Azadirachta indica* leaves as an adsorbent. Processes, 9(3), 559.

Fabre, E., Lopes, C. B., Vale, C., Pereira, E., & Silva, C. M. (2020). Valuation of banana peels as an effective biosorbent for mercury removal under low environmental concentrations. Science of the Total Environment, 709, 135883.

Fabre, E., Vale, C., Pereira, E., & Silva, C. M. (2021). Sustainable water treatment: use of agricultural and industrial wastes to remove mercury by biosorption. Water, Air, & Soil Pollution, 232(7), 284.

Feng, L., & Wang, E. (2025). Adsorption and recovery studies of cadmium and lead ions using biowaste adsorbents from aqueous solution. Separations, 12(1), 16.

Flora, S. J., Flora, G., & Saxena, G. (2006). Environmental occurrence, health effects and management of lead poisoning. In: Lead (pp. 158–228). Elsevier Science BV.

Garg, R., Garg, R., Sillanpää, M., Alimuddin, Khan, M. A., Mubarak, N. M., & Tan, Y. H. (2023). Rapid adsorptive removal of chromium from wastewater using walnut-derived biosorbents. Scientific Reports, 13(1), 6859.

Garg, U., Kaur, M. P., Jawa, G. K., Sud, D., & Garg, V. K. (2008). Removal of cadmium (II) from aqueous solutions by adsorption on agricultural waste biomass. Journal of Hazardous Materials, 154 (1–3), 1149–1157.

Gautam, K., Gupta, N., Patel, D. K., Singh, R. P., & Singh, A. (2023). Comparative study of simple and modified rice husk biochar for cadmium removal: Adsorption performance and possible mechanisms. Aqua – Water Infrastructure, Ecosystems and Society, 72(7), 1269–1287.

Gelaw, Y. B., Dagne, H., Adane, B., Yirdaw, G., Moges, M., Aneley, Z., Kumlachew, L., Aschale, A., Deml, Y. A., Tegegne, E., & Birhan, T. A. (2024). Adsorptive removal of cadmium (II) from wastewater using activated carbon synthesized from stem of *Catha edulis*. Heliyon, 10(22), e40389.

Genchi, G., Sinicropi, M. S., Lauria, G., Carocci, A., & Catalano, A. (2020). The effects of cadmium toxicity. International Journal of Environmental Research and Public Health, 17(11), 3782.

Giraldo, S., Robles, I., Ramirez, A., Flórez, E., & Acelas, N. (2020). Mercury removal from wastewater using agroindustrial waste adsorbents. SN Applied Sciences, 2(6), 1029.

Giri, D. D., Jha, J. M., Srivastava, N., Hashem, A., Abd Allah, E. F., Shah, M., & Pal, D. B. (2022). Sustainable removal of arsenic from simulated wastewater using solid waste seed pods biosorbents of Cassia fistula L. Chemosphere, 287, 132308.

Haider, B., Imran, M., Naeem, A., El-Beltagi, H. S., Arshad, H., Hussain, A., Zulfiqar, U., Rebouh, N. Y., Prasad, P. V., & Djalovic, I. (2025). Adsorptive removal of lead from wastewater using pressmud with evaluation of kinetics and adsorption isotherms. Scientific Reports, 15, 22823.

Hoareau, C. E., Hadibarata, T., & Yılmaz, M. (2022). Occurrence of cadmium in groundwater in China: a review. Arabian Journal of Geosciences, 15(17), 1455.

Hussain, M. K., Khatoon, S., Nizami, G., Fatma, U. K., Ali, M., Singh, B., Quraishi, A., Assiri, M. A., Ahamad, S., & Saquib, M. (2024). Unleashing the power of bio-adsorbents: efficient heavy metal removal for sustainable water purification. Journal of Water Process Engineering, 64, 105705.

Islam, M. M., Mohana, A. A., Rahman, M. A., Rahman, M., Naidu, R., & Rahman, M. M. (2023). A comprehensive review of the current progress of chromium removal methods from aqueous solution. Toxics, 11(3), 252.

Ji, C., Xu, M., Yu, H., Lv, L., & Zhang, W. (2022). Mechanistic insight into selective adsorption and easy regeneration of carboxyl-functionalized MOFs towards heavy metals. Journal of Hazardous Materials, 424, 127684.

Joshi, H. K., Vishwakarma, M. C., Kumar, R., Sharma, H., Joshi, S. K. (2022). Adsorption of Cd^{2+} from synthetic wastewater by modified leaves of *Eupatorium adenophorum* and *Acer oblongum*: thermodynamics, kinetics and equilibrium studies. Discover Water 2, 9.

Kaprara, E. A., Zouboulis, A. I., Simeonidis, K. T., & Mitrakas, M. G. (2015). Potential application of inorganic sulfur reductants for Cr (VI) removal at sub-ppb level. Desalination and Water Treatment, 54(8), 2067–2074.

Kaur, J., Sengupta, P., & Mukhopadhyay, S. (2022). Critical review of bioadsorption on modified cellulose and removal of divalent heavy metals (Cd, Pb, and Cu). Industrial & Engineering Chemistry Research, 61(5), 1921–1954.

Kayranli, B., Gok, O., Yilmaz, T., Gok, G., Celebi, H., Seckin, I. Y., & Mesutoglu, O. C. (2022). Low-cost organic adsorbent usage for removing Ni^{2+} and Pb^{2+} from aqueous solution and adsorption mechanisms. International Journal of Environmental Science and Technology, 19(5), 3547–3564.

Khatoon, A., Uddin, M. K., & Rao, R. A. K. (2018). Adsorptive remediation of Pb(II) from aqueous media using *Schleichera oleosa* bark. Environmental Technology & Innovation, 11, 1–14.

Khoramzadeh, E., Nasernejad, B., & Halladj, R. (2013). Mercury biosorption from aqueous solutions by sugarcane bagasse. Journal of the Taiwan Institute of Chemical Engineers, 44(2), 266–269.

Krishnani, K. K., Meng, X., Christodoulatos, C., & Boddu, V. M. (2008). Biosorption mechanism of nine different heavy metals onto biomatrix from rice husk. Journal of Hazardous Materials, 153(3), 1222–1234.

Kwikima, M. M., Mateso, S., & Chebude, Y. (2021). Potentials of agricultural wastes as the ultimate alternative adsorbent for cadmium removal from wastewater. A review. Scientific African, 13, e00934.

Lata, S., Singh, P. K., & Samadder, S. R. (2015). Regeneration of adsorbents and recovery of heavy metals: a review. International Journal of Environmental Science and Technology, 12(4), 1461–1478.

Liu, B., Xin, Y. N., Zou, J., Khoso, F. M., Liu, Y. P., Jiang, X. Y., Peng, S., & Yu, J. G. (2023). Removal of chromium species by adsorption: fundamental principles, newly developed adsorbents and future perspectives. Molecules, 28, 639.

Moubarik, A., & Grimi, N. (2015). Valorization of olive stone and sugar cane bagasse by-products as biosorbents for the removal of cadmium from aqueous solution. Food Research International, 73, 169–175.

Mudassar, H., Hina, K., Ghani, U., Afzaal, Q., Shah, A. A., Shaffique, S., & Elansary, H. O. (2025). Adsorptive removal of Pb^{2+} from wastewater using ZnO-Biochar nanocomposite. Scientific Reports, 15(1), 29517.

Negi, A. (2024). Adsorptive removal of Cd(II) ions from wastewater using biosorbent of *Tectona grandis* leaves: kinetic, equilibrium and thermodynamic studies (Version 1). Research Square.

Netpae, W., Chaiya, R., Phalakornkule, C., & Sompornpisut, P. (2018). Biosorption efficiency of chromium (VI) from aqueous solution by NaOH-pretreated biomass of *Humicola phialophoroides* fungus. Applied Water Science, 8, 1–11.

Nguyen, T. P., Nguyen, V. D., Trinh, M. T., Han, P. L., Nguyen, T. H., Nguyen, M. T., & Vu, A.-T. (2025). Effective biosorptive removal of Pb^{2+} ions from wastewater using modified lettuce leaves: a novel sustainable and eco-friendly biosorbent. Journal of Hazardous Materials Advances, 19, 100770.

Omar, S., Muhamad, M. S., Te Chuan, L., Hadibarata, T., & Teh, Z. C. (2019). A review on lead sources, occurrences, health effects, and treatment using hydroxyapatite (HAp) adsorbent made from fish waste. Water, Air, & Soil Pollution, 230(12), 275.

Pal, D. B., Saini, R., Srivastava, N., Ahmad, I., Alshahrani, M. Y., & Gupta, V. K. (2022). Waste biomass based potential bioadsorbent for lead removal from simulated wastewater. Bioresource Technology, 349, 126843.

Pal, P., Ghosh, S. K., Mondal, S., & Maiti, T. K. (2025). Lead (Pb^{2+}) biosorption and bioaccumulation efficiency of *Enterobacter chuandaensis* DGI-2: isotherm, kinetics and mechanistic study for bioremediation. Journal of Hazardous Materials, 492, 138017.

Pang, Y., Zhao, C., Li, Y., Li, Q., Bayongzhong, X., Peng, D., & Huang, T. (2022). Cadmium adsorption performance and mechanism from aqueous solution using red mud modified with amorphous MnO2. Scientific Reports, 12, 4424.

Pant, B. D., Neupane, D., Paudel, D. R., Chandra Lohani, P., Gautam, S. K., Pokhrel, M. R., & Poudel, B. R. (2022). Efficient biosorption of hexavalent chromium from water by modified arecanut leaf sheath. Heliyon, 8(4), e09283.

Parlayici, Ş., & Pehlivan, E. (2019). Comparative study of Cr(VI) removal by bio-waste adsorbents: equilibrium, kinetics, and thermodynamic. Journal of Analytical Science and Technology, 10, 15.

Patel, K. S., Pandey, P. K., Martín-Ramos, P., Corns, W. T., Varol, S., Bhattacharya, P., & Zhu, Y. (2023). A review on arsenic in the environment: contamination, mobility, sources, and exposure. RSC Advances, 13(13), 8803–8821.

Poonam, Bharti, S. K., & Kumar, N. (2018). Kinetic study of lead (Pb^{2+}) removal from battery manufacturing wastewater using bagasse biochar as biosorbent. Applied Water Science, 8, 119.

Priya, A. K., Gnanasekaran, L., Dutta, K., Rajendran, S., Balakrishnan, D., & Soto-Moscoso, M. (2022). Biosorption of heavy metals by microorganisms: evaluation of different underlying mechanisms. Chemosphere, 307, 135957.

Qasem, N. A., Mohammed, R. H., & Lawal, D. U. (2021). Removal of heavy metal ions from wastewater: a comprehensive and critical review. Npj Clean Water, 4(1), 36.

Qiu, L., Suo, C., Zhang, N., Yuan, R., Chen, H., & Zhou, B. (2022). Adsorption of heavy metals by activated carbon: effect of natural organic matter and regeneration methods of the adsorbent. Desalination and Water Treatment, 252, 148–166.

Raji, Z., Karim, A., Karam, A., & Khalloufi, S. (2023, September). Adsorption of heavy metals: mechanisms, kinetics, and applications of various adsorbents in wastewater remediation—a review. Waste, 1(3), 775–805.

Rezaei, M., Pourang, N., & Moradi, A. M. (2022). Removal of lead from aqueous solutions using three biosorbents of aquatic origin with the emphasis on the affective factors. Scientific Reports, 12(1), 751.

Rzig, B., Guesmi, F., Sillanpää, M., & Hamrouni, B. (2021). Modelling and optimization of hexavalent chromium removal from aqueous solution by adsorption on low-cost agricultural waste biomass using a response surface methodological approach. Water Science and Technology, 84(3), 552–575.

Saeed, A., Akhter, M. W., & Iqbal, M. (2005). Removal and recovery of heavy metals from aqueous solution using papaya wood as a new biosorbent. Separation and Purification Technology, 45(1), 25–31.

Saha, R., Nandi, R., & Saha, B. (2011). Sources and toxicity of hexavalent chromium. Journal of Coordination Chemistry, 64(10), 1782–1806.

Sahmoune, M. N. (2016). The role of biosorbents in the removal of arsenic from water. Chemical Engineering & Technology, 39(9), 1617–1628.

Saikaew, W., & Kaewsarn, P. (2009). Cadmium ion removal using biosorbents derived from fruit peel wastes. Songklanakarin Journal of Science & Technology, 31(5), 547–554.

Sekaran, M., & Thiagarajan, K. (2025). Green strategies for chromium detoxification in tannery effluents: adsorption and bioconversion with *Portieria hornemannii*. Environmental Technology & Innovation, 38, 104151.

Shabbir, Z., Shahid, M., Natasha, Khalid, S., Khalid, S., Imran, M., Qureshi, M.I., & Niazi, N. K. (2023). Use of agricultural bio-wastes to remove arsenic from contaminated water. Environmental Geochemistry and Health, 45(8), 5703–5712.

Shakoor, M. B., Niazi, N. K., Bibi, I., Shahid, M., Saqib, Z. A., Nawaz, M. F., Shaheen, S. M., Wang, H., Tsang, D. C., Bundschuh, J., & Rinklebe, J. (2019). Exploring the arsenic removal potential of various biosorbents from water. Environment International, 123, 567–579.

Shrivastava, A., Ghosh, D., Dash, A., & Bose, S. (2015). Arsenic contamination in soil and sediment in India: sources, effects, and remediation. Current Pollution Reports, 1(1), 35–46.

Singh, S., Kapoor, D., Khasnabis, S., Singh, J., & Ramamurthy, P. C. (2021). Mechanism and kinetics of adsorption and removal of heavy metals from wastewater using nanomaterials. Environmental Chemistry Letters, 19(3), 2351–2381.

Tafakori, V., Zadmard, R., Tabandeh, F., Amoozegar, M. A., & Ahmadian, G. (2017). Equilibrium isotherm, kinetic modeling, optimization, and characterization studies of cadmium adsorption by surface-engineered *Escherichia coli*. Iranian Biomedical Journal, 21(6), 380–391.

Taha, M. F., Shuib, A. S., Shaharun, M. S., & Borhan, A. (2014). Removal of Ni(II), Zn(II) and Pb(II) ions from single metal aqueous solution using rice husk-based activated carbon. AIP Conference Proceedings, 1621(1), 210–217.

Teshale, F., Karthikeyan, R., & Sahu, O. (2020). Synthesized bioadsorbent from fish scale for chromium (III) removal. Micron, 130, 102817.

Wang, B., Sun, Y. C., & Sun, R. C. (2019). Fractionational and structural characterization of lignin and its modification as biosorbents for efficient removal of chromium from wastewater: a review. Journal of Leather Science and Engineering, 1(1), 1–25.

Wang, G., Zhang, S., Yao, P., Chen, Y., Xu, X., Li, T., & Gong, G. (2015). Removal of Pb(II) from aqueous solutions by *Phytolacca americana* L. biomass as a low-cost biosorbent. Arabian Journal of Chemistry, 11(1), 99–110.

Wang, J., & Chen, C. (2009). Biosorbents for heavy metals removal and their future. Biotechnology Advances, 27(2), 195–226.

Worku, Z., Tibebu, S., Nure, J. F., Tibebu, S., Moyo, W., Ambaye, A. D., & Nkambule, T. T. (2023). Adsorption of chromium from electroplating wastewater using activated carbon developed from water hyacinth. BMC Chemistry, 17, 85.

Wysokowski, M., Klapiszewski, Ł., Moszyński, D., Bartczak, P., Szatkowski, T., Majchrzak, I., Siwińska-Stefańska, K., Bazhenov, V. V., & Jesionowski, T. (2014). Modification of chitin with kraft lignin and development of new biosorbents for removal of cadmium (II) and nickel (II) ions. Marine Drugs, 12(4), 2245–2268.

Zhang, L., & Wong, M. H. (2007). Environmental mercury contamination in China: sources and impacts. Environment International, 33(1), 108–121.

Zhang, Y., Hou, D., O'Connor, D., Shen, Z., Shi, P., Ok, Y. S., Tsang, D., Wen, Y., & Luo, M. (2019). Lead contamination in Chinese surface soils: Source identification, spatial-temporal distribution and associated health risks. Critical Reviews in Environmental Science and Technology, 49(15), 1386–1423.

Zhao, M., Xu, Y., Zhang, C., Rong, H., & Zeng, G. (2016). New trends in removing heavy metals from wastewater. Applied Microbiology and Biotechnology, 100(15), 6509–6518.

Zhou, Y., Zhang, Z., Zhang, J., & Xia, S. (2016). New insight into adsorption characteristics and mechanisms of the biosorbent from waste activated sludge for heavy metals. Journal of Environmental Sciences, 45, 248–256.

9

Bioadsorbents in Emerging Contaminants Removal

9.1 Introduction

Water pollution is a global issue that harms ecosystems and human health. Industrialization, urbanization, and agriculture all emit pollutants into bodies of water, damaging both surface and groundwater. Pharmaceutically active chemicals (PACs), such as antibiotics and hormones, enter the environment through industrial, hospital, and domestic sources. These pollutants can reach humans directly through drinking water or indirectly through crops, livestock, and aquatic food systems. PACs are hazardous to aquatic organisms and endanger environmental stability. Their long lifespan in wastewater treatment plants demonstrates the limitations of conventional treatment approaches. Antibiotics and hormones, which are now designated as pollutants, offer considerable dangers due to the unknown long-term effects. Effective and sustainable solutions, such as biosorbents, are urgently required for their removal (Sahay et al., 2023).

Pesticides, plasticizers, antimicrobials, and flame retardants are all examples of endocrine-disrupting chemicals (EDCs), which are extensively used but pose substantial health risks. These substances disrupt hormonal balance, leading to reproductive and developmental issues (Casals-Casas and Desvergne, 2011). In vitro, in vivo, and epidemiological studies have found correlations between EDC exposure and obesity, metabolic syndrome, and type 2 diabetes. Understanding population-level dangers necessitates prediction approaches based on individual effects. Growing research reveals endocrine disruption mechanisms and their links to environmental and human health consequences (Kabir et al., 2015). The growing issue highlights the critical need for EDC-related risk assessment, monitoring, and mitigation techniques to ensure long-term health and environmental safety.

Disinfectants, perfumes, insect repellents, preservatives, and UV filters are all common household items. Unlike pharmaceuticals, they are applied externally and frequently enter the environment unaltered. Many are persistent, bioactive, and can bioaccumulate, raising environmental issues (Brausch and Rand, 2011). Similarly, per- and polyfluoroalkyl substances (PFAS) are highly stable compounds employed in a variety of industries. Their fluorinated structures last for geologic timescales, dispersing through soil, water, and air before accumulating to dangerous levels. Although remediation is possible, it is still costly and difficult, necessitating safer alternatives and the mitigation of previously released toxins (Evich et al., 2022). Microplastics (MPs) pose new environmental and health hazards. Smaller fragments are more numerous and easily adsorbed, with particles as small as 20 μm capable of passing through cell membranes. Their presence has an impact on organisms' eating, metabolism, reproduction, and behavior, but more research is needed to draw reliable inferences. Humans are exposed through seafood, drinking water, and even indoor spaces, where polymeric materials are prevalent. Sampling, characterization, and measuring long-term ecological repercussions continue to be significant issues. To address the issue, global pollution prevention, advances in degradable polymers and additives, and reduced plastic usage through enhanced recycling and reuse are required (Hale et al., 2020).

Current approaches for detecting new pollutants rely on complex laboratory techniques like HPLC and GC-MS, which are costly and require specialized workers. Immunological assays, such as ELISA, are also utilized, although only in laboratory settings. These limits limit large-scale monitoring. Portable, sensitive, and field-deployable technologies could overcome these constraints, allowing for rapid on-site detection of new contaminants in water, air, and food. Affordable technologies would especially help resource-constrained areas, enhancing public health and environmental safety (Hassan et al., 2022).

Conventional wastewater treatment systems are not designed to remove emerging pollutants, which frequently appear in trace amounts within complex mixes (Morin-Crini et al., 2022). Sophisticated procedures such as ozonation remove drugs like carbamazepine and caffeine by more than 90%, while membrane filtration and sophisticated oxidation processes are also highly efficient. However, these procedures are expensive and may generate dangerous byproducts. Sorption-based technologies provide safer, low-energy options, and activated carbon is frequently employed for developing contaminant removal. However, its fossil origin and cost remain negatives. Biosorbents generated from lignocellulosic biomass and bio-waste offer cost-effective and sustainable alternatives (Vasić et al., 2023).

Cork and biomass wastes have demonstrated excellent potential as low-cost biosorbents for developing pollutants. Granulated cork effectively eliminated phenols, medicines, and personal care items, with 100% efficiency for pentachlorophenol, diclofenac, and triclosan. Adsorption was quick, reaching equilibrium in 30 minutes, and followed the Freundlich and Langmuir models, with efficiency connected to substituent electronegativity (Mallek et al., 2018). Similarly, biomass wastes, including invasive plant roots and tannin-rich residues (coffee and green tea grounds), have a high UV-filter, insecticide, and herbicide biosorption rate. Removal was greater than 88% within 2 hours and remained efficient in saline water. Tannins and pollutant solubility were important performance indicators (Deyris et al., 2023).

The growing presence of emerging contaminants (ECs) such as pharmaceuticals, endocrine disruptors, personal care products, PFAS, and MPs poses severe environmental and health challenges due to their persistence, bioaccumulation, and toxicity. Conventional treatment methods are often inadequate, costly, or generate harmful byproducts. Bioadsorbents, derived from natural or waste biomass, offer sustainable, low-cost, and eco-friendly alternatives for EC removal. This chapter reviews major contaminant classes, their impacts, and challenges in detection and treatment. The novelty lies in linking bioadsorbent performance with specific contaminant groups, providing updated insights into recent advances, limitations, and future directions for scalable, sustainable water treatment solutions.

9.2 Emerging Contaminants

ECs have become a major environmental concern due to their widespread use in medications, insecticides, surfactants, plasticizers, and personal care products. These chemicals are difficult to remove in the short term and remain generally uncontrolled, while efforts in the EU, Canada, and Switzerland have attempted to minimize their release from wastewater treatment facilities (WWTPs). Continuous discharge creates "pseudo-persistence," even in the absence of inherent persistence, as WWTP residues infiltrate water, soil, and sediments. Their low concentrations, complex matrices, and variable chemical characteristics make detection, fate prediction, and ecotoxicological risk assessment challenging (Khan et al., 2022). Figure 9.1 gives the classification of ECs (pharmaceuticals, EDCs, PCPs, PFAS, MPs).

9.2.1 Pharmaceutical and Personal Care Products

Pharmaceuticals are chemicals or mixtures of chemicals that are given to people or animals in order to alter physiological and metabolic processes by means of pharmacological or immunological effects. The biologically active ingredients in medications are known as active pharmaceutical ingredients (APIs), and they are frequently mixed with excipients to improve solubility and adsorption.

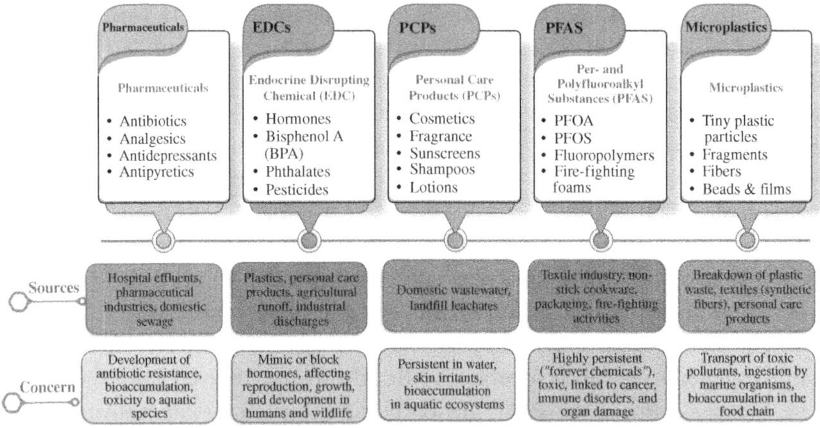

Figure 9.1 Classification of emerging contaminants (pharmaceuticals, EDCs, PCPs, PFAS, microplastics).

When administered, both medications and APIs cause biological effects that alter biological processes. Pharmaceuticals are often made up of organic compounds with hydrocarbon groups and molecular architectures that have both hydrophilic and hydrophobic areas. Their behavior, persistence, and effects on biological systems and the environment are all influenced by these chemical properties (Kayode-Afolayan et al., 2022).

In aquatic habitats, researchers from ten Latin American countries discovered 159 pharmacological active chemicals (PhACs), predominantly analgesics and anti-inflammatories (Valdez-Carrillo et al., 2020). High levels of carbamazepine (830 μg/L) in Ecuador and ethinylestradiol (6.8 μg/L) in Brazil exceed global averages, suggesting significant environmental hazards. Ecotoxicological tests show that these concentrations frequently exceed safe environmental thresholds, and there is no substantial difference between untreated and treated wastewater. Wastewater discharge is still the predominant source of PhAC contamination, highlighting the critical need for effective, low-cost treatment methods. Improved wastewater management strategies and regulatory frameworks are required to reduce pharmaceutical pollution in Latin America's aquatic environments.

Pharmaceuticals may penetrate aquatic systems in a number of ways, but the main one is through wastewater discharge from treatment facilities that are unable to fully eliminate these substances. Human and animal waste products end up in waterways, where they bioaccumulate in fish and other aquatic life. Common medications like ciprofloxacin, sulfamethoxazole, and carbamazepine have been found in surface waters and sediments, according to studies. Untreated wastewater increases bacterial resistance to antibiotics, endangering ecosystems and public health. Leachates from landfills and agricultural runoff that contains veterinary medications are two more sources. To avoid pharmaceutical contamination and protect aquatic ecosystems, efficient wastewater management and ongoing monitoring are essential (Khan and Barros, 2023).

9.2.2 Endocrine Disruptors

Emerging pollutants known as EDCs are present in aquatic habitats at concentrations that can have an impact on both people and wildlife. These include synthetic substances like pesticides, medications, and compounds produced from plastics (phthalates, bisphenol A), as well as natural hormones like estrogen, 17β-estradiol, and estriol. Agricultural runoff and wastewater discharge are the main ways that EDCs get into water systems. Advanced technologies are required since conventional treatment plants are unable to remove them properly. Although some studies only reveal weak associations, exposure to EDCs is connected to conditions like diabetes, infertility, thyroid dysfunction, and neurodegenerative diseases; this highlights the need for more research (Pironti et al., 2021).

Exogenous substances known as EDCs interfere with hormone function and impair vital physiological functions, such as development, reproduction, metabolism, and homeostasis. They work by changing hormone synthesis, metabolism, or receptor activation, or by imitating or inhibiting natural hormones. EDCs vary in structure from phenolic substances (like flavonoids, bisphenols, and PAHs) to non-phenolic substances (like phthalates and perfluorinated chemicals). Pharmaceuticals, insecticides, plasticizers, and industrial waste are some of their origins. Receptor binding, signal transduction interference, and epigenetic modifications are among the ten main disruptive characteristics of EDCs. Interestingly, two prevalent aquatic EDCs that are closely associated with endocrine and reproductive diseases are nonylphenol and bisphenol A (Gałązka and Jankiewicz, 2022).

9.2.3 Polyfluoroalkyl Substances

A broad class of synthetic chemicals known as PFAS is distinguished by a carbon–fluorine backbone and functional groups like amines, carboxylic or sulfonic acids, or both. The structure of certain PFAS, such as ether-types, also includes oxygen atoms. Originally referred to as perfluorinated chemicals (PFCs), the acronym "PFAS" was used to cover polyfluorinated compounds and prevent confusion with perfluorinated carbons. Although they persist in the environment and present health risks, PFAS are frequently utilized as fluorosurfactants and fluorochemicals because of their stability and hydrophobicity

(Gaines, 2023). PFAS have been used extensively in more than 200 commercial and consumer applications, such as surfactants, waterproof textiles, firefighting foams, nonstick cookware, food packaging, and semiconductors, due to their remarkable chemical and thermal durability. Because of this widespread use, thousands of PFAS variations are now in circulation around the world and can be found in water, soil, sediments, plants, and animals. Through the food chain, PFAS bioaccumulate and harm humans and wildlife by causing endocrine disruption, liver damage, immune system malfunction, and cancer. Screening initiatives and mounting data point to PFAS as important pollutants of growing concern (Zhang et al., 2022).

To comprehend the fate, transportation, and transformation of historical and next-generation PFAS in the environment, ongoing monitoring is necessary. The interactions of legacy PFAS, notably PFCAs and PFSAs, with environmental variables like pH, temperature, and co-contaminants are clarified. Their worldwide expansion is facilitated by atmospheric transport; larger molecules exhibit greater bioaccumulation but less mobility. Environmental conditions have a significant impact on PFAS degradation, with anaerobic and marine environments exhibiting slower breakdown. Although incomplete destruction can produce more PFAS byproducts, destructive treatment approaches, whether thermal or nonthermal, are necessary since natural degradation frequently produces new PFAS (Evich et al., 2022).

9.2.4 Microplastics

MPs, which are tiny plastic particles with a size range of 1 μm to less than 5 mm, can be classified as primary (made small) or secondary (produced by degradation). Only 7% of the 360 million tons of plastic that are manufactured each year are recycled, which causes a significant buildup in freshwater habitats. These particles are harmful to the environment and human health because they contaminate aquatic systems and get into food chains. MPs endanger aquatic life and water quality by serving as transporters of harmful chemicals and contaminants. Microplastic pollution affects all facets of freshwater habitats and is a major worldwide concern due to its tenacity and universality (Thacharodi et al., 2024). With concentrations ranging from less than 1 to more than 1,000 MPs/m^3 outside and up to 1,583 $\pm$ 1,181 MPs/m^3 indoors, MPs are increasingly acknowledged as airborne pollutants. They are mostly made of polyethylene terephthalate, polyethylene, and polypropylene. The proximity to sources is shown by the higher deposition rates indoors (475–19,600 MPs/m^2/day) compared to outdoors (0.5–1,357 MPs/m^2/day). Variable microplastic loads are also present in snow and road dust (2–477 MPs/g and 0.1–30,000 MPs/L, respectively). Evidence supports widespread air microplastic pollution with substantial indoor accumulation and the possibility of inhalation exposure, despite methodological differences amongst research (O'Brien et al., 2023).

Both natural and man-made sources contribute to the production of MPs. Secondary MPs come from the breakdown of bigger plastics, but primary MPs, such as microbeads and synthetic fibers, are purposefully created. These pollutants can now be detected in both ocean trenches and mountain peaks. MPs damage aquatic organisms and disturb aquatic ecosystems when paired with chemicals like phthalates and bisphenol A. MPs hinder human reproduction, digestion, and enzyme function while also fostering antibiotic resistance. Therefore, it is imperative to address MP pollution in aquatic environments (Ali et al., 2024).

9.3 Impact of Emerging Contaminants

Pharmaceuticals can persist in aquatic environments because they are very water-soluble and resistant to dissociation. Their ability to penetrate biological barriers and stay stable in organisms, despite their intended physiological effects, poses ecotoxicological dangers to nontarget species. Human medication orthologs have been found to have an impact on 35% of green algae, 61% of daphnids, and 86% of zebrafish. In aquatic products, antibiotics like erythromycin, oxytetracycline, and sarafloxacin have been found; in vegetables, ciprofloxacin and enrofloxacin range from 2.5 to 27.5 and 2.0 to 32.3 mg/kg, respectively. The immune and reproductive systems are harmed, hormone control is disturbed, and these residues bioaccumulate. Significant risks to the environment and human health are further highlighted by the fact that nanoparticles associated with pharmaceutical pollutants cause oxidative stress, cytotoxicity, and organ dysfunction (Kumar et al., 2022).

ECs can have a negative impact on animal and human health by causing endocrine and reproductive problems. Antibiotics and antibiotic-resistant genes, which are produced by bacteria that are resistant to them, are among the main ECs that pose a major risk to public health and ecosystems (Puri et al., 2023). Because contaminants of emerging concern (CECs) readily accumulate in plant roots, shoots, fruits, and grains, human exposure mostly happens through the ingestion of contaminated plant-based foods. Furthermore, CECs—which include medications and MPs—cause genotoxicity, oxidative stress, and growth suppression in aquatic and plant species. The majority of CECs are still unregulated by environmental policies, despite their significant ecological dangers (Maddela et al., 2022).

Because of their overuse in medical and animal husbandry, antibiotic contamination is becoming a greater global concern (Okoye et al., 2022). Aquatic system contamination disturbs microbial ecosystems and encourages the development of genes resistant to antibiotics. Even with advancements in removal methods, problems like secondary pollution and exorbitant expenses still exist. A One Health strategy that combines environmental, animal, and human health via interdisciplinary cooperation is necessary to address new pollutants (Wang et al., 2024). Sustainable solutions can be strengthened by incorporating Indigenous worldviews that emphasize interdependence. Innovations in green chemistry, machine learning, and education are crucial to reducing environmental and public health hazards in addition to addressing legacy pollutants.

Because they interfere with wildlife's hormonal systems, ECs, particularly EDCs, pose a threat to ecosystems. These chemicals evade sewage treatment and continue to exist in water bodies across the globe. While industrial byproducts cause mortality and decrease population stability, pharmaceuticals interfere with the reproduction of aquatic animals. Even in far-flung Arctic regions, persistent pollutants like PBDEs, PFAS, DDT, bisphenol A, and heavy metals are now found. An example of EC toxicity is the sharp drop in the vulture population in India. To preserve ecosystem resilience and safeguard biodiversity, detection and removal methods must be strengthened (Li et al., 2024).

Both human health and aquatic ecosystems are significantly impacted by pharmaceutical contaminants. Even at trace levels, persistent substances like hormones, antidepressants, and antibiotics harm aquatic life. Research has indicated that progestin levonorgestrel lowers rainbow trout fertility, whereas antidepressants slow down frog growth. These pollutants reduce fish populations, cause genetic mutations, and change endocrine functions. Antibiotic residues disrupt nutrient cycling and ecological balance by increasing bacterial resistance. Furthermore, higher creatures, including humans, are at risk for chronic toxicity due to the bioaccumulation of persistent pharmaceutical residues throughout the food chain. Water quality, ecosystem health, and biodiversity are all under risk due to the ongoing release of pharmaceuticals and personal care products (PPCPs). To lessen their long-term environmental effects, sustainable production practices and cutting-edge wastewater treatment technologies must be developed (Samal et al., 2022).

An important factor in the morbidity and mortality rates worldwide is pharmaceutical resistance. Antibiotic-resistant genes are inserted into mobile genetic components, including integrons, plasmids, and transposons, in a large number of bacterial strains found in aquatic systems. This allows for horizontal gene transfer and mutations in natural microbial communities. Microbial resistance is further increased by the presence of efflux pumps, enzymatic changes, and strains that produce extended-spectrum beta-lactamases. Persistent pollutants are introduced into aquatic habitats, landfill leachates, and sediments by the anthropogenic discharge of PPCPs from industrial, municipal, and medical sources. Even at low μg/L levels, chronic exposure causes oxidative stress, endocrine disruption, and carcinogenic consequences in living things. Global detection of persistent PPCPs such as bisphenol A, atrazine, and carbamazepine highlights the urgent need for monitoring and long-term mitigation (Chaturvedi et al., 2021).

9.4 Challenges in Detection and Treatment

The persistence, bioaccumulation, and potential toxicity of ECs in wastewater pose an increasing threat to the environment and public health. The development of sophisticated and sustainable treatment technologies is required because conventional wastewater treatment methods are mainly inefficient in totally eliminating these contaminants (Arman et al., 2021). Notwithstanding significant advancements in recent years, issues still exist, such as the requirement for economical and energy-efficient solutions, regulatory restrictions, and inadequate monitoring systems. To get beyond these obstacles, interdisciplinary cooperation, consistent research, and strong policy execution are necessary. Such initiatives will protect long-term water resource sustainability and environmental health while guaranteeing efficient control of developing pollutants (Corpus et al., 2024). Because of its limited ability to remove ECs and its reliance on socioeconomic factors, conventional wastewater treatment is still insufficient. Numerous physical, chemical, and biological techniques have been investigated; nevertheless, no single strategy guarantees total contamination elimination while preserving economic viability and environmental sustainability. To improve water quality, hybrid systems that combine several treatment techniques are being created (Mahmood et al., 2022).

Analytical detection of ECs in complex matrices faces multiple challenges, including low concentration levels, matrix interferences, and transformation products that hinder quantification (Figure 9.2). The accurate identification of ECs, which represent major risks to ecosystems and human health, is essential to maintaining water quality and safety. Low-cost, field-deployable sensors provide a practical way to monitor huge amounts of water and detect contaminants on-site. However, their high prices, poor reliability, and difficulties in practical application prevent their widespread acceptance. According to Ateia et al. (2024), inadequate stakeholder involvement and a lack of standardized validation are two major technical and nontechnical obstacles that prevent advancement past proof-of-concept. Improving sensor sensitivity, selectivity, robustness, and usability while incorporating user requirements, financial considerations, and data-sharing frameworks is necessary to meet these difficulties. Enhancing workforce development and academia-industry collaboration is crucial to the advancement of useful, scalable, and significant water quality sensor technology.

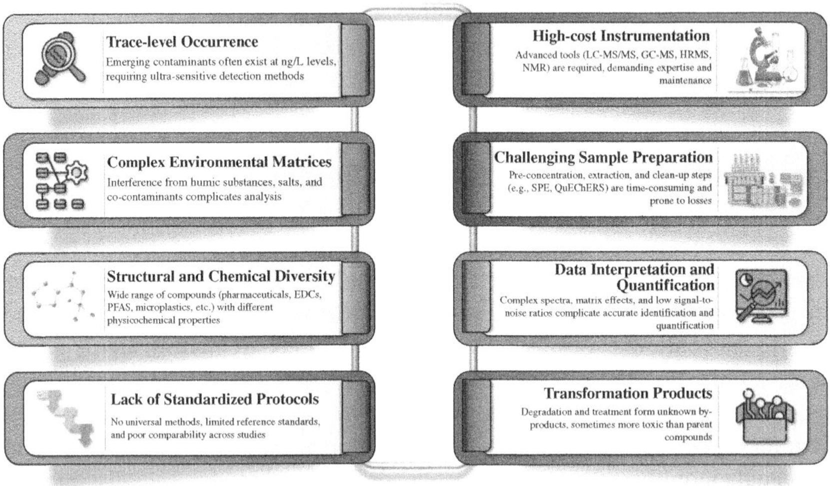

Figure 9.2 Analytical challenges in detection of emerging contaminants in complex matrices.

9.5 Removal of Emerging Contaminants Using Bioadsorbent

Advanced bioadsorbents can be made from agricultural waste to remove ECs from water. High porosity, large surface area, and partial graphitization are characteristics of biochars made by thermal or microwave-assisted pyrolysis that improve π–π interactions with pollutants like diclofenac and chloroquine. Surface acidity and catalytic activity are enhanced by metal oxide functionalization (ZnO, FeO_4, CuO) and chemical activation using $ZnCl_2$, KOH, or H_3PO_4. Through Fenton-type processes, these hybrid adsorbents allow for the partial breakdown of contaminants like ciprofloxacin in addition to their adsorption. With capacities greater than 300 mg/g, adsorption processes include electrostatic forces, ion exchange, and hydrogen bonding. There are still issues with stability, regeneration, and large-scale application despite success in the lab. These bioadsorbents provide affordable, environmentally friendly solutions that are consistent with the ideas of the circular economy (Gonçalves et al., 2025).

9.5.1 Pharmaceutical and Personal Care Products

The health of people and the environment is seriously threatened by ECs, such as pharmaceuticals, personal care items, and EDCs. Up to 100% of EDCs, pesticides, and medications can be eliminated by advanced oxidation processes, while biological therapies, such as anaerobic membrane bioreactors, fungi, and microalgae, have the potential to remove PPCPs. Adsorption is still a viable and efficient method, particularly when low-cost bio-based adsorbents are used. Produced by hydrothermally carbonizing or pyrolyzing agricultural wastes such as sugarcane bagasse, biomass-derived activated carbon provides an economical alternative with a high adsorption effectiveness. Because of their high surface area and reactivity, nano-adsorbents further improve the removal of contaminants (Sethi, 2024).

Research has examined into using tomato waste (TW) as an affordable biosorbent to remove pharmaceuticals from water, including tylosin, febantel, procaine, praziquantel, and dexamethasone. The porous, heterogeneous nature of TW, which is rich in lipids, phosphates, sugars, and amides, was validated by chemical and microstructural analysis using FTIR, EDS, SEM, and porosimetry. Linear, Freundlich, and Dubinin–Radushkevich (D–R) isotherms were used to model sorption studies at 298–308 K, showing spontaneous and primarily physical adsorption. While ΔH° values demonstrated exothermic sorption for the majority of medicines, with the exception of febantel, negative ΔG° values verified spontaneity. Diffusion analysis (Weber–Morris and Boyd models) showed integrated intraparticle and film diffusion control, while kinetic data best matched the pseudo-second-order model (Mutavdžić Pavlović et al., 2021).

The widely accessible jamun seed (*Syzygium cumini*) was carbonized and chemically activated to create an affordable bioadsorbent that can remove the pharmaceutical contaminant ofloxacin from aqueous solutions. According to BET analysis, the activated jamun seed carbon had a high surface area of 366.39 m^2/g, while CHNS examination showed that it contained 55.68% carbon. Particle size analysis, FTIR, and FESEM characterization verified its porous structure and adsorption-promoting functional groups. At pH 5, ACJS had a maximum adsorption capacity of 39.97 mg/g and a 96% removal efficiency after 3.5 hours. A pseudo-second-order model ($R^2 = 0.9999$) was used in the kinetic studies, and monolayer adsorption was suggested by Langmuir isotherm fitting. An exothermic process was found to be possible using thermodynamic analysis. Over five reuse cycles, the adsorbent maintained its effectiveness (Mallik, 2023).

In order to remove the pharmaceutical contaminant carnitine from aqueous media, this study examined the potential of jujube (*Ziziphus mauritiana*) as an environmentally friendly biosorbent. Adsorption-friendly crystallinity, porous shape, and a large surface area (156 m^2/g) were all displayed by the Jujube biosorbent. Box-Behnken Design (BBD) and artificial neural network (ANN) models were optimized, and the results showed that ANN offered better prediction accuracy with higher R^2 and lower error values. With a maximal capacity of 162 mg/g, the Langmuir isotherm ($R^2 = 0.9667$) suggested monolayer adsorption, and the pseudo-second-order kinetic model ($R^2 = 0.9993$) verified chemisorption. According to thermodynamic study, adsorption was both spontaneous and exothermic ($\Delta H^\circ = -3.52$ kJ/mol). Jujube's cost-effectiveness and reusability for sustainable water cleanup were confirmed by the material's continued efficiency after six cycles (Nagappan et al., 2025). The adsorption efficiency of different bioadsorbents reported in recent studies is summarized in Table 9.1.

When an appropriate biosorbent is developed, the biosorption process provides an effective and environmentally responsible way to remove pharmaceutical residues from wastewater. Because of their efficacy and sustainability, the use of renewable biomaterials—like microbial biomass and natural polymers—has drawn a lot of interest in the last 10 years. According to recent research, these biosorbents, which have been created or altered by a variety of processes, have great adsorption capabilities and a high level of contamination tolerance, which makes them more cost-effective than traditional techniques for treating wastewater that is rich in organic materials. However, research is still being done to improve porosity and create composite materials that combine natural polymers with inorganic or microbial components to increase their adsorptive efficiency (Rusu et al., 2023).

9.5.2 Endocrine Disruptors

When discharged into aquatic environments, EDCs, which are found in medications, cosmetics, and pesticides, pose serious dangers to human health and the environment. Research into sustainable biological alternatives is prompted by the fact that conventional treatment approaches are frequently insufficient for their elimination. Through processes like photodegradation, bioadsorption, bioaccumulation, and biodegradation, photosynthetic microorganisms have effectively removed EDCs due to their high tolerance to harsh environments, which makes them suitable for large-scale wastewater treatment applications (Narindri Rara Winayu et al., 2024). Furthermore, with adsorption controlled by pseudo-second-order kinetics and exothermic behavior, natural waste materials such as Mytella falcata shells have demonstrated encouraging adsorption performance, reaching up to 80% removal efficiency for 17-α-methyltestosterone (Henrique et al., 2021). These environmentally friendly methods show promise for long-term EDC cleanup.

Agricultural waste offers economical and environmentally beneficial options for decontaminating water. Using batch and fixed-bed techniques,

Table 9.1 Adsorption performance of bioadsorbents for pharmaceutical chemical removal.

Bioadsorbent source	Emerging contaminant	Modification/ Treatment	Experimental conditions	Adsorption capacity (mg/g)	Adsorption isotherm	Adsorption kinetics	References
Multiwalled carbon nanotube	Ciprofloxacin	Oxidation treatment using sodium hypochlorite	Initial concentration (IC)—150 mg/L Time—60 min pH—4 Temp.—298 K Adsorbent dosage (AD)—20 mg	206	Dubinin–Radushkevich and Langmuir isotherm	Pseudo-second-order kinetic model	Yu et al. (2016)
Wheat grains	Amoxicillin	Chemically modified with tartaric acid	IC—4 mg/L Time—5 min pH—7 Temp.—25 °C AD—4 g	31.3	Temkin isotherm	Pseudo-second-order kinetic model	Boukhelkhal et al. (2016)
Multiwalled carbon nanotube	Tetracycline	Oxidation (2–5.9% O_2 content) using sodium hypochlorite	IC—20 mg/L Time—120 min pH—6.5 Temp.—25 °C AD—0.05 g/L	217.8–269.25	Langmuir isotherm	Not specified	Yu et al. (2014)
Pistachio shell powder	Ciprofloxacin	Dried at 105 °C overnight, grinding, and sieving Coated with ZnO nanoparticles	IC—60 mg/L Time—120 min pH—5 Temp.—25 °C AD—0.1 g/100 mL	92.45	Freundlich isotherm	Pseudo-second-order kinetic model	Mohammed et al. (2019)

(Continued)

Table 9.1 *(Continued)*

Bioadsorbent source	Emerging contaminant	Modification/ Treatment	Experimental conditions	Adsorption capacity (mg/g)	Adsorption isotherm	Adsorption kinetics	References
Pistachio shell powder coated with ZnO nanoparticles (CPS)	Tetracycline (TEC)	Dried at 105 °C overnight, grinding, and sieving Coated with ZnO nanoparticles	IC– 60 mg/L Time—120 min pH—5 Temp.—25 °C AD– 0.1 g/100 mL	98.72	Freundlich isotherm	Pseudo-second-order kinetic model	Mohammed et al. (2019)
Pistachio shell powder coated with ZnO nanoparticles (CPS)	Amoxicillin (AMO)	Dried at 105 °C overnight, grinding, and sieving Coated with ZnO nanoparticles	IC—60 mg/L Time—120 min pH—5 Temp.—25 °C AD—0.1 g/100 mL	132.24	Langmuir isotherm	Pseudo-second-order kinetic model	Mohammed et al. (2019)
Activated carbon from walnut wood	Cephalexin	Chopped, Soaked (85% phosphoric acid for 24 hour at room temperature) Dried at 650 °C for 2 h Coated with cobalt ferrite nanoparticles	IC—50 mg/L Time—30 min pH—3 Temp.—25 °C AD—2 g/L	42.37	Freundlich isotherm	Pseudo-second-order kinetic model	Askari et al. (2021)

Carbon nanoparticles from vine wood	Amoxicillin	Activated using NaOH	IC—20 mg/L Time—8 h pH—2 Temp.—45°C AD—0.4 g/L	2.69	Langmuir isotherm	Pseudo-second-order kinetic model	Pouretedal and Sadegh (2014)
Carbon nanoparticles from vine wood	Cephalexin	Activated using NaOH	IC—20 mg/L Time—8 h pH—2 Temp.—45°C AD– 0.4 g/L	7.08	Langmuir isotherm	Pseudo-second-order kinetic model	Pouretedal and Sadegh (2014)
Carbon nanoparticles from vine wood	Tetracycline	Activated using NaOH	IC—20 mg/L Time—8 h pH—2 Temp.—45°C AD—0.4 g/L	1.98	Langmuir isotherm	Pseudo-second-order kinetic model	Pouretedal and Sadegh (2014)
Carbon nanoparticles from vine wood	Penicillin G	Activated using NaOH	IC—20 mg/L Time—8 h pH—2 Temp.—45°C AD—0.4 g/L	8.41	Langmuir isotherm	Pseudo-second-order kinetic model	Pouretedal and Sadegh (2014)
Biochar from waste sludge/leaf	—	Pyrolysis at 200–500°C Dried, grinded and sieved	IC—10 mg/L Time—4.5 h pH—6.5 Temp.—25°C AD—0.005 g	877	Temkin isotherm	pseudo-second-order kinetic model	Zhang et al. (2020)

(Continued)

Table 9.1 *(Continued)*

Bioadsorbent source	Emerging contaminant	Modification/ Treatment	Experimental conditions	Adsorption capacity (mg/g)	Adsorption isotherm	Adsorption kinetics	References
Multiwall carbon nanotube	Ibuprofen (Ibu)	Acid-functionalized using H_2O_2 (30% v/v)	IC—20 mg/L Time—30 min pH—4 Temp.—25 °C AD—0.002 g	1.15	Langmuir isotherm	Pseudo-second-order kinetic model	Hanbali et al. (2020)
Activated carbon from *Albizia lebbeck* seed pods	Cephalexin (CFX)	Microwave-assisted chemical activation with KOH	IC—20 mg/L Time—90 min pH—7 Temp.—303 K Adsorbent Dose—0.5 g/L	137.02	Langmuir isotherm	Pseudo-second-order kinetic model	Ahmed and Theydan (2012)
Alternanthera philoxeroides Biochars	Metformin (MF)	Heated at 300 °C in N2 environment for 2 h Surface modified using 15% H_2O_2	IC—0.5 mmol/L pH—7 Temp.—45 °C	56.17	Freundlich isotherm	Pseudo-second-order kinetic model	Shearer et al. (2022)

Vieira et al. (2022) created a biochar from wasted mushroom substrate (SMS) and assessed how well it removed endocrine disruptors like progesterone and 17α-ethinylestradiol (EE2) from water. With a removal effectiveness of more than 80%, biochar that was pyrolyzed at 600 °C demonstrated the highest porosity and the best adsorption capabilities. With maximum adsorption capacities of 232.64 mg/g for progesterone and 138.98 mg/g for EE2, the adsorption was conducted according to the Freundlich model and the Langmuir model, respectively. According to these results, SMS-derived biochar shows great efficacy and reusability in water purification applications, making it a feasible and sustainable material for hormone elimination (Vieira et al., 2022).

According to Mathew and Saravanakumar (2023), the functional groups, surface area, and porosity of green-synthesized nanomaterials have superior adsorption capabilities. An environmentally acceptable substitute for traditional treatments is the adsorption process, which is controlled by electrostatic, hydrogen bonding, and π–π interactions. Research also emphasizes how variables like temperature, adsorbent dosage, and pH affect adsorption efficiency. By turning waste into useful adsorbents for pollutant remediation, biosorption's incorporation within a circular bioeconomy framework fosters sustainability. Loffredo et al. (2021) looked at the possibility of digestate, a solid byproduct of anaerobic digestion, as a biosorbent for removing EDCs from water. The study looked at four pollutants: metribuzin, boscalid, bisphenol A, and 4-tert-octylphenol. SEM and FTIR analysis indicated a porous surface with functional groups that promote adsorption. Sorption kinetics showed pseudo-first-order and pseudo-second-order fits depending on the chemical, but isotherm modeling used Freundlich and Langmuir equations. The digestate has significant sorption affinity, particularly for hydrophobic substances, with KOC values up to 2,180 L/kg. Slow and incomplete desorption suggested a significant adsorbate binding. Overall, digestate displayed promising efficiency as a low-cost, long-lasting biosorbent for EDC elimination in water treatment. The adsorption efficiency of various bioadsorbents for endocrine disruptor removal by biosorbent has been summarized in Table 9.2.

Water treatment sludge is a common byproduct that is often discarded without being valorized. Martins et al. (2022) investigated its effectiveness as an adsorbent for removing endocrine disruptors, including 17β-estradiol and 17α-ethinylestradiol. The sludge was physically and chemically modified with phosphoric acid and potassium hydroxide to create adsorbents with increased surface area and porosity. The primary ingredients were silica, quartz, and kaolinite, with a pHpzc of around 6. The optimal adsorption conditions were pH 5.5, 0.5 g dose, and 240 minutes of contact duration. The adsorption

Table 9.2 Adsorption performance of bioadsorbents for endocrine disruptors removal.

Bioadsorbent source	Emerging contaminant	Modification/ Treatment	Experimental conditions (pH, contact time, temp.)	Adsorption capacity (mg/g)	Adsorption isotherm	Adsorption kinetics	References
Spent mushroom substrate (biochar)	17α-ethinylestradiol (EE2)	Pyrolyzed at 600 °C (20 min)	pH 9.56, 24 h, 25 °C	138.98	Freundlich isotherm	Pseudo-second-order kinetic model	Vieira et al. (2022)
Spent mushroom substrate (biochar)	Progesterone	Pyrolyzed at 600 °C (20 min)	pH 9.56, 24 h, 25 °C	232.64	Langmuir isotherm	Pseudo-first-order kinetic model	Vieira et al. (2022)
Lotus seedpod biochar	17β-estradiol (E2)	KOH-treated	pH 10, 20 h, 27 °C	100.6	Langmuir isotherm	Pseudo-second-order kinetic model	Liu et al. (2019)
Graphene-like magnetic sawdust biochar	17β-estradiol (E2)	Pyrolyzed at 500 °C for 2 hour (nitrogen atmosphere) Activated using 0.1 M K2FeO4	pH 5, 15 h, 298 K	133.45	Langmuir isotherm	Pseudo-first-order kinetic model	Zhou et al. (2020b)
Eucalyptus globulus wood (biochar)	Bisphenol A (BPA)	Phosphoric acid activation @ 600 °C, functionalized	pH 3.0–3.5, 42–48 h, 25 °C	4764	Langmuir isotherm	Pseudo-second-order kinetic model	Ahmed et al. (2018)
Eucalyptus globulus wood (biochar)	4-tert-butylphenol (4tBP)	Phosphoric acid activation @ 600 °C, functionalized	pH 3.0–3.5, 42–48 h, 25 °C	4567	Langmuir isotherm	Pseudo-second-order kinetic model	Ahmed et al. (2018)

Eucalyptus globulus wood (biochar)	Estrone (E1)	Phosphoric acid activation @ 600 °C, functionalized	pH 3.0–3.5 (max.), 42–48 h, 25 °C	7588	Langmuir isotherm	Pseudo-second-order kinetic model	Ahmed et al. (2018)
Paper mill sludge (biochar)	Pentachlorophenol (PCP)	Ni-ZVI impregnation (0.5 wt% Ni, ZVI immobilized biochar, 700 °C pyrolysis)	pH 6, 240 min 25 °C	Up to 50	Langmuir isotherm	Pseudo-first-order kinetic model	Devi and Saroha (2015)
Rice husk silica	Estrone (E1)	Heated at 800 °C for 4 h, Grinded and sieved Acid leaching with hydrochloric and sulfuric acid mixture	pH 4, 60 min, 25 °C	56.82	Langmuir isotherm	Pseudo-first-order kinetic model	Zarghi et al. (2019)
Rice husk silica	17β-estradiol (E2)	Heated at 800 °C for 4 h, Grinded and sieved Acid leaching with hydrochloric and sulfuric acid mixture	pH 4, 60 min, 25 °C	62.5	Langmuir isotherm	Pseudo-first-order kinetic model	Zarghi et al. (2019)
Chitosan nanoparticles	Estrogen	None	pH 7.3, 220 min, room temp	5.79	Langmuir isotherm	Pseudo-second-order kinetic model	Davarnejad et al. (2023)
Chitosan/Al_2O_3-*hydroxyapatite* nanocomposite beads	Estradiol	Composite beads: *chitosan* + Al_2O_3 + hydroxyapatite	pH 6, 4 h, *room temp*	39.78	Langmuir *isotherm*	Pseudo-second-*order kinetic* model	Li et al. *(2020)*

(Continued)

Table 9.2 *(Continued)*

Bioadsorbent source	Emerging contaminant	Modification/ Treatment	Experimental conditions (pH, contact time, temp.)	Adsorption capacity (mg/g)	Adsorption isotherm	Adsorption kinetics	References
Chitosan/Al_2O_3-hydroxyapatite nanocomposite beads	Chrysoidine	Composite beads: chitosan + Al_2O_3 + hydroxyapatite	pH 6, 4 h, room temp	23.26	Langmuir isotherm	Pseudo-second-order kinetic model	Li et al. (2020)
Bamboo-derived biochar supported with $CuZnFe_2O_4$ nanoparticles (CZF–biochar)	Bisphenol A (BPA)	One-pot hydrothermal synthesis of CZF NPs on biochar, pyrolyzed at 550 °C	pH 7, 150 min, 25 °C	263.2	Freundlich isotherm	Pseudo-second-order kinetic model	Heo et al. (2019)
Spanish broom cellulose fibers (SB fibers)	Bisphenol A (BPA)	Surface functionalized via 4,4'-diphenylmethane diisocyanate (MDI) coating, hydrophobization (solvent-free)	pH 5, 15 min, 25 °C	~4.86	Langmuir isotherm	Pseudo-second-order kinetic model	Tursi et al. (2018)
Magnesium silicate (Mg/Si = 1:6)	4-Nonylphenol (4-NP)	Synthesized by sol–gel method at 60 °C, amorphous mesoporous structure with high surface area (117.7 m^2/g)	pH 7, contact time ~300 min, temperature 25 °C (optimized); dose 0.2 g/L	30.84	Sips model	Pseudo-second-order kinetic model	Yan et al. (2022)

followed chemisorption pathways and corresponded to both Langmuir and Freundlich isotherms. The study found that modified water treatment sludge is a cost-effective and environmentally friendly material for EDC elimination, enabling circular waste usage.

9.5.3 Polyfluoroalkyl Substances

PFAS, also known as "forever chemicals," remain in the environment because of their strong carbon–fluorine linkages and resistance to degradation. Their widespread use and toxicity have caused serious environmental and health issues. Short-chain PFAS are particularly problematic because of their high solubility and inadequate removal by traditional adsorbents such as activated charcoal and ion-exchange resins. Bioadsorbents made from renewable biomass have recently acquired popularity as low-cost and environmentally friendly alternatives. Functionalized bioadsorbents improve PFAS removal effectiveness via electrostatic and hydrophobic interactions, providing a green, efficient, and scalable strategy for water purification (Li et al., 2021).

A recent study found that innovative cationic biopolyamides (C-bioPAs) derived from renewable bisfuran-based monomers may effectively remove perfluoroalkyl and polyfluoroalkyl compounds (PFAS) from water. C-bioPAs produced by polycondensation and quaternization with alkyl bromides were highly biocompatible, stable, and reusable. The structural and thermal properties were confirmed through characterization using NMR, FTIR, GPC, TGA, and DSC. The C-bioPAs obtained up to 99.8% PFAS removal effectiveness, with rapid sorption rates (>2.964 mmol/g/h) and short equilibrium durations (<12 hours). Furthermore, they maintained high effectiveness after five reuse cycles, demonstrating their durability and potential as an environmentally benign alternative to traditional PFAS adsorbents (Shu et al., 2025).

PFAS are extremely persistent, bioaccumulative contaminants that pose health and environmental hazards. Omo-Okoro et al. (2021) studied the removal of PFAS from aqueous medium with silver nanocomposite-activated carbons derived from maize tassel biomass. Two adsorbents, physically activated maize tassel silver (PAMTAg) and chemically activated maize tassel silver (CAMTAg), were created and characterized using SEM, FTIR, and BET analysis. Following the Langmuir and Freundlich isotherms, CAMTAg had the maximum adsorption capabilities for PFOS at 454.1 mg/g and PFOA at 321.2 mg/g, respectively. The adsorption was spontaneous and largely exothermic, regulated by electrostatic and hydrophobic interactions, indicating that maize-tassel-based nanocomposites are potential, environmentally benign alternatives for PFAS remediation (Omo-Okoro et al., 2021).

9.5.4 Microplastics

MPs are emerging pollutants that endanger aquatic ecosystems and human health. Moghri et al. (2025) explored the use of untreated pine (*Pinus eldarica*) pollen grains as a long-lasting, chemical-free biosorbent for MP removal from water. The porous and functionalized surface of the pollen allowed for effective adsorption through hydrophobic and electrostatic interactions. The biosorbent had removal efficiency of 66.2% for polystyrene (PS), 87% for PVC, and 95.2% for PET MPs. Adsorption occurred along the Freundlich isotherm ($R^2 = 0.977$), showing diverse active sites. Surfactants increased removal efficiency to about 98%, and the material preserved over 50% efficiency after four reuse cycles, indicating high reusability and environmental potential (Moghri et al., 2025).

Anthropogenic activities, particularly the improper disposal of post-consumer plastics, are mostly responsible for the increasing contamination of soil and water. Bhagwat et al. (2024) investigated the potential of natural hydrophobic cattail (*Typha latifolia*) fibers (CFs) as sustainable bioadsorbents for the removal of microplastic particles (MPPs) from wastewater. The study used batch adsorption tests to investigate the spontaneous sorption of five MPPs—PVC, PP, LDPE, HDPE, and Nylon 6—in two distinct adsorption environments: industrial wastewater (PW) and distilled water. The results indicated greater removal efficiency (89–100%) in PW for PVC, PP, LDPE, and HDPE, with Nylon 6 adsorption improved by 50%. The study concluded that hydrophobic interactions predominantly cause MPP adsorption, establishing CFs as environmentally acceptable, efficient, and low-cost wastewater treatment solutions.

Oliveira et al. (2023) found that surfactant-modified bioadsorbents can effectively remove MPs (<0.5 mm) from wastewater. The study included coconut bagasse, sugarcane bagasse, banana peel bagasse, bentonite, and diatomite as adsorbents, which were modified with a Tween 80-based microemulsion system. The microemulsion zone was characterized using the pseudoternary phase diagram, and surface characterization validated stability using surface tension, droplet size, and pH measurements. Batch adsorption investigations demonstrated that surfactant-modified adsorbents removed MPs with up to 80% effectiveness, beating unmodified adsorbents. The proposed mechanism involves enhanced hydrophobic interactions and surface modification that increase adsorption efficiency (Figure 9.3). The comparative adsorption performance of various bioadsorbents used for personal care products and PFAS removal is summarized in Table 9.3, highlighting the potential of eco-friendly and low-cost materials in microplastic remediation.

Table 9.3 Adsorption performance of bioadsorbents for removal of Personal care products and polyfluoroalkyl substances.

Bioadsorbent source	Emerging contaminant (EC)	Application of EC	Modification/ Treatment	Experimental conditions (pH, contact time, temp.)	Adsorption capacity (mg/g)	Adsorption isotherm	Adsorption kinetics	References
MIL-101-OH (functionalized MOF)	PPCPs: naproxen, ibuprofen, oxybenzone	Removal from aqueous solution	Hydroxyl functionalization (—OH groups)	pH—5.4 Time 30 min—12 h Temp.—25 °C	Up to ~365 mg/g	Langmuir isotherm	Pseudo-second-order kinetic model	Seo et al. (2016)
Raw and chemically modified agricultural waste (e.g., rice straw, coconut shell, banana peel)	PPCPs, dyes, pharmaceuticals	Wastewater treatment	Acid, alkali, salt treatment; pyrolysis	pH range varies (2–9), contact times mostly 1–24 h, temp 25–55 °C	16.7–441 mg/g	Langmuir isotherm	Pseudo-second-order kinetic model	Tang (2024)
Carbonized lignin bio-based adsorbent	PFAS: PFBA, PFBS, PFOA	Removal from water	Carbonization to increase surface area	Conditions typical for aqueous batch: details vary	1.6–5.3 ng/g	Freundlich isotherm	—	Roman et al. (2025)
Biochars from pinewood and date seeds	PFAS (PFHxA, GenX, PFOA, PFOS, PFDA, PFTeDA)	Removal from aqueous solutions	Raw and modified biochars	pH not always specified, contact times standard batch	Up to ~188 mg/g for PFAS	Langmuir/ piecewise isotherms	—	Papini et al. (2024)

(Continued)

Table 9.3 *(Continued)*

Bioadsorbent source	Emerging contaminant (EC)	Application of EC	Modification/ Treatment	Experimental conditions (pH, contact time, temp.)	Adsorption capacity (mg/g)	Adsorption isotherm	Adsorption kinetics	References
Miscanthus lignin (raw)	PFAS: PFBA, PFBS, PFOA, PFOS	PFAS removal from water	None (raw lignin)	pH ~5.3 Time ~100 hours	~0.009 (PFBA) to 0.082 (PFOS) L/g	Mainly linear isotherm	—	Roman et al. (2025)
Carbonized miscanthus lignin	PFAS: PFBA, PFBS, PFOA, PFOS	PFAS removal from water	Carbonization at 600 °C in N2 atmosphere	pH ~5.3 Time ~100 hours	3.6 L/g (PFOS)	Langmuir isotherm	—	Roman et al. (2025)
Activated carbon	PFAS: PFBA, PFBS, PFOA, PFOS	PFAS removal from water	Commercial granular activated carbon	pH ~5.3 Time ~100 hours	160 L/g (PFAS)	Langmuir isotherm	—	Roman et al. (2025)
Banana peel activated carbon (BPAC)	Polycyclic aromatic hydrocarbons (naphthalene, fluorene, phenanthrene)	Removal of PAHs from aqueous systems, industrial and wastewater treatment	Chemical activation with phosphoric acid, carbonization at 200 °C and activation at 500 °C	pH 7, Time 80 min, Temp. 25 ± 2 °C	Naphthalene: 333.33 Fluorene: 285.71 Phenanthrene: 217.39	Freundlich isotherm	—	Gupta and Gupta (2016)
OP with DIC Treatment (OP-DIC)	Phenol	Removal of phenol from aqueous solutions	Physical treatment by Instant Controlled Pressure Drop (DIC)	pH 7 Temp. 25 °C	15.23	Langmuir isotherm	—	Romero-Cano et al. (2016)

Waste fish scales	Personal care products, dyes	Wastewater treatment from seafood processing	Fe_3O_4 nanoparticles modification	pH not specified, 90 min, ambient temp	Up to 291 mg/g	Sips, Langmuir	Pseudo-second-order kinetic model	Tang (2024)
Chitosan beads (CS)	Naphthalene (NAP)	Removal from aqueous solutions	Impregnation with surfactants TX100, CTAB, SDS	pH 6 Time 24 h Temp. 30 °C	CS bead alone: 3.56 mg/g	Chapman sigmoidal	—	Chatterjee et al. (2010)
TX100-impregnated chitosan beads (TCB)	Naphthalene (NAP)	Removal from aqueous solutions	Impregnated with 0.5 g/L bead volume (BV) Triton X-100	pH 6 Time 24 h Temp. 30 °C	12.06	Chapman sigmoidal	—	Chatterjee et al. (2010)
CTAB-impregnated chitosan beads (CCB)	Naphthalene (NAP)	Removal from aqueous solutions	Impregnated with 0.5 g/L BV cetyltrimethyl ammonium bromide	pH 6 Time 24 h Temp. 30 °C	12.52	Chapman sigmoidal	—	Chatterjee et al. (2010)
SDS-impregnated chitosan beads (SCB)	Naphthalene (NAP)	Removal from aqueous solutions	Impregnated with 0.8 g/L BV sodium dodecyl sulfate	pH 6 Time 24 h Temp. 30 °C	8.32	Chapman sigmoidal	—	Chatterjee et al. (2010)

(Continued)

Table 9.3 *(Continued)*

Bioadsorbent source	Emerging contaminant (EC)	Application of EC	Modification/ Treatment	Experimental conditions (pH, contact time, temp.)	Adsorption capacity (mg/g)	Adsorption isotherm	Adsorption kinetics	References
UiO-66	Diflunisal, salicylic acid, carbamazepine	Removal from aqueous solutions	Zirconium-based MOF synthesized by solvothermal method, no functionalization	pH 5 (diflunisal), contact time 1–10 h, Temp. 26 °C	Diflunisal: 109.5 Salicylic acid: (not specified) Carbamazepine: (less than UiO-66-NH2)	Langmuir isotherm	Pseudo-second-order kinetic model	Zhou et al. (2020a)
UiO-66-NH2	Diflunisal, Salicylic acid, Carbamazepine	Removal from aqueous solutions	Amino-functionalized UiO-66 MOF synthesized by solvothermal method	pH 5 (diflunisal), contact time 1–10 h, Temp. 26 °C	Diflunisal: 136.6 Salicylic acid: (higher than UiO-66) Carbamazepine: (lower than UiO-66)	Langmuir isotherm	Pseudo-second-order kinetic model	Zhou et al. (2020a)

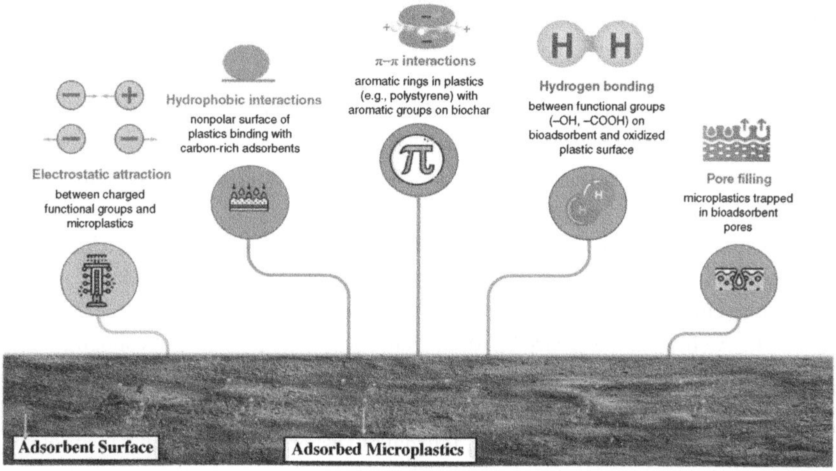

Figure 9.3 Adsorption mechanism of microplastics on bioadsorbent surfaces.

9.6 Conclusion

The persistence of ECs such as pharmaceuticals, endocrine disruptors, personal care products, PFAS, and MPs poses significant threats to aquatic ecosystems and human health. Conventional treatment technologies often fail to eliminate these pollutants effectively due to their complex chemical structures and resistance to degradation. Bioadsorbents derived from natural materials, including agricultural residues, plant fibers, biopolymers, and biochar, have shown great potential as sustainable, low-cost, and environmentally friendly alternatives for contaminant removal. Their high surface area, tunable functional groups, and biodegradability enhance adsorption efficiency through electrostatic, hydrophobic, and π–π interactions. Despite substantial progress, further research is required to optimize bioadsorbent modification, regeneration, and large-scale application. Future advancements integrating bioadsorption with hybrid systems and predictive modeling could revolutionize water purification technologies, promoting circular and green economy principles for sustainable environmental management.

References

Ahmed, M. B., Zhou, J. L., Ngo, H. H., Johir, M. A. H., & Sornalingam, K. (2018). Sorptive removal of phenolic endocrine disruptors by functionalized biochar: competitive interaction mechanism, removal efficacy and application in wastewater. Chemical Engineering Journal, 335, 801–811.

Ahmed, M. J., & Theydan, S. K. (2012). Adsorption of cephalexin onto activated carbons from Albizia lebbek seed pods by microwave-induced KOH and K_2CO_3 activations. Chemical Engineering Journal, 211–212, 200–207.

Ali, N., Khan, M. H., Ali, M., Ahmad, S., Khan, A., Nabi, G., Ali, F., Bououdina, M., & Kyzas, G. Z. (2024). Insight into microplastics in the aquatic ecosystem: properties, sources, threats and mitigation strategies. Science of the Total Environment, 913, 169489.

Arman, N. Z., Salmiati, S., Aris, A., Salim, M. R., Nazifa, T. H., Muhamad, M. S., & Marpongahtun, M. (2021). A review on emerging pollutants in the water environment: existences, health effects and treatment processes. Water, 13(22), 3258.

Askari, R., Afshin, S., Rashtbari, Y., Moharrami, A., Mohammadi, F., Vosuoghi, M., & Dargahi, A. (2021). Synthesis of activated carbon from walnut wood and magnetized with cobalt ferrite ($CoFe_2O_4$) and its application in removal of cephalexin from aqueous solutions. Journal of Dispersion Science and Technology, 44(7), 1183–1194.

Ateia, M., Wei, H., & Andreescu, S. (2024). Sensors for emerging water contaminants: overcoming roadblocks to innovation. Environmental Science & Technology, 58(6), 2636–2651.

Bhagwat, K. P., Rodrigue, D., & Romero-Zerón, L. (2024). Effective removal of microplastic particles from wastewater using hydrophobic bio-substrates. Pollutants, 4(2), 231–250.

Boukhelkhal, A., Benkortbi, O., Hamadache, M., Ghalem, N., Hanini, S., & Amrane, A. (2016). Adsorptive removal of amoxicillin from wastewater using wheat grains: equilibrium, kinetic, thermodynamic studies and mass transfer. Desalination and Water Treatment, 57(56), 27035–27047.

Brausch, J. M., & Rand, G. M. (2011). A review of personal care products in the aquatic environment: environmental concentrations and toxicity. Chemosphere, 82(11), 1518–1532.

Casals-Casas, C., & Desvergne, B. (2011). Endocrine disruptors: from endocrine to metabolic disruption. Annual Review of Physiology, 73(1), 135–162.

Chatterjee, S., Lee, D. S., Lee, M. W., & Woo, S. H. (2010). Enhanced molar sorption ratio for naphthalene through the impregnation of surfactant into chitosan hydrogel beads. Bioresource Technology, 101(12), 4315–4321.

Chaturvedi, P., Shukla, P., Giri, B. S., Chowdhary, P., Chandra, R., Gupta, P., & Pandey, A. (2021). Prevalence and hazardous impact of pharmaceutical and personal care products and antibiotics in environment: a review on emerging contaminants. Environmental Research, 194, 110664.

Corpus, R. M. B., Bayani, M. S., Aguilar, J. L., Aguilar, J. C. B., & Aguilar, H. B. (2024, July). Emerging pollutants in waste water: challenges and advancements in treatment technology. IOP Conference Series: Earth and Environmental Science, 1372(1), 012037.

Davarnejad, R., Sarvmeili, K., Safari, Z., & Kennedy, J. F. (2023). Estrogen adsorption from an aqueous solution on the chitosan nanoparticles. International Journal of Biological Macromolecules, 237, 124224.

Devi, P., & Saroha, A. K. (2015). Simultaneous adsorption and dechlorination of pentachlorophenol from effluent by Ni–ZVI magnetic biochar composites synthesized from paper mill sludge. Chemical Engineering Journal, 271, 195–203.

Deyris, P. A., Pelissier, F., Grison, C. M., Hesemann, P., Petit, E., & Grison, C. (2023). Efficient removal of persistent and emerging organic pollutants by biosorption using abundant biomass wastes. Chemosphere, 313, 137307.

Evich, M. G., Davis, M. J., McCord, J. P., Acrey, B., Awkerman, J. A., Knappe, D. R., Lindstrom, A. B., Speth, T. F., Tebes-Stevens, C., Strynar, M. J., Wang, Z., Weber, E.J., Henderson, W. M., & Washington, J. W. (2022). Per-and polyfluoroalkyl substances in the environment. Science, 375(6580), eabg9065.

Gaines, L. G. (2023). Historical and current usage of per-and polyfluoroalkyl substances (PFAS): a literature review. American Journal of Industrial Medicine, 66(5), 353–378.

Gałązka, A., & Jankiewicz, U. (2022). Endocrine disrupting compounds (nonylphenol and bisphenol A)–sources, harmfulness and laccase-assisted degradation in the aquatic environment. Microorganisms, 10(11), 2236.

Gonçalves, J. O., Leones, A. R., de Farias, B. S., da Silva, M. D., Jaeschke, D. P., Fernandes, S. S., Ribeiro, A. C., Cadaval Jr, T. R. S., & Pinto, L. A. D. A. (2025). A comprehensive review of agricultural residue-derived bioadsorbents for emerging contaminant removal. Water, 17(14), 2141.

Gupta, H., & Gupta, B. (2016). Adsorption of polycyclic aromatic hydrocarbons on banana peel activated carbon. Desalination and Water Treatment, 57(20), 9498–9509.

Hale, R. C., Seeley, M. E., La Guardia, M. J., Mai, L., & Zeng, E. Y. (2020). A global perspective on microplastics. Journal of Geophysical Research: Oceans, 125(1), e2018JC014719.

Hanbali, G., Jodeh, S., Hamed, O., Bol, R., Khalaf, B., Qdemat, A., & Samhan, S. (2020). Enhanced ibuprofen adsorption and desorption on synthesized functionalized magnetic multiwall carbon nanotubes from aqueous solution. Materials, 13(15), 3329.

Hassan, M. H., Khan, R., & Andreescu, S. (2022). Advances in electrochemical detection methods for measuring contaminants of emerging concerns. Electrochemical Science Advances, 2(6), e2100184.

Henrique, D. C., Quitela, D. U., Ide, A. H., Lins, P. V., Perazzini, M. T., Perazzini, H., Oliveira, L. M., Duarte, J. L., & Meili, L. (2021). Mollusk shells as adsorbent for removal of endocrine disruptor in different water matrix. Journal of Environmental Chemical Engineering, 9(4), 105704.

Heo, J., Yoon, Y., Lee, G., Kim, Y., Han, J., & Park, C. M. (2019). Enhanced adsorption of bisphenol A and sulfamethoxazole by a novel magnetic $CuZnFe_2O_4$– biochar composite. Bioresource Technology, 281, 179–187.

Kabir, E. R., Rahman, M. S., & Rahman, I. (2015). A review on endocrine disruptors and their possible impacts on human health. Environmental Toxicology and Pharmacology, 40(1), 241–258.

Kayode-Afolayan, S. D., Ahuekwe, E. F., & Nwinyi, O. C. (2022). Impacts of pharmaceutical effluents on aquatic ecosystems. Scientific African, 17, e01288.

Khan, A. H. A., & Barros, R. (2023). Pharmaceuticals in water: risks to aquatic life and remediation strategies. Hydrobiology, 2(2), 395–409.

Khan, S., Naushad, M., Govarthanan, M., Iqbal, J., & Alfadul, S. M. (2022). Emerging contaminants of high concern for the environment: current trends and future research. Environmental Research, 207, 112609.

Kumar, R., Qureshi, M., Vishwakarma, D. K., Al-Ansari, N., Kuriqi, A., Elbeltagi, A., & Saraswat, A. (2022). A review on emerging water contaminants and the application of sustainable removal technologies. Case Studies in Chemical and Environmental Engineering, 6, 100219.

Li, D., Londhe, K., Chi, K., Lee, C. S., Venkatesan, A. K., & Hsiao, B. S. (2021). Functionalized bio-adsorbents for removal of perfluoroalkyl substances: a perspective. AWWA Water Science, 3(6), e1258.

Li, L., Iqbal, J., Zhu, Y., Wang, F., Zhang, F., Chen, W., Wu, T., & Du, Y. (2020). Chitosan/Al_2O_3-HA nanocomposite beads for efficient removal of estradiol and chrysoidin from aqueous solution. International Journal of Biological Macromolecules, 145, 686–693.

Li, X., Shen, X., Jiang, W., Xi, Y., & Li, S. (2024). Comprehensive review of emerging contaminants: detection technologies, environmental impact, and management strategies. Ecotoxicology and Environmental Safety, 278, 116420.

Liu, S., Li, M., Liu, Y., Liu, N., Tan, X., Jiang, L., Wen, J., Hu, X., & Yin, Z. (2019). Removal of 17β-estradiol from aqueous solution by graphene oxide supported activated magnetic biochar: adsorption behavior and mechanism. Journal of the Taiwan Institute of Chemical Engineers, 102, 330–339.

Loffredo, E., Carnimeo, C., Silletti, R., & Summo, C. (2021). Use of the solid by-product of anaerobic digestion of biomass to remove anthropogenic organic pollutants with endocrine disruptive activity. Processes, 9(11), 2018.

Maddela, N. R., Ramakrishnan, B., Kakarla, D., Venkateswarlu, K., & Megharaj, M. (2022). Major contaminants of emerging concern in soils: a perspective on potential health risks. RSC Advances, 12(20), 12396–12415.

Mahmood, T., Momin, S., Ali, R., Naeem, A., & Khan, A. (2022). Technologies for removal of emerging contaminants from wastewater. In: Wastewater Treatment. (pp. 1–20). IntechOpen.

Mallek, M., Chtourou, M., Portillo, M., Monclus, H., Walha, K., ben Salah, A., & Salvado, V. (2018). Granulated cork as biosorbent for the removal of phenol derivatives and emerging contaminants. Journal of Environmental Management, 223, 576–585.

Mallik, S. (2023). Development of fruit-waste derived bio-adsorbent for removal of pharmaceutical contaminant from aqueous solution.

Martins, D. S., Estevam, B. R., Perez, I. D., Américo-Pinheiro, J. H. P., Isique, W. D., & Boina, R. F. (2022). Sludge from a water treatment plant as an adsorbent of endocrine disruptors. Journal of Environmental Chemical Engineering, 10(4), 108090.

Mathew, A. T., & Saravanakumar, M. P. (2023). Remediation of endocrine disrupting compounds and organic dye pollutants through biosorbents in a circular bioeconomy: prospects and constraints. In: Sustainable and Cleaner Technologies for Environmental Remediation: Avenues in Nano and Biotechnology (pp. 163–175). Cham: Springer International Publishing.

Moghri, P., Sabahi, H., Rezayan, A. H., & Akbari, B. (2025). A green approach: utilizing untreated pine pollen grains as a natural biosorbent for microplastic removal from water systems. Journal of Contaminant Hydrology, 274, 104675.

Mohammed, A. A., Al-Musawi, T. J., Kareem, S. L., & Al-Ma'abreh, A. M. (2019). Simultaneous adsorption of tetracycline, amoxicillin, and ciprofloxacin by pistachio shell powder coated with zinc oxide nanoparticles. Arabian Journal of Chemistry, 13(3), 4629–4643.

Morin-Crini, N., Lichtfouse, E., Fourmentin, M., Ribeiro, A. R. L., Noutsopoulos, C., Mapelli, F., Fenyvesi, É., Vieira, M. G. A., Picos-Corrales, L. A., Moreno-Piraján, J. C., Giraldo, L., Sohajda, T., Huq, M. M., Soltan, J., Torri, G., Magureanu, M., Bradu, C., & Crini, G. (2022). Removal of emerging contaminants from wastewater using advanced treatments. A review. Environmental Chemistry Letters, 20(2), 1333–1375.

Mutavdžić Pavlović, D., Ćurković, L., Mandić, V., Macan, J., Šimić, I., & Blažek, D. (2021). Removal of pharmaceuticals from water by tomato waste as novel promising biosorbent: equilibrium, kinetics, and thermodynamics. Sustainability, 13(21), 11560.

Nagappan, A., Krishnamoorthy, R., & Kui, C. C. (2025). Biosorptive removal of pharmaceutical contaminants using jujuba seeds for sustainable water pollution mitigation: mechanistic and modeling insights. Water, Air, & Soil Pollution, 236(11), 1–21.

Narindri Rara Winayu, B., Chu, F. J., Sutopo, C. C. Y., & Chu, H. (2024). Bioprospecting photosynthetic microorganisms for the removal of endocrine disruptor compounds. World Journal of Microbiology and Biotechnology, 40(4), 120.

O'Brien, S., Rauert, C., Ribeiro, F., Okoffo, E. D., Burrows, S. D., O'Brien, J. W., Wang, X., Wright, S. L., & Thomas, K. V. (2023). There's something in the air: a review of sources, prevalence and behaviour of microplastics in the atmosphere. Science of the Total Environment, 874, 162193.

Okoye, C. O., Nyaruaba, R., Ita, R. E., Okon, S. U., Addey, C. I., Ebido, C. C., Opabunmi, A. O., Okeke, E. S., & Chukwudozie, K. I. (2022). Antibiotic resistance in the aquatic environment: analytical techniques and interactive impact of emerging contaminants. Environmental Toxicology and Pharmacology, 96, 103995.

Oliveira, A. C., Neto, A. D., Moura, M. C. P. A., & Dantas, T. C. (2023). Use of surfactant-modified adsorbents in the removal of microplastics from wastewater. Journal of Environmental Chemical Engineering, 11(5), 110827.

Omo-Okoro, P. N., Curtis, C. J., Marco, A. M., Melymuk, L., & Okonkwo, J. O. (2021). Removal of per-and polyfluoroalkyl substances from aqueous media using synthesized silver nanocomposite-activated carbons. Journal of Environmental Health Science and Engineering, 19(1), 217–236.

Papini, M. P., Senofonte, M., Cuzzola, R. A., Remmani, R., Pettiti, I., Riccardi, C., & Simonetti, G. (2024). Adsorption technology for PFAS removal in water: comparison between novel carbonaceous materials. Materials, 17(17), 4169.

Pironti, C., Ricciardi, M., Proto, A., Bianco, P. M., Montano, L., & Motta, O. (2021). Endocrine-disrupting compounds: an overview on their occurrence in the aquatic environment and human exposure. Water, 13(10), 1347.

Pouretedal, H. R., & Sadegh, N. (2014). Effective removal of amoxicillin, cephalexin, tetracycline and penicillin G from aqueous solutions using activated carbon nanoparticles prepared from vine wood. Journal of Water Process Engineering, 1, 64–73.

Puri, M., Gandhi, K., & Kumar, M. S. (2023). Emerging environmental contaminants: a global perspective on policies and regulations. Journal of Environmental Management, 332, 117344.

Roman, J. B., Kemperman, A. J. B., van der Meer, W. G. J., & Wood, J. A. (2025). Assessing the viability of bio-based adsorbents for PFAS removal. Chemical Engineering Science, 306, 121215.

Romero-Cano, L. A., Gonzalez-Gutierrez, L. V., Baldenegro-Perez, L. A., & Medina-Montes, M. I. (2016). Preparation of orange peels by instant controlled pressure drop and chemical modification for its use as biosorbent of organic pollutants. Revista Mexicana de Ingeniería Química, 15(2), 481–491.

Rusu, L., Suceveanu, E. M., Blaga, A. C., Nedeff, F. M., & Şuteu, D. (2023). Insights into recent advances of biomaterials based on microbial biomass and natural polymers for sustainable removal of pharmaceuticals residues. Polymers, 15(13), 2923.

Sahay, P., Mohite, D., Arya, S., Dalmia, K., Khan, Z., & Kumar, A. (2023). Removal of the emergent pollutants (hormones and antibiotics) from wastewater using different kinds of biosorbent—a review. Emergent Materials, 6(2), 373–404.

Samal, K., Mahapatra, S., & Ali, M. H. (2022). Pharmaceutical wastewater as Emerging Contaminants (EC): treatment technologies, impact on environment and human health. Energy Nexus, 6, 100076.

Seo, P., Bhadra, B., Ahmed, I., Khan, N. A., & Jhung, S. H. (2016). Adsorptive removal of pharmaceuticals and personal care products from water with functionalized metal-organic frameworks: remarkable adsorbents with hydrogen-bonding abilities. Scientific Reports, 6, 34462.

Sethi, S. (2024). Agro waste based adsorbents: application in cleaning of pharmaceuticals and personal care products from wastewater. In: Development in Wastewater Treatment Research and Processes (pp. 115–140). Elsevier.

Shearer, L., Pap, S., & Gibb, S. W. (2022). Removal of pharmaceuticals from wastewater: a review of adsorptive approaches, modelling and mechanisms for metformin and macrolides. Journal of Environmental Chemical Engineering, 10(4), 108106.

Shu, H. H., Shi, L. Q., Zhang, H. Q., Liu, Y., Han, S. L., Rao, J. Y., & Liu, C. M. (2025). Synthesis of furan-based cationic biopolyamides and their removal abilities for perfluoroalkyl and polyfluoroalkyl substances. Chemical Engineering Journal, 507, 160306.

Tang, K. H. D. (2024). Valorization of organic waste as biosorbents for wastewater treatment. Water Emerging Contaminants & Nanoplastics, 3, 25.

Thacharodi, A., Meenatchi, R., Hassan, S., Hussain, N., Bhat, M. A., Arockiaraj, J., Ngo, H. H., Le, Q. H., & Pugazhendhi, A. (2024). Microplastics in the environment: a critical overview on its fate, toxicity, implications, management, and bioremediation strategies. Journal of Environmental Management, 349, 119433.

Tursi, A., Chatzisymeon, E., Chidichimo, F., Beneduci, A., & Chidichimo, G. (2018). Removal of endocrine disrupting chemicals from water: adsorption of bisphenol-a by biobased hydrophobic functionalized cellulose. International Journal of Environmental Research and Public Health, 15(11), 2419.

Valdez-Carrillo, M., Abrell, L., Ramírez-Hernández, J., Reyes-López, J. A., & Carreón-Diazconti, C. (2020). Pharmaceuticals as emerging contaminants in the aquatic environment of Latin America: a review. Environmental Science and Pollution Research, 27(36), 44863–44891.

Vasić, V., Kukić, D., Šćiban, M., Đurišić-Mladenović, N., Velić, N., Pajin, B., Crespo, J., Farre, M., & Šereš, Z. (2023). Lignocellulose-based biosorbents for the removal of contaminants of emerging concern (CECs) from water: a review. Water, 15(10), 1853.

Vieira, R. A. L., Pickler, T. B., Segato, T. C. M., Jozala, A. F., & Grotto, D. (2022). Biochar from fungiculture waste for adsorption of endocrine disruptors in water. Scientific Reports, 12(1), 6507.

Wang, F., Xiang, L., Leung, K. S. Y., Elsner, M., Zhang, Y., Guo, Y., Pan, B., Sun, H., An, T., Ying, G., Brooks, B. W., Hou, D., Helbling, D. E., Sun, J., Qiu, H., Vogel, T. M., Zhang, W., Gao, Y., Simpson, M. J., Luo, Y., & Tiedje, J. M. (2024). Emerging contaminants: a one health perspective. The Innovation, 5(4), 100612.

Yan, X., Zhang, Q., Rao, Q., Chen, S., Wang, X., Song, W., Cheng, L., Guan, S., & Song, W. (2022). Efficient removal of nonylphenol contamination from water using optimized magnesium silicate. Materials (Basel, Switzerland), 15(13), 4445.

Yu, F., Ma, J., & Han, S. (2014). Adsorption of tetracycline from aqueous solutions onto multi-walled carbon nanotubes with different oxygen contents. Science Reports 4, 5326.

Yu, F., Sun, S., Han, S., Zheng, J., & Ma, J. (2016). Adsorption removal of ciprofloxacin by multi-walled carbon nanotubes with different oxygen contents from aqueous solutions. Chemical Engineering Journal, 285, 588–595.

Zarghi, M., Roodbari, A., Jorfi, S., & Fard, N. (2019). Removal of estrogen hormones (17β-estradiol and estrone) from aqueous solutions using rice husk silica. Chemical and Biochemical Engineering Quarterly, 33(3), 281–293.

Zhang, H., Tu, Y.-J., Duan, Y.-P., Liu, J., Zhi, W., Tang, Y., Xiao, L.-S., & Meng, L. (2020). Production of biochar from waste sludge/leaf for fast and efficient removal of diclofenac. Journal of Molecular Liquids, 299, 112193.

Zhang, Z., Sarkar, D., Biswas, J. K., & Datta, R. (2022). Biodegradation of per- and polyfluoroalkyl substances (PFAS): a review. Bioresource Technology, 344, 126223.

Zhou, W., Dai, Q., Zhan, T., Wang, L., Bian, X., Fan, S., Xiong, P., Xia, Y., & Chen, J. (2020a). Adsorption removal of pharmaceutical and personal care products with functionalized metal-organic framework: adsorptive selectivity and mechanism. Desalination and Water Treatment, 191, 231–238.

Zhou, Y., Liu, S., Liu, Y., Tan, X., Liu, N., & Wen, J. (2020b). Efficient removal 17-estradiol by graphene-like magnetic sawdust biochar: preparation condition and adsorption mechanism. International Journal of Environmental Research and Public Health, 17(22), 8377.

10

Economics and Environmental Aspects of Bioadsorbents

10.1 Introduction

The growing scarcity of freshwater resources has increased the demand for efficient wastewater treatment methods. Each water treatment process has unique advantages and disadvantages that are impacted by water composition, pollutant type, local conditions, and treatment objectives. To achieve sustainability, the most appropriate technique must be chosen while keeping technical, economic, and environmental factors in mind. Among the available options, adsorption is the most efficient and cost-effective method for heavy metal removal. Although commercial adsorbents are commonly employed, their exorbitant cost prevents large-scale usage (Hamad and Idrus, 2022). As a result, research has shifted toward low-cost and sustainable bioadsorbents for the effective removal of heavy metal ions and dyes from contaminated water. These bioadsorbents, made from natural or waste materials, are inexpensive, widely available, and environmentally beneficial. Their use not only improves pollutant removal efficiency but also encourages waste valorization and circular economy activities, making them a viable alternative to traditional adsorbents (Thanki et al., 2021). It combines cheap operational costs, versatility, and great removal efficiency over multiple pH ranges. Furthermore, bioadsorbents increase their ecological and circular economic value by employing renewable natural resources and lowering waste output (Meftah et al., 2025c; Meftah et al., 2025a).

The experimental setup in this study was extremely low-cost because bioadsorbents were generated from waste materials and no energy was used. All necessary chemicals, filter papers, and laboratory equipment were obtained from existing sources. However, scaling up the process would incur extra expenditures, such as chemicals, electricity, personnel, transportation, and analytical testing. For drinking water applications, additional treatment procedures such as chemical oxygen demand (COD) and total dissolved solids (TDS) reduction, pH adjustment, and heavy metal removal may be necessary. Bioadsorbents provide a cost-effective and environmentally benign alternative to conventional CETP procedures for chromium removal, reducing chemical use and secondary contamination (Jacob and Soni, 2024).

This study provided a complete techno-economic and sensitivity analysis of chitosan-based bioadsorbent manufacturing with raw chitosan and titanium tetraisopropoxide (TTIP). The study found that chitosan microbeads (CM) and TiO_2-modified chitosan microbeads ($CMTiO_2$) can be produced sustainably at costs of \$64.40/kg and \$37.00/kg, respectively. While the CM process yielded a higher ROI, the $CMTiO_2$ process offered greater economic stability in the face of fluctuating operating expenses. Meramo-Hurtado and González-Delgado (2020) found that $CMTiO_2$ has a cheaper price due to increased yield and lower raw material costs, making it a cost-effective industrial alternative.

Agro-waste feedstocks for this techno-economic evaluation (TEA) were sourced from local agro-processing companies within a 50-km radius of Coimbatore. Because these wastes are frequently dumped or used domestically, purchase prices were low. The analysis also examined transportation and electricity costs, notably those incurred during the grid-powered pyrolysis process. The most cost-effective alternative for removing BR09 dye was groundnut shell biochar, which cost ₹0.91 per gram and had a high adsorption capacity of 46.29 mg/g. Coconut shell and rice husk biochars are substantially cheaper than commercial activated carbon (AC), with prices of ₹4.54 and ₹0.97 per gram, respectively. Developing efficient regeneration and disposal solutions for dye-loaded biochars can improve cost-effectiveness while reducing secondary pollution (Praveen et al., 2021).

Biochar has become recognized as a low-cost, environmentally acceptable adsorbent capable of extracting organic and inorganic pollutants from water. Its manufacturing yields carbon-rich materials with well-developed pores, a large surface area, and several functional groups, which improve adsorption efficiency (Huang et al., 2019). Regeneration, desorption, or safe reuse of spent sorbents in applications such as fertilizers, glass and cement fillers, and photocatalytic processes are examples of environmentally friendly waste management practices. Agricultural waste-based adsorbents are desirable due to their availability, low cost, and regeneration potential. Furthermore, the synthesis and modification of biomass-derived biochar can be enhanced to yield diverse carbon compounds with improved surface properties for effective wastewater treatment (Mihai et al., 2025).

This chapter uniquely integrates economic and environmental perspectives on bioadsorbents, emphasizing their sustainability, cost-effectiveness, and industrial applicability. Unlike previous studies focusing only on adsorption efficiency, it connects TEA to assess real-world feasibility. The novelty lies in linking production cost, processing methods, and environmental performance within a circular economy framework, offering a comprehensive understanding for promoting large-scale, eco-friendly, and economically viable wastewater treatment technologies.

10.2 Economic Aspects of Bioadsorbents

TEA is critical in determining both the technical efficiency and economic viability of any process or product. It bridges the gap between laboratory research and practical application by combining process modeling, design engineering, energy balance, and cost analysis. In the case of biochar and hydrochar production, TEA involves a thorough life cycle assessment (LCA) that includes carbon footprint estimation, energy consumption analysis, and a detailed cost-benefit analysis of the entire process. This method assures that the manufacturing system is both environmentally sustainable and economically viable for large-scale adoption and commercialization, easing the transition to cleaner technologies and circular bioeconomy models (Kumar et al., 2020).

10.2.1 Cost of Raw Materials

The economic feasibility of adsorbent production is determined by the cost per unit mass of raw materials utilized in the synthesis process. The costs vary based on the activation process, with chemical activation often being more expensive than physical activation. Raw material prices, chemicals used for modification, and operational fees are all major cost drivers. Other key variables are adsorption capacity, selectivity, and material stability during reuse. In some circumstances, low-cost or waste-derived components can help down overall costs. Additionally, energy usage, transportation, and equipment operation all play important factors in determining total production costs. Understanding these characteristics allows for effective process design and optimization, guaranteeing that large-scale adsorbent manufacturing is economically viable and competitive in industrial applications (Ighalo et al., 2022). Table 10.1 shows the cost comparison of various raw materials used for bioadsorbents, including agricultural waste, algae, and other biomass sources.

The cost of purchased biomass can be determined using the annual biomass flow rate and the feedstock's specific purchased cost. It is computed by multiplying the annual biomass input by the unit cost, resulting in the overall cost of purchasing biomass for biochar production. This cost varies greatly according to the type and source of feedstock. Agricultural residues, forestry debris, and industrial byproducts have varying price ranges depending on availability, transportation, and processing needs. Understanding these cost relationships is

Table 10.1 Cost comparison of various raw materials used for bioadsorbents (agricultural waste, algae, biomass, etc.).

Raw material source	Availability/ Abundance	Collection cost	Processing cost	Adsorption capacity (mg/g)	Surface area (m^2/g)	Functional groups	Waste management benefit	References
Rice husk	Readily available as a major agricultural waste in rice-producing regions. Rice husk collected free from a paddy field for research purposes.	Negligible cost—often provided for free by rice milling companies for academic use	Low (primarily energy for oven/furnace)	6–9 mg/g for phenol	Thermally treated: 24–201 m^2/g; Chemically treated: 3.2 m^2/g	—OH, C=O, Si—O—Si, C—C, C—O—C (varies with treatment)	Reduces open burning/landfill; valorizes waste to value-added adsorbent	Daffalla et al. (2020)
Banana peel powder	Highly abundant, worldwide food/agro-industrial waste.	Not specified, negligible (food/agro waste stream)	Sun-drying, grinding; low, as no chemical activation needed	~33 (methylene blue); 3.26 (atrazine); 3.02 (glyphosate)	Microporous, agglomerated micro-particle structure; numerical value not specified	Hydroxyl, carboxyl, other O-containing groups	Converts problematic agro/food waste into value-added material; supports circular economy, reduces landfill/methane	Farias et al. (2023)
Orange peel	Highly abundant; ~32 million tons annual global waste. Easily sourced from juice industries and street vendors.	Negligible cost-waste	Low (primarily energy for oven/furnace)	Raw: 3–5 mg/g for dyes/heavy metals; modified up to 289 mg/g depending on treatment.	Raw: Smooth, nonporous; modified: increased surface area, more microporous	Pectin, —OH, —COOH, carbonyl, amine	Converts massive, high-moisture food waste into useful product; avoids landfilling and methane emission	Michael-Igolima et al. (2023)

Sawdust/ wood waste	Byproduct of timber industry	Low	Moderate (carbonization)	50–150 (dyes, Pb^{2+}, Cu^{2+})	100–300	—OH, —COOH, lignocellulosic	Reduces sawmill waste disposal	Habib and Mohamed (2023)
Coconut shell	Easily available, abundant	Low—byproduct cost, usually ₹1–3/kg	₹15–50/kg	156.25	747	—OH, C=C, C=O	Converts waste shells, reduces landfill mass, creates high-value adsorbent	Karaer Yağmur and Kaya (2021)
Coffee grounds	Grows with coffee consumption	Negligible (waste stream)	₹3–8 ($0.04–0.11)	43–232 (Cr, Pb, Cu, dyes)	Up to 193 (biochar)	—OH, phenolics, proteins	Reduces solid waste from cafes	Sulyman et al. (2017)
Tea waste	High (tea industry waste, ~2 M tons/year in China)	Very low/ near zero	Moderate (washing, pyrolysis ~ ₹15–25/kg)	4.7–8.9	~50–150	—OH, —COOH, aromatic C=C, C—O	Utilizes agro-industrial waste, reduces landfill burden	Fan and Zhang (2021)
Peanut shell	Available seasonally	₹1–2	₹3–8 ($0.04–0.10)	18–56 (Cu, Cr, Pb, Ni)	35–150	Cellulose, lignin, —COOH	Waste valorization, landfill avoidance	Sen (2023)
Water hyacinth (Eichhornia crassipes)	Overabundant invasive aquatic weed	Negligible (easy to harvest, but transport cost high)	Low–moderate (drying, carbonization)	40–70	150–250	—OH, —COOH, C=O	Eco-friendly removal of nuisance weed	Habib and Mohamed (2023)
Sewage sludge	Large volumes from wastewater treatment plants	Very low (waste)	Moderate–high (drying, pyrolysis, safety handling)	30–60	100–250	—OH, —COOH, amine groups	Converts hazardous waste to value-added adsorbent	Fan and Zhang (2021)

(Continued)

Table 10.1 *(Continued)*

Raw material source	Availability/ Abundance	Collection cost	Processing cost	Adsorption capacity (mg/g)	Surface area (m^2/g)	Functional groups	Waste management benefit	References
Bamboo/ wood waste	Widely available forestry/ agricultural residue	Low	Moderate	20–40	200–600	—OH, C=O, aromatic structures	Reduces biomass disposal problems	Fan and Zhang (2021)
Spent mushroom substrate	Byproduct of mushroom cultivation	Very low	Low–moderate	10–25	50–150	—OH, —COOH	Utilizes nutrient-rich agricultural residue	Fan and Zhang (2021)
Bael (*Aegle marmelos*) shell carbon	Available in fruit-processing regions	Very low, <₹2/kg	₹10–15/kg (pyrolysis, activation)	95.1 (MB dye)	282	Lignocellulose, —OH, —COOH	Valorizes fruit waste, reduces disposal load	Hamad and Idrus (2022)
Neem leaf powder (raw, not activated)	Widespread in India, highly accessible	₹0.5–2/kg	Minimal, sun drying/ pulverization	83.33 (MB dye)	46	Lignin, —OH, —COOH	Utilizes fallen leaves, easy disposal	Hamad and Idrus (2022)
Banana stem	Plantation/ agro-waste	Very low	Low–moderate (activation)	Reported up to 194.6 mg/g	~200–500	—OH, —COOH	Reduces plantation residue disposal	Misran et al. (2022)
Algae (Brown, *Sargassum*)	Widely available coastal biomass	Very low/ near zero	Moderate (pretreatment, modification)	Up to 279.14 mg/g (crystal violet); 193.8 mg/g (acridine orange); 110.4 mg/g (malachite green)	~50–200	Amino, amine, hydroxyl, phosphate, sulfate	Eco-utilization of algal blooms and invasive seaweed	Shen et al. (2019)

critical for determining the overall economic feasibility of biochar production, as feedstock expenses frequently account for a significant portion of total production costs, affecting the profitability and scalability of biochar-based industries (Ahmed et al., 2016).

10.2.2 Processing and Modification Costs

The cost of purchased biomass can be determined using the annual biomass flow rate and the feedstock's specific purchased cost. It is computed by multiplying the annual biomass input by the unit cost, resulting in the overall cost of purchasing biomass for biochar production. This cost varies greatly according to the type and source of feedstock. Agricultural residues, forestry debris, and industrial byproducts have varying price ranges depending on availability, transportation, and processing needs. Understanding these cost relationships is critical for determining the overall economic feasibility of biochar production, as feedstock expenses frequently account for a significant portion of total production costs, affecting the profitability and scalability of biochar-based industries (Ahmed et al., 2016).

The economic assessment of polymeric adsorbents was conducted using data gathered mostly through in-person interviews with managers from two Greek chemistry laboratories with experience in the sector. These unstructured interviews gave useful qualitative insights into the cost characteristics of commonly used adsorbents, highlighting the significance of raw material and energy costs. A recipe analysis of chitosan-based adsorbents was used to further evaluate costs. According to the survey, labor accounted for about 70% of total manufacturing costs, with raw materials and energy accounting for ~23% and 45%, respectively. High energy costs were mostly linked with the pyrolysis stage. Overall, operational costs—particularly labor and energy—were discovered to be the most important elements determining total recipe cost (Gkika et al. 2019).

10.2.3 Comparison with Conventional Adsorbents

Commercial AC is commonly used to remove contaminants from liquid solutions, but its high cost prevents large-scale utilization (Gupta et al., 2015). To solve this issue, researchers investigated nontraditional, low-cost adsorbents produced from natural, industrial, and agricultural waste. Materials including bagasse flyash, pine cones, grape debris, pistachio shells, and cherry kernels have been demonstrated to effectively remove heavy metals. Biochar and activated agricultural carbon stand out for their abundance and inexpensive manufacturing costs. Waste cherry kernels, for example, are a sustainable and cost-effective precursor for adsorbent manufacture. Integrating eco-design concepts, which emphasize

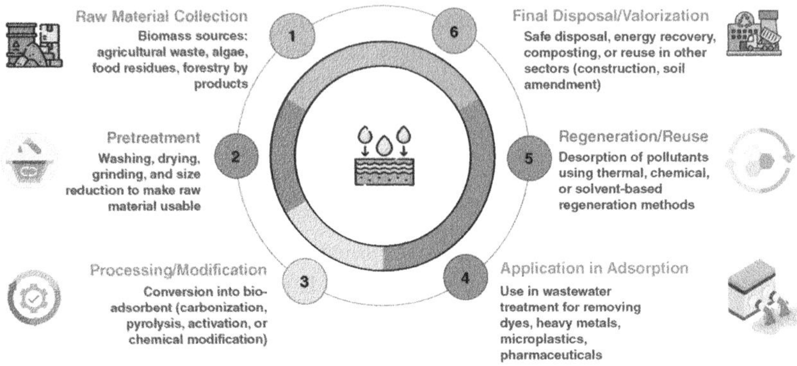

Figure 10.1 Flow diagram of life cycle stages of bioadsorbents.

minimizing environmental impact throughout the product's life cycle, improves the sustainability of adsorbent manufacture. Figure 10.1 presents a flow diagram illustrating the life cycle stages of bioadsorbents. LCA remains an important method for optimizing environmentally friendly activities (Vukelic et al., 2018). Table 10.2 compares the costs of conventional adsorbents such as AC, synthetic resins, and zeolites.

A variety of nontraditional adsorbents generated from agricultural and forest wastes, including tea trash, coffee husks, sawdust, corncobs, and fruit peels, have emerged as intriguing alternatives to conventional materials. Compared to commercial AC and synthetic ion-exchange resins, these biosorbents are inexpensive, renewable, and environmentally beneficial (Ali, 2010). However, their adsorption efficiency, selectivity, and regeneration capacity differ greatly depending on the surface chemistry and pollutant type. Conventional adsorbents have higher removal efficiency and faster kinetics, whereas nonconventional ones are more sustainable and economically viable. Despite numerous investigations, systematic comparisons are limited due to varying experimental settings and a lack of cost data. As a result, hybrid techniques combining conventional and bio-based adsorbents are being investigated to balance efficiency, cost, and environmental effect (Crini et al., 2019).

10.2.4 Scale-up and Commercialization Challenges

Recent advances in adsorption-based water cleanup have attempted to address both economic and ethical concerns by prioritizing efficiency, sustainability, and affordability (AlAqad et al., 2025). The cost-effectiveness of adsorbents is substantially determined by synthesis and operating parameters. Green chemistry methodologies currently allow for the synthesis of low-cost, high-performance

Table 10.2 Cost comparison of various conventional adsorbents (activated carbon, synthetic resins, zeolites).

Conventional adsorbents	Availability/ Abundance	Cost (₹/kg or $/kg)	Processing cost (₹/kg or $/kg)	Adsorption capacity (mg/g)	Surface area (m^2/g)	Functional groups	Waste management benefit	References
Activated carbon	Widely available; produced from biomass, coal, and waste chars	₹300–900/kg	₹200–600/kg	35.7 (trimethylamine) 1,930 mg/g (methylene blue dye)	500–2,400 (2,276 m^2/g typical)	Carboxyl, hydroxyl, phenol, ether groups	Converts waste biomass; reduces landfill; reusable via regeneration	Park et al. (2023)
Synthetic resins	Industrially manufactured, widely available	$10–20/kg (₹800–1600/kg)	High (polymerization, functionalization)	100–400 mg/g (heavy metals, dyes)	50–800	$-SO_3H$, $-NH_2$, $-COOH$, quaternary amines	Can be regenerated, but non-biodegradable, disposal issue	Pérez-Botella et al. (2022)
Zeolites	Natural and synthetic varieties widely mined and produced	$5–20/kg	Moderate cost for modification and synthesis	50–200 mg/g (CO_2, NH_3, heavy metals depending on type)	300–800	Al—O—Si frameworks, ion-exchange sites	Natural mineral, regenerable, selective separation; minimal waste	Pérez-Botella et al. (2022)
Activated carbon	Widely available; derived from coal, peat, wood, coconut shells; commercially produced	High (₹300–1000/kg; depending on grade)	High Involves pyrolysis and activation (steam/chemical)	200–1,200 mg/g for dyes/organics	500–2,000 m^2/g	—OH, —COOH, —C=O	Effective for broad spectrum pollutants; regenerable, but regeneration costly	Crini et al. (2019)

(Continued)

Table 10.2 *(Continued)*

Conventional adsorbents	Availability/ Abundance	Cost (₹/kg or $/kg)	Processing cost (₹/kg or $/kg)	Adsorption capacity (mg/g)	Surface area (m^2/g)	Functional groups	Waste management benefit	References
Ion-exchange resins	Manufactured; synthetic polymers (styrene-divinylbenzene, acrylic resins); commercially abundant	Expensive (₹500–2000/kg; $10–25/kg)	High – polymerization and crosslinking processes	50–500 mg/g (depending on pollutant)	100–1,000 m^2/g	$-SO_3H$, $-NH_2$, —COOH (depending on type)	No adsorbent loss on regeneration, solvent recovery possible; reduces wastewater toxicity	Crini et al. (2019)
Zeolites	Naturally occurring and synthetic aluminosilicates; abundant minerals but synthesis costly	Moderate to high (₹100–500/kg natural; $2–10/kg; synthetics higher)	Moderate to high – hydrothermal synthesis or mining and processing	50–400 mg/g for dyes/heavy metals	300–1,000 m^2/g	—Si—OH, —Al—OH, framework negative charges for ion exchange	Environmentally stable; regenerable; widely used in water softening and gas separation	Crini et al. (2019)
Natural clinoptilolite	Naturally mined worldwide	$5–15/kg	$2–4/kg	~100 mg/g for heavy metals	300–350	Al—O—Si framework	Regenerable, selectively adsorbs ions	Pérez-Botella et al. (2022)
Synthetic Faujasite	Synthetic, industrial scale	$8–20/kg	$3–6/kg	Up to 150 mg/g CO_2	350–800	Si—O, Al—O, ion-exchange sites	Regenerable, high selectivity	

Synthetic zeolites	Naturally abundant aluminosilicates; also synthesized hydrothermally	Moderate (₹100–500/kg natural; \$2–10/kg; higher for synthetics)	Moderate to high – mining or hydrothermal synthesis	89.3 mg/g (Ni^{2+} removal)	300–925 m^2/g (depending on type)	—Si—OH, —Al—OH, framework negative sites	Stable, regenerable, eco-friendly, widely used in ion exchange & softening	Hashmi et al. (2024)
Organic/ Synthetic Resins	Commercially manufactured (styrene-divinylbenzene, acrylic resins)	Expensive (₹500–2000/kg; \$10–25/kg)	High—polymerization, crosslinking	182 mg/g (TOC removal)	100–1,000 m^2/g	$—SO_3H$, $—NH_2$, —COOH	Efficient, reusable, minimizes secondary waste	Hashmi et al. (2024)
Clinoptilolite (zeolite)	Natural mined zeolite	\$5–15/kg	\$2–4/kg	~100 mg/g (heavy metals)	300–350	Al—O—Si framework	Regenerable, selective ion adsorbent	Welsch and Laird (2024)
Synthetic polymer adsorbent (TOC)	Laboratory synthesized polymer	Low cost (lab-scale preparation)	Moderate (chemical synthesis)	Pb (II): 122.7 mg/g; Cu (II): 95.9 mg/g	Not specified	Sulfur and oxygen (thiophene, carbonyl)	High adsorption capacity, excellent reusability, chemical stability	Chen and Zhou (2024)
Synthetic Resin (macroporous high adsorbent resin [MHAR])	Synthesized from acrylic acid, sodium carboxymethyl cellulose (CMC), and sodium alginate (SA)	Low-cost laboratory synthesis (not exact provided)	Moderate chemical synthesis cost	Cu^{2+} adsorption capacity ~80–100	Not quantitatively specified	Carboxyl (—COOH), hydroxyl (—OH)	Biodegradable, renewable materials; high adsorption selectivity	Shi et al. (2021)

(Continued)

Table 10.2 *(Continued)*

Conventional adsorbents	Availability/ Abundance	Cost (₹/kg or $/kg)	Processing cost (₹/kg or $/kg)	Adsorption capacity (mg/g)	Surface area (m^2/g)	Functional groups	Waste management benefit	References
Synthetic resin from waste polystyrene foam (RPSF)	Waste polystyrene foam widely available	Low cost (waste material)	Moderate (sulfonation, neutralization)	Acid orange 67: 82.64–113.6 (Langmuir model)	23	Sulfonic groups ($-SO_3Na$)	Converts plastic waste to value-added adsorbent	Hilal et al. (2019)
Zeolites (NH4BETA, NH4ZSM-5, NaY)	Naturally abundant and synthetically produced; ~1.1 million tons mined/year	$0.5–2/kg (₹40–150/kg)	Low–Moderate (mining, ion exchange, calcination)	27–105 mg/g (dyes, MB/RB)	200–700 m^2/g (NH4BETA > ZSM-5 > NaY)	—Si—O—Al— lattice, Lewis and Brønsted acid sites	Environmentally safe, reusable, no toxic byproducts; stable under regeneration	Rakanović et al. (2022)

adsorbents by environmentally friendly routes such as sol-gel, hydrothermal, and bio-fabrication methods. These technologies reduce energy use and eliminate harmful precursors, aligning with sustainability objectives. Furthermore, ethical considerations have prompted the adoption of circular economy concepts, which emphasize biodegradability, reusability, and longer adsorbent lifespans via surface modification and composite synthesis. Integration with hybrid technologies, such as membrane filtration and photocatalysis, improves pollutant removal efficiency while lowering operating costs, indicating substantial progress toward scalable, ethical, and economically viable water treatment options (Satyam and Patra, 2024).

Despite their promising laboratory-scale performance, green adsorbents encounter considerable obstacles when scaling up and commercializing. The majority of investigations are limited to batch trials, with just a few number of pilot- or industrial-scale applications being investigated. Process optimization, raw material unpredictability, regeneration cost, and maintaining adsorption efficiency all impede the shift from lab to large-scale production. Furthermore, large-scale synthesis frequently necessitates uniform material quality, sustained adsorption performance, and economic viability, which are all difficult to achieve with biomass-derived adsorbents. Regeneration and disposal of exhausted sorbents raise environmental and financial concerns. Furthermore, the absence of standardized production processes, limited industrial collaborations, and insufficient government incentives impede the economic viability of these environmentally beneficial adsorbents (Hashmi et al., 2024).

Bioadsorbents' scaling and commercialization face significant technological and economic challenges. The sophisticated mathematical modeling of adsorption columns in multicomponent systems hinders practical application because the models are difficult to use and interpret. Simplified models and artificial intelligence tools help speed up process design. Another critical issue is the disposal and reuse of used bioadsorbents, which influences both sustainability and cost-effectiveness (Das et al., 2024). Technologies without practical reuse or low-cost disposal are less appealing for commercialization. Optimizing operational parameters, such as the mass transfer zone and bed use, is critical for increasing efficiency. However, few studies have focused on large-scale or continuous adsorption systems. As a result, life cycle and techno-economic assessments are critical for moving bioadsorbent technologies forward toward greater readiness and industrial acceptance (Coelho et al., 2025).

A diverse range of biomaterials generated from agricultural sources are widely available around the world, including wastes, byproducts, and coproducts from farming and processing industries. If these wastes are not managed properly, they might have negative environmental and economic consequences. A circular economy approach reduces waste output while increasing the value of these resources through cost-effective utilization (Abou Taleb and Al Farooque, 2021). Small and medium-sized companies in developing nations, particularly in rural areas, frequently discharge untreated wastewater containing harmful dyes and heavy metals due to financial constraints and a lack of treatment facilities. Using agricultural waste as low-cost adsorbents provides a sustainable option for these sectors. Although these materials may have lower adsorption capacity than synthetic adsorbents, their low cost and availability make them suitable for wastewater cleanup in small-scale applications (Das et al., 2020).

10.2.5 Techno-economic Evaluation

Cost is an important factor in establishing the viability of any water treatment technology, with the adsorbent cost being the most significant component of the adsorption process (Gupta et al., 2009). As a result, using locally available, low-cost materials with high adsorption effectiveness is necessary. Walnut shells, an agricultural byproduct, are an attractive choice because of their availability and low cost—~$385 per metric ton in India. Their vast surface area and metal-binding functional groups make them effective at removing chromium (VI). According to economic research, small and medium-sized enterprises can implement this approach at a low cost because the adsorbent can be reused several times before being replaced. Thus, walnut shells are an efficient, sustainable, and cost-effective green adsorbent for industrial wastewater treatment (Banerjee et al., 2018).

The economic feasibility of CC-NaOH bioadsorbent production was assessed using common profitability metrics such as rate of return on investment (ROI), payout time (POT), percent profit on sales (POS), and break-even point (BEP). The facility, located in Pasang Kayu Regency, South Sulawesi, Indonesia, runs 330 days a year and has a capacity of 7,200 tons. The key assumptions are 60 : 40 owned capital, 5.75% interest rate, and 3.18% inflation rate. Labor expenditures cover both expert and unskilled labor involved in biomass operations. The analysis, based on USD 1 = IDR 14,340, demonstrated that the procedure was economically viable, similar to prior studies on BE processing that showed lucrative returns at $100/ton (Ifa et al., 2022).

The work assesses the techno-economic viability of biochar production from orchard waste in California's Central Valley, which accounts for 8% of US agricultural output, 25% of national food production, and nearly 100% of almond supply. Monte Carlo simulations indicated biochar production costs between $300 and 700 per tonne, whereas IMPLAN modeling demonstrated the potential to create $25 million in gross output and 150–200 jobs each year. A suggested dual-column adsorption-based water treatment system is commercially viable, with an estimated unit cost of €0.89/m^3. The findings support biochar's promise as a cost-effective alternative for sustainable water treatment and resource recovery.

Future study should aim to improve treatment efficiency by incorporating additional advanced purification methods (Nematian et al., 2021).

Liu et al. (2021) conducted a techno-economic analysis comparing two biomass processing scenarios using three indicators: ROI, payback period, and net present value (NPV). Scenario 2 demonstrated more economic feasibility, with a ROI of 0.2 (twice that of Scenario 1) and a shorter payback period of 3.04 versus 4.78 years. Nonetheless, Scenario 2 required careful temperature control during activation to minimize charcoal over-oxidation, meaning a higher investment in automation. Environmentally, both scenarios used wasted mushroom substrate (SMS) as a feedstock to generate AC and fuel gas. Scenario 2 has higher CO_2 emissions but higher energy recovery due to increased gas flow. This suggests more profitability but raises larger operational and environmental challenges.

The simulated results for processing 38.9 thousand tonnes of sugarcane bagasse with different NaOH:PB ratios revealed considerable variances in AC yields. The output varied between a maximum of 8.3 thousand tonnes per year for scenario S0.5 and a minimum of 1.0 thousand tonnes per year for scenario S3. For perspective, Brazil imported 5.0 thousand tonnes of AC in 2016 and 5.8 thousand tonnes in 2017, implying that a single integrated manufacturing unit might significantly reduce import dependence. Furthermore, the required bagasse quantity represents only a small percentage of the biomass produced by a typical sugarcane mill. As a result, large-scale AC production from sugarcane bagasse is an economically viable and sustainable option, particularly for NaOH:PB ratios of 1 to 2.5 (Fingolo et al., 2020).

Kalengyo et al. (2024) conducted a TEA of three scenarios for using orange peel powder (OPP) and biochar to treat textile wastewater (TWW). Scenario 1 utilized simple adsorption, which had the lowest capital cost ($320) and operating simplicity, but was economically unfeasible, with a monthly profit of only $48. Scenario 2 combined adsorption with anaerobic digestion, raising the capital cost to $1,220 but increasing earnings to $452 per month, with an 8.9-year payback period. Scenario 3 (adsorption/digestion/pyrolysis) had the highest capital ($1,460) and operational expenditures ($370/month), but it yielded the best return of $564/month and a shorter payback period of 7.5 years. Thus, scenario 3 emerged as the most economically and environmentally sustainable option (Kalengyo et al., 2024).

The techno-economic analysis of biogas production from barley straw revealed that it is a viable but currently unprofitable option under current cost and revenue conditions. Two scenarios were compared: Scenario 1, which featured untreated and mechanically pretreated barley straw, and Scenario 2, which included both mechanical and chemical preparation using ZnO nanoparticles. The results showed that Scenario 2 had larger cumulative biogas and methane production, resulting in more energy generation and revenue. However, the high prices of field preparation, fertilizer, and pretreatment chemicals made profitability difficult. To improve economic viability, the study recommended reducing cultivation expenses by employing nonfood or perennial crops and increasing biogas generation through optimized digester types, co-digestion, and effective pretreatment procedures (Hassaan et al., 2020).

10.3 Environmental Aspects of Bioadsorbents

The environmental benefits of bioadsorbents underline their significance in sustainable wastewater treatment, particularly for dye-laden industrial effluents (Sayed et al., 2025). Bioadsorbents, which are generally obtained from agricultural or plant waste, offer a low-cost and environmentally acceptable alternative to traditional adsorbents such as AC. Their efficiency is determined by adsorbent-adsorbate features such as surface area, porosity, and surface charge, as well as environmental variables such as pH, ionic strength, and competing ions. Regeneration and proper disposal of wasted adsorbents are critical for reducing secondary pollution. To ensure practical and long-term uses, investigations should go beyond laboratory-scale batch tests and include pilot-scale and genuine industrial effluents. Comprehensive characterization and mechanistic assessments improve both efficiency and environmental safety (Bello et al., 2015).

Bioadsorbents derived from lignocellulosic biomass are environmentally friendly and contribute to the circular bioeconomy by recycling waste and lowering pollution. AC manufacturing from biomass necessitates environmentally friendly processes, as chemical activation with strong acids or bases can produce harmful byproducts. Safer, less expensive self-activation methods are favored. Bioadsorbents can also help photocatalysts remove pollutants and regenerate themselves. Scaling up is difficult due to technological constraints such as light penetration and biomass oxidation. Mixed municipal biomass is more difficult to process but improves resource usage. Standardized methods, pilot-scale testing, and techno-economic assessments are required to ensure that large-scale applications are safe, effective, and sustainable (Mergbi et al., 2023).

The regeneration of bioadsorbents is critical for lowering operational costs and providing environmentally friendly water treatment. Thermal, chemical, and microbiological regeneration procedures are commonly used, and they employ physical waves, electrical currents, chemical reagents, or microbes. These work by desorption, pH shifts, extraction, or degradation, although the majority do not allow for adsorbate recovery. In some circumstances, regeneration may not be feasible, particularly if it generates more trash or involves more effort than creating new adsorbent, making incineration or landfill disposal desirable. Adsorption in real water is a challenge. Most research employs synthetic solutions, frequently

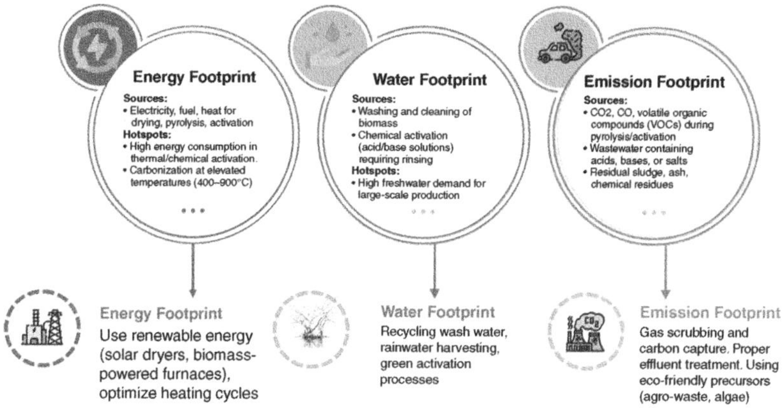

Figure 10.2 Environmental footprint analysis of bioadsorbent production.

containing a single contaminant, but real wastewater comprises complicated mixes. Improving adsorbent selectivity and employing magnetic adsorbents can simplify separation while increasing practical efficiency (Dotto and McKay, 2020). Figure 10.2 displays the environmental footprint analysis of bioadsorbent production.

The regeneration of bioadsorbents is determined by their source, technique, and dosage. Chemical regeneration, such as NaOH or pH correction, is effective but expensive and may result in secondary pollution. Thermal regeneration increases efficiency by forming new pores, but carbon loss is higher. Microwave and supercritical fluid technologies show promise for rapid regeneration with low biomass loss, but they are still limited to the laboratory scale. Chemical modification improves adsorption capacity and industrial usability; however, environmental concerns must be addressed. Spent adsorbents should be burnt at high temperatures (1,100–1,200 °C) using gas collection devices. Adegoke et al. (2023) emphasize the need for life cycle studies in determining adsorbent reuse cycles and ensuring sustainable management.

10.4 Environmental Benefits and Circular Economy

The increased emphasis on environmental sustainability has prompted study into biomass as an important alternative resource. Plant biomass, such as rubber seeds, eucalyptus bark, rice husks, fruit peels, and animal bones, can be converted into high-value products. These materials have applications in wastewater treatment, energy, building, and fertilizer production. According to studies, biomass-derived products can compete with conventional materials like adsorbents, biofuels, concrete additives, and biofertilizers. Using biomass waste not only reduces pollution but also promotes the circular economy by transforming low-cost, underutilized materials into useful products, emphasizing the dual environmental and economic benefits of biomass valorization (Martini et al., 2023). Table 10.3 summarizes the environmental benefits of bioadsorbents, including carbon footprint reduction, waste valorization, and reduced eco-toxicity.

Because of their biodegradability, renewability, and low ecological imprint, bioadsorbents outperform conventional adsorbents in the environment. They assist in reducing environmental pollution by converting low-value leftovers into valuable adsorptive materials. The use of bioadsorbents promotes waste valorization and resource recovery, which helps to create a circular economy. Unlike synthetic adsorbents, bioadsorbents do not emit toxic byproducts during production or disposal, making them environmentally friendly. Furthermore, after use, discarded bioadsorbents can be regenerated, burnt for energy recovery, or converted into biochar for soil amendment, ensuring a sustainable end-of-life pathway. Bioadsorbents are a viable alternative for sustainable wastewater treatment and environmental protection due to their eco-friendliness and efficient pollutant removal (Kumar and Balasubramani, 2025).

Efficient utilization of biomass waste as a precursor promotes both environmental and economic sustainability, in accordance with circular economy principles. A study created a new adsorbent, Zn_2SiO_4/activated carbon (Zn_2SiO_4/AC), by carbonizing wasted coffee grounds, TEOS, and $ZnCl_2$ in a single pot. The composite's high surface area (625.54 m^2/g) and porous structure resulted in a maximum crystal violet adsorption capacity of 163.39 mg/g under optimal conditions (pH 5, 2 g/L dose, 45 °C, 105 min). Adsorption was shown to follow pseudo-second-order kinetics and Freundlich isotherm behavior, with major interactions involving hydrogen bonding, π–π interactions, coordination, and electrostatic attractions. After five cycles, Zn_2SiO_4/AC maintained high capacity (77.96 mg/g), indicating reusability and promise for sustainable wastewater treatment (Hoang et al., 2025).

Using waste biomass to make carbon electrodes for electrochemical water treatment promotes sustainable waste management and contributes to the circular economy. Recent research has focused on the use of biomass-derived carbon sources in electrochemical advanced oxidation processes (e-AOPs), capacitive deionization, and bioelectrochemical systems (BES). These carbon compounds have high surface areas, electrical conductivity, and variable hydrophobicity, making them ideal for a variety of electrochemical applications. Incorporating waste-derived carbons into BES can increase electrode efficiency while decreasing production costs (Ramirez et al., 2024).

Table 10.3 Summary of environmental benefits of bioadsorbents (carbon footprint reduction, waste valorization, eco-toxicity).

Bioadsorbent source	Waste valorization benefit	Carbon footprint reduction	Eco-toxicity impact	End-of-life disposal/ Reusability	Sustainability advantage	References
Rice husk	Converts agro-residue into value-added product	Low energy requirement, reduces open burning	Nontoxic	Can be regenerated multiple cycles	Reduces agricultural waste burden	Yefremova et al. (2023)
Agricultural and organic waste-derived biochar	Converts agricultural/ organic waste into value-added adsorbents for pollutant removal	Lowers CO_2 emissions by utilizing waste biomass and renewable energy for production	Generally low eco-toxicity, biodegradable, reduces pollutants effectively	Can be regenerated, reused, or disposed with minimal environmental impact	Sustainable waste recycling and cost-effective water treatment alternative	Tang (2024)
Biomass from plants, fungi, algae, microbes	Uses nonliving biomass from abundant natural sources for dye and heavy metal removal	Bio-based materials reduce reliance on synthetic chemicals and fossil fuels	Low eco-toxicity, eco-friendly, and often biodegradable	Can be reused after regeneration; some compostable	Low-cost, abundant, and renewable resource with high adsorption efficiency	Aragaw and Bogale (2021)
Modified biochar (magnetic, chemical modification)	Upcycles biomass with enhanced adsorption capacity for microplastics and heavy metals	Improved pollutant removal reduces indirect emissions from contamination	Surface modifications designed to minimize additional toxicity	Magnetic modification allows easy recovery and reuse	Enhanced removal efficiency with retained or improved sustainability	Mulindwa et al. (2024)
Waste rubber tires (chemically modified)	Valorizes waste tires as adsorbents for pesticide removal	Less environmental burden than conventional disposal or incineration	Potentially lower toxicity compared to raw waste; treatment reduces toxic release	Can be reused after treatment cycles	Provides a circular economy approach to solid waste management	Vievard et al. (2023)

(Continued)

Table 10.3 *(Continued)*

Bioadsorbent source	Waste valorization benefit	Carbon footprint reduction	Eco-toxicity impact	End-of-life disposal/ Reusability	Sustainability advantage	References
Aquatic plants and microbes	Use of invasive or abundant biomass for heavy metal biosorption	Avoids chemical synthetics, carbon neutral absorption process	Minimal eco-toxicity; biodegradation possible	Biomass can be disposed safely or composted	Bioremediation with minimal environmental footprint	Mathew et al. (2016)
Agricultural residues (walnut)	Utilizing abundant agro-waste to adsorb heavy metals and organic pollutants	Reduces reliance on synthetic adsorbents, low embodied carbon	Low eco-toxicity, biodegradable	Reusable after regeneration; biowaste compatible	Circular economy approach, cost-effective wastewater treatment	Jha (2024)
Bamboo charcoal biochar	Valorizes bamboo waste as porous adsorbent for CO_2 capture	Significant potential for carbon capture and storage	Biochar is generally low toxicity	Stable and reusable in carbon capture systems	Supports carbon neutrality and renewable bio-based adsorbent production	Zhang et al. (2023)
Activated carbon from food waste	Converts food waste (peels, shells) into high surface area adsorbents	Uses waste resources, avoiding new synthetic carbon production	Minimal eco-toxicity, biodegradable	Reusable and thermally stable	Waste valorization contributing to landfill reduction and pollution control	Schwarz et al. (2024)
Sugarcane bagasse and rice husk	Agricultural byproducts used for toxic metal adsorption	Low carbon footprint related to biowaste reuse	Nontoxic adsorption process	Regenerable and reusable	Low-cost treatment with waste minimization	Mathew et al. (2016)

Lignocellulosic biomass (wood, bark)	Valorizes forestry residues and wood waste for pollutant uptake	Renewable carbon source with lower emissions	Low environmental toxicity	Can be composted or reused	Renewable, abundant, and eco-friendly material source	Teklemedhin et al. (2025)
Shellfish waste-derived bioadsorbents	Uses discarded shells as adsorbents for heavy metals and dyes	Employs waste biomaterials, reducing environmental burden	Nontoxic, biodegradable	Biodegradable and safe disposal	Converts waste to resource in circular economy	Meftah et al. (2025b)
Citrus fruit peels (orange, lemon)	Agricultural food waste converted to adsorb pollutants	Low carbon footprint due to waste utilization	Low toxicity, biodegradable	Compostable, reusable after treatment	Readily available, low-cost renewable	Schwarz et al. (2024)
Microbial biomass (bacteria, fungi)	Uses microbial byproducts/biosynthesis residues as adsorbents	Bio-based, low carbon emissions in production	Typically, biodegradable and low toxicity	Often biodegradable and regenerable	Fast-growing and sustainable source of adsorbents	Aragaw and Bogale (2021)
Fibrous plant-based food waste (lentil husk, leaves)	Low-cost bioadsorbents from food industry waste fibers	Utilizes food waste, reducing landfill methane emissions	Eco-friendly and biodegradable	Regenerable in water treatment cycles	Sustainable waste recycling and pollutant uptake	Karim et al. (2023)
Magnetic bio-nanocomposites with natural polymers	Enables easy recovery and multiple reuse cycles	Reduced secondary pollution and energy use	Low toxicity with biopolymer matrices	Strong mechanical stability for reuse	High-efficiency sustainable adsorbents with reusability	Satyam and Patra (2024)

10.5 Conclusion

The increasing need for sustainable and cost-effective wastewater treatment has highlighted the significance of bioadsorbents as viable alternatives to conventional adsorbents. Bioadsorbents, derived from agricultural residues, lignocellulosic biomass, algae, and other organic wastes, offer multiple advantages, including low production costs, high adsorption efficiency, and eco-friendly characteristics. Economic assessments show that raw material selection, processing, and modification costs significantly influence feasibility, while scale-up and commercialization remain critical challenges. TEAs suggest that integrating bioadsorbent production with existing industrial processes, such as biorefineries, can enhance profitability and reduce operational costs. From an environmental perspective, bioadsorbents contribute to carbon footprint reduction, waste valorization, and minimized eco-toxicity, aligning with circular economy principles. LCAs indicate that careful selection of biomass, sustainable activation methods, and efficient regeneration strategies can maximize environmental benefits while maintaining economic viability. Moreover, advances in regeneration techniques, including thermal, chemical, and microwave-assisted methods, enhance reusability and resource efficiency. Overall, bioadsorbents present a promising approach for sustainable wastewater treatment, combining economic and environmental advantages. Future research should focus on real-world applications, pilot-scale studies, and development of standardized protocols to facilitate large-scale adoption and integration into circular economy frameworks.

References

Abou Taleb, M., & Al Farooque, O. (2021). Towards a circular economy for sustainable development: an application of full cost accounting to municipal waste recyclables. Journal of Cleaner Production, 280, 124047.

Adegoke, K. A., Akinnawo, S. O., Adebusuyi, T. A., Ajala, O. A., Adegoke, R. O., Maxakato, N. W., & Bello, O. S. (2023). Modified biomass adsorbents for removal of organic pollutants: a review of batch and optimization studies. International Journal of Environmental Science and Technology, 20(10), 11615–11644.

Ahmed, M. B., Zhou, J. L., Ngo, H. H., & Guo, W. (2016). Insight into biochar properties and its cost analysis. Biomass and Bioenergy, 84, 76–86.

AlAqad, K. M., Abdelnaby, M. M., Tanimu, A., Abdulazeez, I., & Elsharif, A. M. (2025). Adsorbent materials for water treatment: a review of current trends and future challenges. Environmental Pollution and Management, 2, 1–13.

Ali, I. (2010). The quest for active carbon adsorbent substitutes: inexpensive adsorbents for toxic metal ions removal from wastewater. Separation & Purification Reviews, 39(3–4), 95–171.

Aragaw, T. A., & Bogale, F. M. (2021). Biomass-based adsorbents for removal of dyes from wastewater: a review. Frontiers in Environmental Science, 9, 764958.

Banerjee, M., Basu, R. K., & Das, S. K. (2018). Cr (VI) adsorption by a green adsorbent walnut shell: adsorption studies, regeneration studies, scale-up design and economic feasibility. Process Safety and Environmental Protection, 116, 693–702.

Bello, O. S., Adegoke, K. A., Olaniyan, A. A., & Abdulazeez, H. (2015). Dye adsorption using biomass wastes and natural adsorbents: overview and future prospects. Desalination and Water Treatment, 53(5), 1292–1315.

Chen, D., & Zhou, W. (2024). Synthesis of a novel polymer adsorbent and its adsorption of Pb (II) and Cu (II) ions in water. Processes, 12(2), 2901.

Coelho, R. S., Soares, L. C., Adarme, O. F. H., Maia, L. C., Costa, C. S. D., Guibal, E., & Gurgel, L. V. A. (2025). A review on advances in the use of raw and modified agricultural lignocellulosic residues in mono-and multicomponent continuous adsorption of inorganic pollutants for upscaling technologies. Polymers, 17(7), 953.

Crini, G., Lichtfouse, E., Wilson, L. D., & Morin-Crini, N. (2019). Conventional and non-conventional adsorbents for wastewater treatment. Environmental Chemistry Letters, 17(1), 195–213.

Daffalla, S. B., Mukhtar, H., & Shaharun, M. S. (2020). Preparation and characterization of rice husk adsorbents for phenol removal from aqueous systems. PLoS One, 15(12), e0243540.

Das, A., Bar, N., & Das, S. K. (2020). Pb (II) adsorption from aqueous solution by nutshells, green adsorbent: adsorption studies, regeneration studies, scale-up design, its effect on biological indicator and MLR modeling. Journal of Colloid and Interface Science, 580, 245–255.

Das, K., Sukul, U., Chen, J. S., Sharma, R. K., Banerjee, P., Dey, G., Taharia, M., Wijaya, C. J., Lee, C. I., Wang, S. L., Nuong, N. H. K., & Chen, C. Y. (2024). Transformative and sustainable insights of agricultural waste-based adsorbents for water defluoridation: biosorption dynamics, economic viability, and spent adsorbent management. Heliyon, 10(8), e29747.

Dotto, G. L., & McKay, G. (2020). Current scenario and challenges in adsorption for water treatment. Journal of Environmental Chemical Engineering, 8(4), 103988.

Fan, S., & Zhang, L. (2021). Production and characterization of tea waste–based biochar and its application in treatment of Cd-containing wastewater. Biomass Conversion and Biorefinery, 11, 1719–1732.

Farias, K. C., Guimarães, R. C., Oliveira, K. R., Nazário, C. E., Ferencz, J. A., & Wender, H. (2023). Banana peel powder biosorbent for removal of hazardous organic pollutants from wastewater. Toxics, 11(8), 664.

Fingolo, A. C., Klein, B. C., Rezende, M. C., Silva e Souza, C. A., Yuan, J., Yin, G., Bonomi, A., Martinez, D. S., & Strauss, M. (2020). Techno-economic assessment and critical properties tuning of activated carbons from pyrolyzed sugarcane bagasse. Waste and Biomass Valorization, 11(8), 1–13.

Gkika, D. A., Liakos, E. V., Vordos, N., Kontogoulidou, C., Magafas, L., Bikiaris, D. N., Bandekas, D. V., Mitropoulos, A. C., & Kyzas, G. Z. (2019). Cost estimation of polymeric adsorbents. Polymers, 11(5), 925.

Gupta, V. K., Carrott, P. J. M., Ribeiro Carrott, M. M. L., & Suhas. (2009). Low-cost adsorbents: growing approach to wastewater treatment—a review. Critical Reviews in Environmental Science and Technology, 39(10), 783–842.

Gupta, V. K., Nayak, A., Bhushan, B., & Agarwal, S. (2015). A critical analysis on the efficiency of activated carbons from low-cost precursors for heavy metals remediation. Critical Reviews in Environmental Science and Technology, 45(6), 613–668.

Habib, M. A., & Mohamed, I. E. (2023). Agriculture by -products wastes as low-cost bio-sorbents for heavy metals removal from industrial waste stream, short review. Preprints.

Hamad, H. N., & Idrus, S. (2022). Recent developments in the application of bio-waste-derived adsorbents for the removal of methylene blue from wastewater: a review. Polymers, 14(4), 783.

Hashmi, Z., Jatoi, A. S., Nadeem, S., Anjum, A., Imam, S. M., & Jangda, H. (2024). Comparative analysis of conventional to biomass-derived adsorbent for wastewater treatment: a review. Biomass Conversion and Biorefinery, 14(1), 45–76.

Hassaan, M. A., Pantaleo, A., Santoro, F., Elkatory, M. R., De Mastro, G., Sikaily, A. E., Ragab, S., & Nemr, A. E. (2020). Techno-economic analysis of ZnO nanoparticles pretreatments for biogas production from barley straw. Energies, 13(19), 5001.

Hilal, N. M., Badawy, N. A., Mostafa, O. I., & Elrefay, H. M. (2019). Synthetic and application of a novel resin from waste foam packing for adsorption of acid orange 67 from aqueous solution. Bulletin of the National Research Centre, 43, 58.

Hoang, V. H., Hoang, H. Y., Nguyen, A. T., Doan, V. D., & Tran, V. A. (2025). Synthesis and adsorption behavior of Zn_2SiO_4 nanoparticles incorporated with biomass-derived activated carbon as a novel adsorbent in a circular economy framework. Biomass and Bioenergy, 193, 107554.

Huang, Q., Song, S., Chen, Z., Hu, B., Chen, J., & Wang, X. (2019). Biochar-based materials and their applications in removal of organic contaminants from wastewater: state-of-the-art review. Biochar, 1(1), 45–73.

Ifa, L., Syarif, T., Sartia, S., Juliani, J., Nurdjannah, N., & Kusuma, H. S. (2022). Techno-economics of coconut coir bioadsorbent utilization on free fatty acid level reduction in crude palm oil. Heliyon, 8(3), e09146.

Ighalo, J. O., Omoarukhe, F. O., Ojukwu, V. E., Iwuozor, K. O., & Igwegbe, C. A. (2022). Cost of adsorbent preparation and usage in wastewater treatment: a review. Cleaner Chemical Engineering, 3, 100042.

Jacob, A. P., & Soni, P. (2024). Chromium (VI) removal from water using different bio-adsorbents: comparison and scope in the treatment of local water bodies with economic analysis. Desalination and Water Treatment, 318, 100314.

Jha, P. (2024). Agricultural waste assessment as adsorbents for the removal of pollutants from industrial effluents. In: Abstracts of the International Conference on Separation and Purification Technologies 2023 (AIJR Abstracts). AIJR Publisher, 6.

Kalengyo, R. B., Ibrahim, M. G., Fujii, M., & Nasr, M. (2024). Utilizing orange peel waste biomass in textile wastewater treatment and its recyclability for dual biogas and biochar production: a techno-economic sustainable approach. Biomass Conversion and Biorefinery, 14(16), 19875–19888.

Karaer Yağmur, H., & Kaya, İ. (2021). Synthesis and characterization of magnetic $ZnCl_2$-activated carbon produced from coconut shell for the adsorption of methylene blue. Journal of Molecular Structure, 1232, 130071.

Karim, A., Raji, Z., Karam, A., & Khalloufi, S. (2023). Valorization of fibrous plant-based food waste as biosorbents for remediation of heavy metals from wastewater-a review. Molecules (Basel, Switzerland), 28(10), 4205.

Kumar, A., Saini, K., & Bhaskar, T. (2020). Hydochar and biochar: production, physicochemical properties and techno-economic analysis. Bioresource Technology, 310, 123442.

Kumar, P. S., & Balasubramani, S. R. (2025). Bio-adsorbents from biomass for environmental remediation. In: Biomass for Environmental Remediation (pp. 41–54). Elsevier.

Liu, L., Qian, H., Mu, L., Wu, J., Feng, X., Lu, X., & Zhu, J. (2021). Techno-economic analysis of biomass processing with dual outputs of energy and activated carbon. Bioresource Technology, 319, 124108.

Martini, S., Kharismadewi, D., & Ginting, Y. R. (2023, May). Biomass potential as an alternative resource for valuable products in the perspective of environmental sustainability and a circular economy system. IOP Conference Series: Earth and Environmental Science 1175(1), 012012.

Mathew, B. B., Jaishankar, M., Biju, V. G., & Beeregowda, K. N. (2016). Role of bioadsorbents in reducing toxic metals. Journal of Toxicology, 2016, 4369604.

Meftah, S., Meftah, K., Babassa, N., Malous, K., Drissi, M., Amahrous, A., Jabri, N., Fathi, J., Aguerchi, K., Laababid, Y., Adnouss, O., Taib, M., Zinaoui, O., ElHassan, O., & Bouyazza, L. (2025a). The versatility and effectiveness of bio-adsorbents in the removal of chemical pollutants from water: adsorption mechanisms, optimization by ANN and RSM, SWOT analysis, and contribution to the 3rd and 6th sustainable development goals. Discover Sustainability, 6(1), 971.

Meftah, S., Meftah, K., Drissi, M., Radah, I., Malous, K., Amahrous, A., Chahid, A., Tamri, T., Rayyad, A., Darkaoui, B., Hanine, S., El-Hassan, O., & Bouyazza, L. (2025b). Heavy metal polluted water: effects and sustainable treatment solutions using bio-adsorbents aligned with the SDGs. Discover Sustainability, 6(1), 137.

Meftah, S., Meftah, K., El Hassan, O., & Bouyazza, L. (2025c). Eco-friendly adsorbents for the removal of heavy metals. Materials International, 7(1), 8.

Meramo-Hurtado, S. I., & González-Delgado, A. D. (2020). Application of techno-economic and sensitivity analyses as decision-making tools for assessing emerging large-scale technologies for production of chitosan-based adsorbents. ACS Omega, 5(28), 17601–17610.
Mergbi, M., Galloni, M. G., Aboagye, D., Elimian, E., Su, P., Ikram, B. M., Nabgan, W., Bedia, J., Amor, H. B., Contreras, S., Medina, F., & Djellabi, R. (2023). Valorization of lignocellulosic biomass into sustainable materials for adsorption and photocatalytic applications in water and air remediation. Environmental Science and Pollution Research, 30(30), 74544–74574.
Michael-Igolima, U., Abbey, S. J., Ifelebuegu, A. O., & Eyo, E. U. (2023). Modified orange peel waste as a sustainable material for adsorption of contaminants. Materials, 16(3), 1092.
Mihai, S., Bondarev, A., & Necula, M. (2025). The potential of biogenic materials as sustainable and environmentally benign alternatives to conventional adsorbents for dyes removal: a review. Processes, 13(2), 589.
Misran, E., Bani, O., Situmeang, E. M., & Purba, A. S. (2022). Banana stem based activated carbon as a low-cost adsorbent for methylene blue removal: isotherm, kinetics, and reusability. Alexandria Engineering Journal, 61(3), 1946–1955.
Mulindwa, P., Kasule, J. S., Nantaba, F., Wasswa, J., & Expósito, A. J. (2024). Bioadsorbents for removal of microplastics from water ecosystems: a review. International Journal of Sustainable Engineering, 17(1), 582–599.
Nematian, M., Keske, C., & Ng'ombe, J. N. (2021). A techno-economic analysis of biochar production and the bioeconomy for orchard biomass. Waste Management, 135, 467–477.
Park, J. E., Jo, E. S., Lee, G. B., Lee, S. E., & Hong, B. U. (2023). Adsorption capacity and desorption efficiency of activated carbon for odors from medical waste. Molecules, 28(2), 785.
Pérez-Botella, E., Valencia, S., & Rey, F. (2022). Zeolites in adsorption processes: state of the art and future prospects. Chemical Reviews, 122(24), 17647–17695.
Praveen, S., Gokulan, R., Pushpa, T. B., & Jegan, J. (2021). Techno-economic feasibility of biochar as biosorbent for basic dye sequestration. Journal of the Indian Chemical Society, 98(8), 100107.
Rakanović, M., Vukojević, A., Savanović, M. M., Armaković, S., Pelemiš, S., Živić, F., Sladojević, S., & Armaković, S. J. (2022). Zeolites as adsorbents and photocatalysts for removal of dyes from the aqueous environment. Molecules, 27(19), 6582.
Ramirez, A., Muñoz-Morales, M., Lopez-Fernandez, E., Fernández-Morales, F. J., & Llanos, J. (2024). Advancing circular economy: Critical insights into waste biomass

derived carbon electrodes for (bio) electrochemical water treatment. Current Opinion in Electrochemistry, 46, 101492.

Satyam, S., & Patra, S. (2024). Innovations and challenges in adsorption-based wastewater remediation: a comprehensive review. Heliyon, 10(9), e29573.

Sayed, A., Behalo, M., Feteha, A., Younis, A., Keshawy, M., Yousry, R., & Abdel-Raouf, M. E. (2025). Unveiling the removal performance of banana peel biochar and rice husk nanoparticles for crystal violet and methylene blue dyes from aqueous media. Discover Environment, 3(1), 1–24.

Schwarz, B. B., Kunst, S. R., Soares, L. G., Morisso, F. D., Oliveira, C. T., & Machado, T. C. (2024). Study of the use of food waste as bioadsorbents to remove methylene blue dye. Matéria (Rio de Janeiro), 29(3), e20240100.

Sen, T. (2023). Agricultural solid wastes based adsorbent materials in the remediation of heavy metal ions from water and wastewater by adsorption: a review. Molecules (Basel, Switzerland), 28(14), 5575.

Shen, X., Huang, P., Li, F., Wang, X., Yuan, T., & Sun, R. (2019). Compressive alginate sponge derived from seaweed biomass resources for methylene blue removal from wastewater. Polymers, 11(6), 961.

Shi, C., Li, Y., Li, X., Zhao, X., Ma, X., Zhou, X., Cui, Y., Ma, S., Xu, W., & Ren, C. (2021). Preparation of macroporous high adsorbent resin and its application for heavy metal ion removal. ChemistrySelect, 6(34), 9038–9045.

Sulyman, M., Namiesnik, J., & Gierak, A. (2017). Low-cost adsorbents derived from agricultural by-products/wastes for enhancing contaminant uptakes from wastewater: a review. Polish Journal of Environmental Studies, 26(2), 479–510.

Tang, K. H. (2024). Valorization of organic waste as biosorbents for wastewater treatment. Water Emerging Contaminants & Nanoplastics, 3, 25.

Teklemedhin, T. B., Wan, Z., Hossain, N., Guo, Y., & Li, H. (2025). Bio-adsorbents derived from carbohydrates, lignin, and protein biomass as potential environmental remediation: a comprehensive review. International Journal of Biological Macromolecules, 320(Pt 2), 145861.

Thanki, A., Thanki, A., Singh, R., & Sohal, K. S. (2021). The application of low-cost natural bio-adsorbents for the removal of heavy metals—A review. Sustainable Development Through Engineering Innovations: Select Proceedings of SDEI 2020, 355–371.

Vievard, J., Alem, A., Pantet, A., Ahfir, N. D., Aellano-Sánchez, M. G., Devouge-Boyer, C., & Mignot, M. (2023). Bio-based adsorption as ecofriendly method for wastewater decontamination: a review. Toxic, 11(5), 404.

Vukelic, D., Boskovic, N., Agarski, B., Radonic, J., Budak, I., Pap, S., & Sekulic, M. T. (2018). Eco-design of a low-cost adsorbent produced from waste cherry kernels. Journal of Cleaner Production, 174, 1620–1628.

Welsch, M. L., & Laird, B. B. (2024). Cation dependence of gas adsorption in clinoptilolite and RHO zeolites using Monte Carlo simulation: a linear model. Journal of Chemical & Engineering Data, 69(5), 2031–2040.

Yefremova, S., Kablanbekov, A., Satbaev, B., & Zharmenov, A. (2023). Rice husk-based adsorbents for removal of metals from aqueous solutions. Materials, 16(23), 7353.

Zhang, C., Ji, Y., Li, C., Zhang, Y., Sun, S., Xu, Y., Jiang, L., & Wu, C. (2023). The application of biochar for CO_2 capture: influence of biochar preparation and CO_2 capture reactors. Industrial & Engineering Chemistry Research, 62(42), 17168–17181.

Index

Printed and bound by CPI Group (UK) Ltd, Croydon, CR0 4YY
24/08/2026
14947832-0002